THIN FILMS FROM FREE ATOMS AND PARTICLES

CONTRIBUTORS

R. AVNI

D. BOUCHIER

G. GAUTHERIN

KENNETH J. KLABUNDE

HIROYUKI MATSUNAMI

MASANORI MURAKAMI

GEORGE C. NIEMAN

C. SCHWEBEL

ARMIN SEGMÜLLER

M. VENUGOPALAN

JAMES B. WEBB

C. WEISSMANTEL

THIN FILMS FROM FREE ATOMS AND PARTICLES

Edited by

KENNETH J. KLABUNDE

Department of Chemistry
Kansas State University
Manhattan, Kansas

1985

ACADEMIC PRESS, INC.
Harcourt Brace Jovanovich, Publishers
Orlando San Diego New York Austin
London Montreal Sydney Tokyo Toronto

ACADEMIC PRESS, INC.
Orlando, Florida 32887

United Kingdom Edition published by
ACADEMIC PRESS INC. (LONDON) LTD.
24–28 Oval Road, London NW1 7DX

Library of Congress Cataloging in Publication Data
Main entry under title:

Thin films from free atoms and particles.

Includes index.
1. Thin films. 2. Layer structure (Solids).
3. Crystals–Growth. 4. Atoms. 5. Particles.
I. Klabunde, Kenneth J.
QC176.83.T47 1983 530.4'1 85-6025
ISBN 0–12–410755–9 (alk. paper)
ISBN 0–12–410756–7 (paperback)

PRINTED IN THE UNITED STATES OF AMERICA

85 86 87 88 9 8 7 6 5 4 3 2 1

To David

CONTENTS

CONTRIBUTORS

Numbers in parentheses indicate the pages on which the authors' contributions begin.

R. AVNI (49), Division of Chemistry, Nuclear Research Center Negev, Beer-Sheva 84190, Israel

D. BOUCHIER (203), Université Paris-Sud, Institut d'Electronique Fondamentale, 91405 Orsay, France

G. GAUTHERIN (203), Université Paris-Sud, Institut d'Electronique Fondamentale, 91405 Orsay, France

KENNETH J. KLABUNDE (1, 23), Department of Chemistry, Kansas State University, Manhattan, Kansas 66506

HIROYUKI MATSUNAMI (301), Department of Electrical Engineering, Faculty of Engineering, Kyoto University, Kyoto 606, Japan

MASANORI MURAKAMI (325), IBM Thomas J. Watson Research Center, Yorktown Heights, New York 10598

GEORGE C. NIEMAN (23), Department of Chemistry, Monmouth College, Monmouth, Illinois 61462

C. SCHWEBEL (203), Université Paris-Sud, Institut d'Electronique Fondamentale, 91405 Orsay, France

ARMIN SEGMÜLLER (325), IBM Thomas J. Watson Research Center, Yorktown Heights, New York 10598

M. VENUGOPALAN (49), Department of Chemistry, Western Illinois University, Macomb, Illinois 61455

JAMES B. WEBB (257), Semiconductor Research Group, National Research Council Canada, Division of Chemistry, Ottawa K1A-OR6, Canada

C. WEISSMANTEL (153), Sektion Physik/EB, Technische Hochschule, 9010 Karl-Marx-Stadt, PSF 964, German Democratic Republic

PREFACE

In 1980 the editor of this volume published a book entitled "Chemistry of Free Atoms and Particles." "Free atoms" are single neutral atoms of the elements. "Free particles" are coordination-deficient molecules that normally exist as condensed or polymerized solids. Examples of reactive particles are BF, CS, C_3, Ni_2, SiF_2, and Si:O. Free atoms and particles are usually prepared by high-energy methods such as high-temperature vaporization using resistive heating or lasers or by discharge plasma methods. Such species have found a unique and important place in the chemistry laboratory as synthons. The possibilities are enormous when the number of species (thousands) that could be formed and studied is considered.

The availability of excellent vacuum technology, cryogenics, high-temperature apparatus, and plasma apparatus now allows the modern laboratory the means to prepare and study many free atoms and particles. Their chemistries with other added molecules are fascinating and are being slowly elucidated.

As one might anticipate, one of the primary reaction modes of atoms or coordination-deficient particles is polymerization—condensation. It is this process with which this book is concerned because of the present technological importance of such phenonema and also because of the tremendous new possibilities that exist if we can work together.

$$\text{new molecules} \xleftarrow[\text{reactant}]{\text{chemical}} \underset{\text{(gas)}}{\text{Ni atoms}} \xrightarrow{\text{surface}} \text{Ni film}$$

$$\text{new molecules} \xleftarrow[\text{reactant}]{\text{chemical}} \underset{\text{(gas)}}{\text{Ti:O molecules}} \xrightarrow{\text{surface}} \text{Ti:O film}$$

Perhaps the one area of science and technology that could benefit most by an interdisciplinary approach is that of thin-film materials. The tremendous impact of thin films on the electronics/semiconductor industry emphasizes the importance of the area to physicists and engineers.

Chemists, however, who are needed if new thin-film materials are to be synthesized, need to be more firmly established in this area. This book is meant to aid this process and to help bring physicists, engineers, materials scientists, and chemists together with a common interest in new thin-film technology.

The coverage of this volume is as follows: First, an introduction to free atoms and particles is followed by a chapter describing the embryonic growth of films (dimers, trimers, and other small telomers formed and detected). Next, the current understanding of discharge processes for forming free atoms and particles is reviewed, and then several chapters dealing with current technology, techniques, and materials follow.

The editor gratefully acknowledges the contributing authors of these chapters, who are world-renowned experts in their respective fields. Their enthusiasm for contributing to such an interdisciplinary venture is most gratifying. Special thanks are given to Dr. W. D. Westwood of Bell Northern Research for many helpful discussions and suggestions. The editor also acknowledges the help of his students in many ways, as well as his family and expecially his wife, Linda, for their patience.

1 INTRODUCTION TO FREE ATOMS AND PARTICLES

Kenneth J. Klabunde

Department of Chemistry
Kansas State University
Manhattan, Kansas

I. FREE ATOMS

The liberation of an atom from its elemental condensed form usually requires a great deal of energy. The atom thus formed, for example, a single atom of titanium (Ti), will be highly chemically reactive with added reagents. Of course, Ti atoms will also be very reactive with themselves, thus reforming titanium metal (or film).

The formation of free atoms can be carried out in a number of ways, all involving some high-temperature method in which the element is vaporized or a compound of the element is decomposed. These methods will be discussed in more detail later. At this point it is sufficient to say that over 90 of the elements can be treated in this way, that is, treatment with high temperature to form the atoms, which can then be used as chemical synthons or as building blocks for films.

II. FREE PARTICLES

Elemental solids are not the only sources of reactive, gaseous building blocks. High-temperature treatment of condensed molecular matter, such as metal oxides or sulfides, can often lead to gaseous coordination-deficient molecules, for example, titanium monoxide (TiO), which can be used as chemical synthons or as building blocks for films.

In theory the possibilities for generation of coordination-deficient reactive free particles are almost limitless. If two to four combined atoms make up the particle, almost all of the permutations of the elements must be considered; Ni_2, TiO, Ti_3, TiS, MoO_3, B_2, BSi_2, SiC, RuC, PtTi, etc. In practice, *current* experimental methods limit the possibilities to hundreds (rather than thousands) of combinations.

To summarize, a free atom is a single, isolated reactive atom in a condensed inert matrix or in the gas phase. A free particle is a reactive coordination-deficient molecule of two to four atoms. These species are almost always formed by a high-temperature/energy process of one kind or another.

III. THE ELEMENTS

Figure 1.1 shows a periodic chart listing heats of formation of the elements [1, 2], which clearly points out that almost all of the elements exist as very stable condensed solids. The exceptions are the diatomic gases (H_2, O_2, N_2, F_2, Cl_2, Br_2), the nonreactive atoms (He, Ne, Ar, Kr, Xe, Ra), and the liquids (Hg, Br_2).

Nearly all of the solid elements vaporize monatomically. The most notable exception is carbon where C_1, C_2, C_3, and C_4 are formed, the

	H	Li	Be											B	(C)	N	O	F	Ne
kcal	52	39	78											134	171	113	59	19	0
		Na	Mg											Al	Si	(P)	(S)	Cl	Ar
kcal		26	35											78	107	79	67	29	0
		K	Ca	Sc	Ti	V	Cr	Mn	Fe	Co	Ni	Cu	Zn	Ga	Ge	(As)	(Se)	Br	Kr
kcal		21	42	82	113	123	95	67	100	101	103	80	31	69	69	72	54	27	0
		Rb	Sr	Y	Zr	Nb	Mo	Te	Ru	Rh	Pd	Ag	Cd	In	(Sn)	(Sb)	(Te)	I	Xe
kcal		20	39	86	145	172	159	(155)	155	133	89	68	27	58	72	63	47	25	0
		Cs	Ba	La	Hf	Ta	W	Re	Os	Ir	Pt	Au	Hg	Tl	Pb	Bi	Po	At	Rn
kcal		19	42	104	145	187	202	186	187	160	135	88	15	43	47	50	34	—	0
		Ce	Pr	Nd	Sm	Eu	Gd	Tb	Dy	Ho	Er	Tm	Yb	Lu					
kcal		97	85	77	50	43	82	87	71	75	75	56	40	102					
		Th	U																
kcal		135	117																

Fig. 1.1. Periodic chart showing heats of formation of the elements (kilocalories per mole). Those elements that yield polyatomics on vaporization of the bulk solid element are circled.

TABLE 1.1

Solid Elements[a] with Physical Properties and Vapor Species[b]

Element	mp (°C)	bp[c] (°C)	Vaporization temperature under vacuum (°C at about 10 μ pressure)	Vaporization method	Vapor composition
Li	180	1347	535	Resistive heating	Li
Na	98	883	289	Resistive heating	Na
K	64	774	208	Resistive heating	K, K_2 (small)
Rb	39	688	173	Resistive heating	Rb
Cs	28	678	145	Resistive heating	Cs
Be	1278	2970	1225	Resistive heating	Be, Be_2 (small)
Mg	649	1090	439	Resistive heating, arc, e-beam	Mg
Ca	839	1484	597	Resistive heating, arc, e-beam	Ca
Sr	769	1384	537	Resistive heating, arc, e-beam	Sr
Ba	725	1640	610	Resistive heating, arc, e-beam	Ba
Sc	1541	2831	—	Arc	Sc
Ti	1660	3287	—	Laser, e-beam, arc	Ti
V	1890	3380	—	E-beam, arc	V
Cr	1857	2672	—	Resistive heating, laser, e-beam, arc	Cr
Mn	1244	1962	—	Arc, resistive heating, levitation	Mn
Y	1552	3338	—	Arc	Y
Zr	1852	4377	—	Laser, e-beam, arc, sublimation by resistive heating	Zr
Nb	2468	4742	—	E-beam, arc	Nb
Mo	2617	4612	—	Laser, e-beam, arc, sublimation by resistive heating	Mo
Hf	2227	4602	—	E-beam	Hf
Ta	2996	5425	—	Laser, e-beam, arc	Ta

(*continues*)

TABLE 1.1 (*continued*)

Element	mp (°C)	bp[c] (°C)	Vaporization temperature under vacuum (°C at about 10 μ pressure)	Vaporization method	Vapor composition
W	3410	5660	—	Laser, e-beam, arc, sublimation by resistive heating	W
Re	3180	—	—	E-beam, arc sublimation by resistive heating	Re
Fe	1535	2750	—	Laser, e-beam induction, resistive heating	Fe
Ru	2310	3900	—	Resistive heating	Ru
Os	3045	5020	—	—	Os
Co	1495	2870	—	E-beam, induction, arc, resistive heating	Co
Rh	1966	3727	—	Arc, resistive heating, e-beam	Rh
Ir	2410	4130	—	E-beam, arc	Ir
Ni	1453	2732	—	Laser, e-beam, induction, arc, resistive heating	Ni
Pd	1552	3140	—	Arc, resistive heating	Pd
Pt	1772	3827	—	E-beam, resistive heating	Pt
Cu	1083	2567	—	E-beam, discharge, induction, laser, resistive heating	Cu Cu_2 (0.0009)
Ag	961	2212	—	E-beam, induction, arc, resistive heating	Ag Ag_2 (0.0005)
Au	1064	2807	—	E-beam, arc, resistive heating	Au Au_2 (0.0007)

TABLE 1.1 *(continued)*

Element	mp (°C)	bp[c] (°C)	Vaporization temperature under vacuum (°C at about 10 μ pressure)	Vaporization method	Vapor composition
Zn	420	907	—	E-beam, arc laser, resistive heating	Zn
Cd	321	765	—	Arc, resistive heating, laser	Cd
Hg	−39	357	—	Arc, resistive heating	Hg
B	2300	2550	—	E-beam, DC arc, induction, laser, resistive heating of C crucible	B, B_2
Al	660	2467	—	Laser, resistive heating from BN or TiB_2 or C crucible, e-beam, levitation	Al
Ga	30	2403	—	Resistive heating	Ga, Ga_2 (small)
In	156	2080	—	Resistive heating	In, In_2 (small)
Tl	304	1457	—	Resistive heating	Tl, Tl_2 (small)
C	3632	4827	—	Sublimation by resistive heating, arc, laser	C, C_2, C_3, C_4, C_5
Si	1410	2355	—	E-beam, laser levitation, resistive heating of Si rod	Si
Ge	937	2830	—	Resistive heating, laser	Ge
Sn	232	2260	—	Resistive heating, laser, e-beam	Sn, Sn_2, Sn_3, Sn_4
Pb	328	1740	—	Resistive heating, laser, e-beam	Pb
As	817	613 (subl)	—	Resistive heating	As, As_6, As_8
Sb	631	1750	—	Resistive heating, e-beam, laser	Sb_3

(continues)

TABLE 1.1 (*continued*)

Element	mp (°C)	bp[c] (°C)	Vaporization temperature under vacuum (°C at about 10 μ pressure)	Vaporization method	Vapor composition
Bi	271	1560	—	Laser, resistive heating	Bi, Bi_2, Bi_4
Se	217	685	—	Resistive heating, laser	Se_2–Se_9
Te	452	1390	—	Resistive heating, laser	Te_5
La	921	3457	—	Resistive heating	La
Ce	799	3426	—	Resistive heating	Ce
Pr	931	3512	—	Resistive heating	Pr
Nd	1021	3068	—	Resistive heating	Nd
Pm	—	—	—	—	—
Sm	1077	1791	650–900	Resistive heating	Sm
Eu	822	1597	—	Resistive heating	Eu
Gd	1313	3266	—	Resistive heating	Gd
Tb	1360	3123	—	Resistive heating	Tb
Dy	1412	2562	—	Resistive heating	Dy
Ho	1474	2695	—	Resistive heating	Ho
Er	1529	2863	—	Resistive heating	Er
Tm	1545	1947	—	Resistive heating	Tm
Yb	819	1194	500–650	Resistive heating	Yb
Lu	1663	3395	—	Resistive heating	Lu
Th	—	—	—	—	—
Pa	< 1600	—	—	—	—
U	1132	3818	—	Resistive heating, e-beam	U
Np	640	3902	—	—	—
Pu	641	3232	—	—	—
Am	994	2607	—	—	—

[a] Arranged by families of elements in the periodic table moving from left to right.
[b] Taken mainly from Klabunde [1].
[c] Atmospheric pressure.

proportion of which depends on the method of graphite vaporization. A few other elements behave similarly, and Table 1.1 lists the polyatomics observed for these materials.

Table 1.1 also lists the melting points, boiling points (atmospheric pressure), and temperatures at which the vapor pressure of the element is 1×10^{-2} torr [3], which is an estimate of the temperature at which convenient vaporization under vacuum occurs. Listed also are the most commonly used methods of vaporization of each of the elements. These high-temperature techniques are discussed in more detail later in this chapter.

It should be emphatically stated here that a great deal of additional work is needed on determining vapor compositions. Most of the data listed are based on Knutsen cell/mass spectrometry experiments, which are fine for systems at thermal equilibrium. However, synthetic applications of high-temperature species requires vaporization from an open crucible or hearth, and the vapor compositions for many of the elements may be quite different under these conditions.

IV. MOLECULAR SOLIDS

Synthetic and natural molecular solids exist everywhere. These include the metal oxides, metal sulfides, metal carbides, metal nitrides, metal phosphides, metal halides, and metal alloys. If ternary or quarternary compounds of the elements are considered a huge number of possible combinations exist.

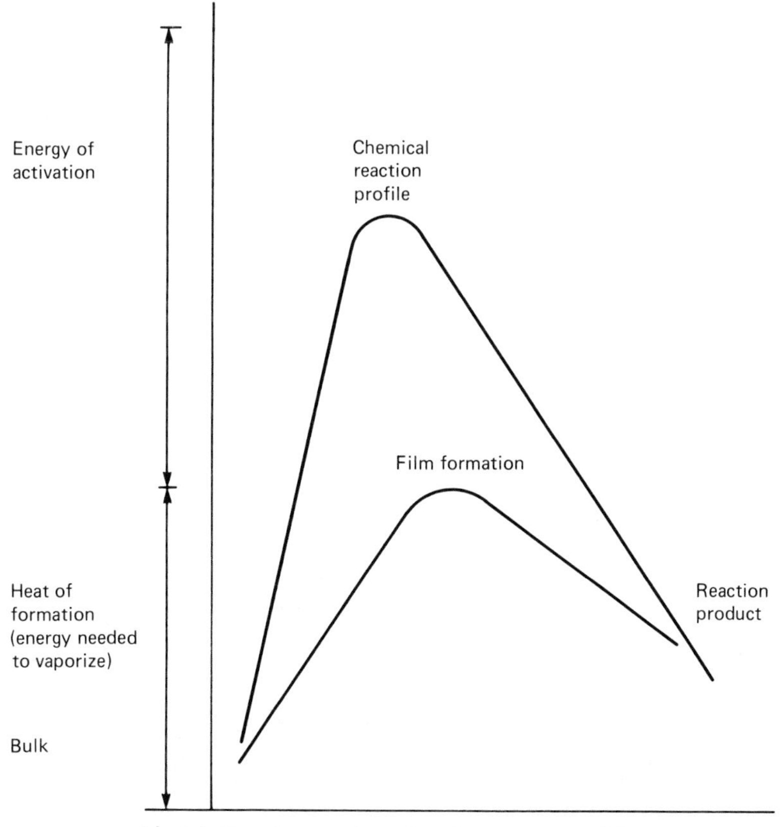

Fig. 1.2. Reaction coordinate for a free atom or particle.

TABLE 1.2

Examples of Molecular Solids[a] Studied under Vaporization Conditions[b]

Substance	mp (°C)	bp[c] (°C)	Vaporization temperature under vacuum	Vapor species and comments
LiF	845	1676	1047	LiF, $(LiF)_2$ (large)
NaCl	801	1413	865	NaCl, $(NaCl)_2$ (large)
KBr	734	1435	795	KBr, $(KBr)_2$
RbBr	693	1340	781	RbBr, $(RbBr)_2$ (18%)
CsCl	645	1290	744	CsCl, $(CsCl)_2$ (20%)
Li_2O	>1700	1200	980	Li_2O
K_2O	d 350	—	450	K_2O, K, O_2
BeF_2	—	—	500–1100	$BeF_2(BeF_2)_2$ (1%)
$BeBr_2$	490	520	—	$BeBr_2(BeBr_2)_2$
$MgBr_2$	711	—	570	$MgBr_2$, $(MgBr_2)_2$ (1%)
SrO	2430	3000	1400–1500	SrO, Sr (trace)
ScF_3	—	—	—	ScF_3 (D_{3h} geometry)
$ScCl_3$	939	—	—	$ScCl_3$, $(ScCl_3)_2$
$TiCl_4$	−25	136	—	$TiCl_4(TiCl_4)_2$
TiO_2	1830	2500–3000	—	Ti_3O_5, Ti_4O_7, TiO, TiO_2
TiO	1750	>3000	—	TiO(mainly), Ti_2O, Ti_2O_3, Ti_3O_5, Ti_4O_7, Ti_5O_9, $Ti_{10}O_{19}$, TiO_2, Ti
TiS	—	—	1900	TiS, Ti, S_2, S
VCl_4	−28	149	—	VCl_4
VS	d	—	—	VS, V, S
MnF_2	856	—	—	MnF_2, Mn_2F_5
MnF_3	d	—	—	MnF_3, Mn_2F_5
YF_3	1387	—	—	YF_3(C_{3v} geometry), YF_2 (C_{2v} geometry)
YO	2410	—	—	YO
ZrF_4	—	subl 903	600	ZrF_2, ZrF_3, ZrF_4
ZrO_2	2700	5000	—	ZrO, ZrO_2
ZrS_2	1550	—	1400	Zr_2S_5, S_2
NbF_5	72	236	—	Nb_3F_{15}, NbF_5
NbO	—	—	1630	Nb, NbO, NbO_2
MoO_3	795	subl 1155	—	Mo_3O_9, Mo_4O_{12}, Mo_5O_{15}
MoS_2	1185	subl 450	—	S_2
WO_3	1473	—	—	W_3O_9
ReO_2	d 1000	vap 350	—	Re_2O_7, ReO_3, $HReO_4$
ReO_3		vap 750	—	Re_2O_7, ReO_3, $HReO_4$
$FeBr_2$	d 684	—	—	$FeBr_2$, $(FeBr_2)_2$
FeI_2	red heat	—	—	FeI_2, $(FeI_2)_2$
Fe_2O_3	1565	—	—	O_2, lower oxides
$CoCl_2$	724	1049	—	$CoCl_2$, $(CoCl_2)_2$
CoO	1935	—	—	Co, O_2, CoO

TABLE 1.2 *(continued)*

Substance	mp (°C)	bp[c] (°C)	Vaporization temperature under vacuum	Vapor species and comments
$NiBr_2$	963	—	—	$NiBr_2$ (linear)
RhO_2	—	—	—	Rh, RhO, RhO_2
$PdBr_2$	d	—	—	$(PdBr_2)_4$, $(PdBr_2)_6$
PdO	870	—	—	Pd, PdO
$PtCl_2$	d 581	—	—	Pt_5Cl_{10}, Pt_6Cl_{12}, Pt_4Cl_8
CuCl	430	1490	—	CuCl, Cu_2Cl_2, Cu_3Cl_3
CuBr	492	1345	—	CuBr, Cu_2Br_2, Cu_3Br_3
CuF_2	d 950	—	—	CuF_2
Cu_2O	1235	d 1800	—	Cu, Cu_2O, CuO
AgCl	455	1550	—	AgCl, Ag_3Cl_2, $(AgCl)_3$
AgBr	432	1505	—	AgBr, $(AgBr)_3$
AgI	558	1506	—	AgI, Ag_3I_2, I, I_2
$ZnBr_2$	394	650	—	$ZnBr_2$
$CdCl_2$	568	960	—	$CdCl_2$, $(CdCl_2)_2$
CdS	—	subl 980	—	Cd, S_2
Hg_2Br_2	236	318	—	Hg, $HgBr_2$
HgO	d 500	—	—	Hg
BF	(BF_3 + B at 2000°C →)		—	BF, BF_3
BCl	(BCl_3 + B at 1100°C →		—	BCl, B_2Cl_4, BCl_3
B_2O_3	460	1860	—	B_2O_3
BS_2	—	—	550–1100	BS_2, $(BS_2)_2$, $(BS_2)_4$
B_2S_3	310	—	300–600	B_2S_3 polymers, BS_2, B_2S_2, B_6S_{12}, B_7S_{14}
AlF	(AlF_3 + Al at 800–1000°C →)		—	AlF, $(AlF)_2$
Al_2O_3	2020	2980	2000–4000	Al_2O_3, AlO, Al_2O, Al, O, O_2
GaF	(CaF_2 + Ga at 800–1300°C →)		—	GaF
Ga_2O	(Ga_2O_3 + Ga →)		—	Ga_2O
Ga_2S_3	1255	—	950	Ga_4S_5, S
GaAs	1238	—	700–900	Ga, As, As_2, As_4
GaP	—	—	741–953	Ga, P_2
InCl	225	608	—	InCl
In_2Cl_3	—	641	340–450	InCl, $InCl_3$
In_2O	(In_2O_3 + In at 600–950°C →)		—	In_2O
TlF	327	826	—	TlF, Tl_2F_2
Tl_2O	300	—	—	Tl_2O
CS[d]	−130	$\left(CS_2 \xrightarrow{\text{discharge}}\right)$ several other methods have also been used	—	CS, S, CS_2
SiF_2	(SiF_4 + Si at 1150°C →)		—	SiF_2, SiF_4
GeF_2	d 350	subl	—	GeF_2, $(GeF_2)_2$
GeO_2	1115	—	—	GeO, GeO_2, O

(continues)

TABLE 1.2 (*continued*)

Substance	mp (°C)	bp[c] (°C)	Vaporization temperature under vacuum	Vapor species and comments
SnS	880	1230	1000–1200	SnS, $(SnS)_2$
$PbBr_2$	373	1166	530–660	$PbBr_2$
PbO	888	1132	—	Pb_4O_4 favored, Pb_nO_n, where $n = 1$–6
Sb_2O_3	656	subl 1550	290–425	SbO, Sb_2O_2, Sb_3O_3, Sb_4O_4, Sb_2O_4, Sb_2O_5, Sb_2O_6
Bi_2O_3	820	1855	—	Bi_4O_4, Bi_4O_2, Bi_3O_3, Bi_2O_2, BiO, Bi, O_2
TeO	d 370	—	—	Te, O_2, TeO
LaF_3	—	—	—	LaF_3, La_2F_6
$YbCl_2$	702	1900	—	$YbCl_2$
ThO_2	3050	4400	1500–2000	—
UO_2	2500	subl 1400–2300	—	UO_2 mainly, U, UO

[a] Arranged by families of the periodic chart moving from left to right.
[b] Taken mainly from Klabunde [1].
[c] Atmospheric pressure.
[d] See Klabunde *et al.* [5] for further reading on the synthetic chemistry of CS.

A severe limitation in the formation of free molecules (particles) from these solids is the fact that high-temperature vaporization often leads to decomposition/disproportionation. Thus, for example, sublimation of MoO_3 solid yields vapor species Mo_3O_9, Mo_4O_{12}, Mo_5O_{15}, and others. Also, HfS_2 yields Hf_2S_5 and S_2, while RhO_2 yields Rh atoms, RhO, and RhO_2 [1]. On the other hand, many molecular solids can be smoothly vaporized. For example, TiO yields mainly TiO vapor [4] and $NiBr_2$ yields $NiBr_2$ vapor. Summaries of these processes are available [1]. Still, a great deal needs to be done regarding vaporization of molecular solids; so much is still not known.

Of course the structure of the solid and the strength of the bonding determines the ease with which vaporization and/or decomposition can occur. Generally high-melting-point/high-boiling-point materials have some ionic bonding which must be broken up. Bridging bonds also need to be broken:

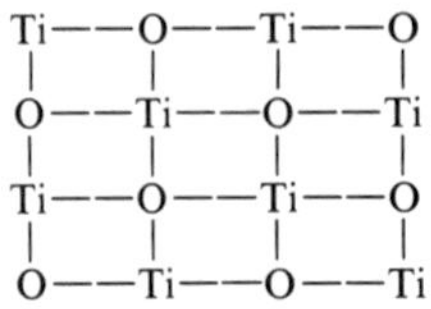

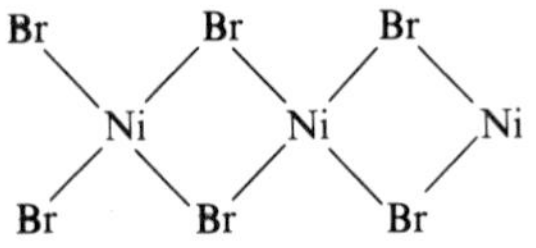

```
Ti——O——Ti——O          Br        Br        Br
|   |   |   |            \      /  \      /  \
O——Ti——O——Ti              Ni        Ni        Ni
|   |   |   |            /      \  /      \  /
Ti——O——Ti——O          Br        Br        Br
|   |   |   |
O——Ti——O——Ti
```

The reactivity of the resultant free particle TiO or $NiBr_2$ will depend on its readily available orbitals, low steric restrictions to reaction, and somewhat on the energy required to form it. Thus, TiO is difficult to form (vaporizes at 1700–2000°C under vacuum; the free TiO molecules are extremely reactive), while $NiBr_2$ readily sublimes at several hundred degrees and the resultant $NiBr_2$ particles have less tendency to react with added reagents, and usually repolymerize to $(NiBr_2)_n$ film. The heat of formation (or energy to vaporize) is a potential energy that puts the particle higher on the energy/reaction coordinate for a chemical reaction (see Fig. 1.2).

Although a more thorough summary is available in an earlier book [1], Table 1.2 summarizes the physical properties of the most important solid substances that have been or could be vaporized and the resultant free particles used as synthons of new compounds or as building blocks for films.

V. EXPERIMENTAL METHODS

The development of high-temperature species as synthons and film precursors has been heavily dependent on the availability of excellent vacuum technology and on high-temperature ceramics. Industrial technology in these areas is highly developed and improving rapidly. Technology is available and is being used for vaporization of large quantities of metals, metal oxides, metal sulfides, and other substances [6]. Techniques will be dealt with in some detail in later chapters. A brief summary here may be helpful, however.

A. Resistive Heating Vaporization

Electrical resistive heating is commonly employed in industry and in the laboratory. Many different crucibles or vaporization sources are available commercially [7]. Usually a tungsten coil is coated with Al_2O_3, and the resultant crucible is very useful for vaporizing Mn, Fe, Co, Ni, Pd, Cu, Ag, Au, Sn, Pb, $NiCl_2$, and many other materials. Another variation is a tungsten coil wrapped around an Al_2O_3 cup, which serves well for low-temperature metals such as Mg, Ca, Zn, and others. Sometimes tungsten, molybdenum, or tantalum boats are used to vaporize metals directly, and these work well for Ti, V, and Cr vaporizations. The limitations are the temperatures Al_2O_3 can withstand (~2000°C) or that tungsten can withstand without alloying with the evaporant. Other variations exist where materials can be vaporized downwards continuously using a wire feed [7, 8].

B. Electron-Beam Vaporization

A beam of electrons can be focused on the sample, and by using voltage differences between the filament emitting the electrons and the water-cooled hearth holding the sample, electrons can be passed through it causing heating and eventual vaporization [9]. The main advantages of this method are that it can be scaled up readily and, more importantly, high-boiling metals such as Pt, Rh, Mo, W, and U can be vaporized without materials problems, since the hearth can be kept cool. Semiconductor substances such as YO, TiO, ThF_3, TaO, SrF_2, Nd_2O_3, and Nb_2O_5 can also be vaporized this way. Both magnetically and electrically focused beams can be fashioned. Also, positive filament or positive hearth (reversed polarity from normal) e-beams can be employed; the positive hearth type is more difficult to engineer but minimizes stray parasitic electrons and x rays.

C. Laser Evaporation

Many metal vaporization studies using a variety of lasers have appeared. Koerner von Gustorf [10] used YAG (yttrium aluminum garnet doped with Nd^{3+}) lasers; CO_2 lasers are also usable but with less efficiency. Small-scale vaporizations can be done with ease and great control [11]. However, large-scale vaporizations are very destructive [12]. Two problems are encountered: molten metals behave as good mirrors for the laser beam and the window allowing the laser beam into the vacuum chamber can become coated with evaporant film. The advantages of laser evaporation are precise control of the vaporization "spot", continuous vaporization can be done either with a cw laser or with continuing laser bursts, and stray radiation is less of a problem than with electron beam methods.

D. Sputtering and Magnetron Sputtering

These methods will be covered in some detail later in this book. For the sake of definition, sputtering is vaporization of a material by bombarding it with high-energy atoms and ions of argon or other inert gases. The force of impact causes atoms or molecules to be ejected from the surface. Sputtering is very important in industry even though it has poor energy efficiency (5–10% compared with 35% for e-beams) [6]. Magnetron sputtering is dealt with in some detail in Chapter 7 and makes use of permanent magnets coupled with rf or dc sputtering.

E. Induction Heating

By using a conducting coil surrounding the sample, a magnetic field can be set up such that electron current in the sample is induced, thereby heating and eventually vaporizing the sample. Under some conditions the sample can even be levitated [13]. Induction heating is limited because of the coupling needed between the coil and the sample. When vacuum apparatus is required, it becomes difficult to position the coil and sample properly for effective coupling.

VI. USES OF FREE ATOMS AND PARTICLES IN CHEMICAL AND FILM-FORMATION PROCESSES

The high-temperature generation of free atoms and particles has been very useful in chemistry [1]. Low-temperature matrix isolation of these species for spectroscopic study has taught us a great deal about electronic states and

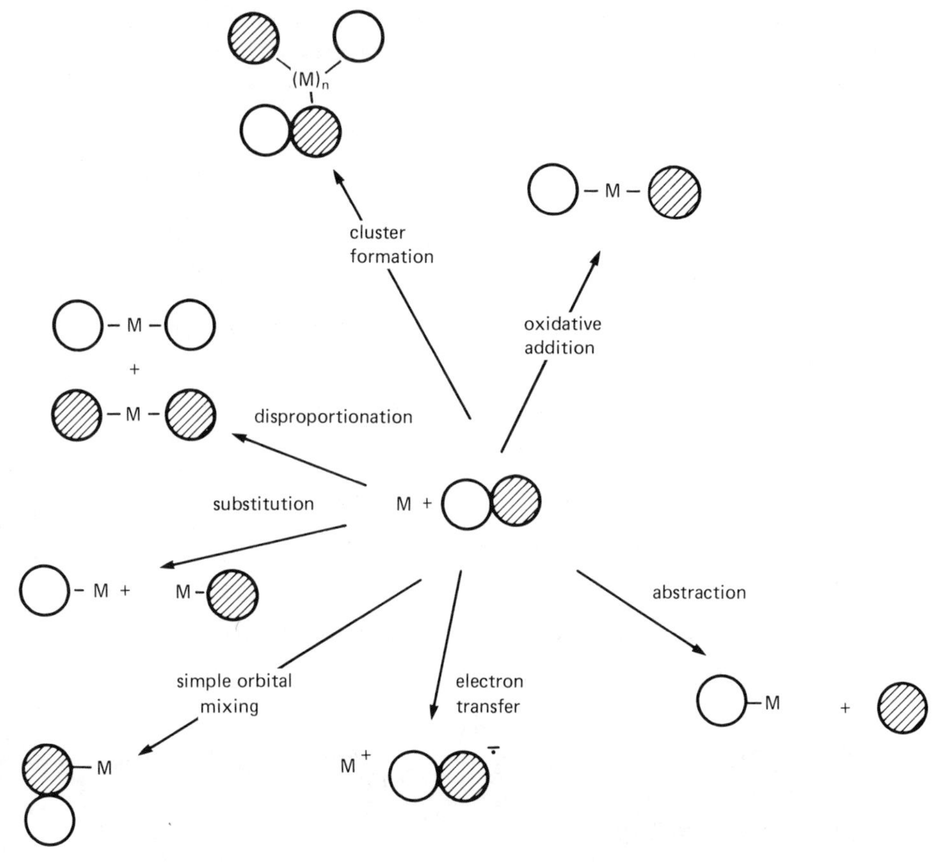

Scheme 1.1. Illustration of low-temperature reaction chemistry of free atoms; (M is a metal atom, and the open circles and cross-hatched circles are organic or inorganic molecules, respectively, with some functional group) [1, 7].

bonding in metal atoms, dimers, and trimers [14]. Particles such as metal halide molecules or fragments have also been frozen and spectroscopically analyzed [15]. Matrix isolation spectroscopy has also taught us much about the electronic/bonding properties of free atoms and particles. In Chapter 2 we will take a closer look at some of these studies.

Reaction chemistry of free atoms and particles has been extremely fruitful and has made "vapor synthesis" or the "metal atom/vapor technique" an established synthetic method in inorganic and organometallic chemistry [1]. The high reactivity of free atoms and particles allows low-temperature reaction chemistry to proceed smoothly, and many unusual new compounds have been prepared by combining metal atoms with organic molecules. These studies have not only led to new molecules but have also contributed to a more fundamental understanding of metal–organic interactions, which of course is important in catalysis and other surface chemistry phenomena.

Basically the low-temperature reaction chemistry of free atoms and particles can be illustrated by Scheme 1.1. Some classic chemical reactions are shown below as examples [1]:

Oxidative addition:

$$Ni + C_6F_5Br \longrightarrow C_6F_5{-}Ni{-}Br$$

$$SiF_2 + C_6F_6 \longrightarrow C_6F_5SiF_3$$

Electron transfer:

$$Li_{atom} + (NC)_2C{=}C(CN)_2 \longrightarrow Li^+(NC)_2C{=}(CN)_2^{\dot{-}}$$

Abstraction:

$$Ag_{atom} + CF_3I \longrightarrow AgI + CF_3$$

Simple orbital mixing:

$$Cr_{atom} + C_6H_6 \longrightarrow (C_6H_6)_2Cr$$

benzene sandwich compound

Substitution:

$$2Ag_{atom} + (CF_3)_2CFI \longrightarrow AgI + AgCF(CF_3)_2$$

Disproportionation and ligand transfer:

$$2Ni_{atom} + (CH_2{=}CHCH_2)_4Sn \longrightarrow H{-}C\begin{matrix}\diagup CH_2 \\ \diagdown CH_2\end{matrix}(-Ni-)\begin{matrix}CH_2\diagdown \\ CH_2\diagup\end{matrix}C{-}H + (Sn)_n$$

Cluster formation [16]:

$$Mg_{atom} + Mg_{atom} \longrightarrow Mg{-}Mg \xrightarrow{CH_3Br} CH_3MgMgBr$$

TABLE 1.3

Materials Normally Vaporized in Industry: Suggested Methods, Remarks, and Uses[a]

Material[b]	mp (°C)	Evaporation temperature	Evaporation method[c]	Remarks	Uses[d]
Al	660	—	Res W boat or wire, boron nitride crucibles	Use stranded W wire, alloys W, corrodes Al_2O_3	Films on plastics and other materials
Al_2O_3	2050	2000	EB or sputtering Res W boat	Recommend rf sputtering	Protective anti-reflective coatings
Sb	630	—	Res alumina crucible, Mo, Ta boats	Toxic, use external heater	—
Sb_2O_3	656	400–500	Res Pt	—	Multilayer coating
Sb_2S_3	550	300–500	Res Mo or Ta boats or wires	—	IR filters, photo-conductors
As	814	—	Res alumina crucible	Sublimes, toxic use external heater	—
B	2000	—	Res C boats or strips, EB	—	—
Ba	717	—	Res Mo, Ta, W boats	Does not alloy	—
Be	1284	—	Res W boat, EB	Alloys refractory metals, toxic	—
Bi	271	—	Res alumina crucible	Toxic, use external heater	—
Bi_2O_3	820	800–1000	Res Pt or alumina crucible	—	Heat-reflecting films support for Au films on glass
Cd	321	—	Res alumina crucible, Mo, Ta, W boats	Use external heater	—
CdS	1750	600–800	Res Mo, Ta, or C boats	—	IR filters, multi-layer coatings, photo-conductors, thin-film transistors

(*continues*)

TABLE 1.3 (*continued*)

Material[b]	mp (°C)	Evaporation temperature	Evaporation method[c]	Remarks	Uses[d]
CdSe	1350	500–700	Res Mo, Ta, or alumina crucible	—	IR filters, multi-layer coatings, photo-conductors, thin-film transistors
CdTe	1041	600–1000	Res Mo, Ta wires or boats	—	IR filters, multi-layer coatings, photo-conductors, thin-film transistors
Ca	810	—	Res alumina crucible	Use external heater	—
CaF_2	1360	1300–1500	Res Mo, Ta wire or boats	—	Multilayer coatings, insulating films in thin-layer circuits
C	3650	—	Carbon arc	—	—
CeF_3	1324	1200–1600	Res Mo, Ta, W wires or boats	—	Multilayer coatings
CeO_2	1950	—	EB, Res W boats, or alumina crucible	EB best; reacts with refractory source	Multilayer coatings
CsI	621	600–800	Res Mo, Ta	—	X-ray fluorescent screens
Cr	1900	—	Res W boat	—	—
CrO_3	196	1900–2000	Res W	—	Absorbent brown films for optical glasses
Co	1478	—	Res alumina crucible	—	—
Cb	2500	—	Res W boat	—	—
Cu	1083	—	Res alumina crucible, Mo, Ta, W boats	—	—
Ga	30	—	Res alumina crucible or coated boat	Use external heater	—
GaP	1348	—	Res W boat	—	—
Ge	959	—	Res alumina crucible, Mo, Ta, W boats	—	—
GeO_2	1115	—	Res Mo, Ta, W boats	—	—

TABLE 1.3 (*continued*)

Material[b]	mp (°C)	Evaporation temperature	Evaporation method[c]	Remarks	Uses[d]
Au	1063	—	Res alumina crucible or coated boat	—	—
In	157	—	Res Mo, W boat	—	—
In_2O_3	850[e]	—	Res Pt wires	—	Transparent heating elements on glass
Fe	1535	—	Res alumina crucibles, W boats	Sputters well	—
Fe_2O_3	1565	—	Res W boat	Reactively sputter	IR interference film and beam splitter
LaF_3		1200–1600	Res Mo, Ta	—	Multilayer coatings
Pb	328	—	Res alumina crucible, Mo, W boats	Use external heater, toxic	—
PbF_2	855	700–1000	Res Pt or alumina crucibles	—	Multilayer coatings
Li	179	—	Res Fe or quartz crucibles	Use external heater	—
LiF	870	800–1000	Res Mo, Ta EB	—	UV reflective films, multilayer coatings
Mg	651	—	Res W boats or alumina crucible	Use external heater	—
MgF_2	1395	—	Res Mo, W boats, C crucible	—	Antireflective films on glass, decorative coatings
MgO	2800	2000	EB	—	Multilayer coatings
Mn	1244	—	Res Mo, W boats or alumina crucible	—	—
Mo	2622	—	EB, Res Mo wire	Sputters well	—
Nd_2O_3	—	1600–2000	EB, Res W	—	Multilayer coatings
Nd_2O_5	1520	1400–1600	EB, Res W	—	Dielectric films, multilayer coatings

(*continues*)

TABLE 1.3 (*continued*)

Material[b]	mp (°C)	Evaporation temperature	Evaporation method[c]	Remarks	Uses[d]
Ni	1455	—	Res alumina crucible, W boats	Sputters very well	—
Ni/Cr	1360	—	Res alumina crucible, W boats	Deposits as chrome alloy	—
Pd	1555	—	Res alumina crucible, W boats	Alloys with refractory metals	—
Pt	1774	—	Res W filament or wire	Alloys and must be evaporated rapidly	—
Se	217	—	Res alumina crucible or W boat	Toxic	—
Se	1420	—	Res beryllium oxide or C crucible, EB	Can use rf sputtering	—
SiO_2	1800	1600	EB	—	Laser coatings
SiO	—	1200–1600	Res Mo, Ta wires or boats	—	Antireflective films, protective dielectric films, decorative coatings, electron microscopy specimens
Ag	961	—	Res alumina boats, Ta, Mo boats	—	—
NaF	1040	1000–1400	Res Mo, Ta	—	Multilayer coatings
SrF_2	1190	1000–1400	EB, Res W	—	Low index layers for IR films
Ta	2996	—	EB	Sputters well	—
Ta_2O_5	1470[e]	2000	EB	—	Dielectric films
Te	452	—	Res alumina crucible, W boat	Use external heater, toxic	—
Th	1827	—	Res W boat	Wets W	—
ThF_4	—	1000–1200	Res Mo, Ta	—	Laser coatings, multilayer coatings

TABLE 1.3 (*continued*)

Material[b]	mp (°C)	Evaporation temperature	Evaporation method[c]	Remarks	Uses[d]
Sn	232	—	Res alumina crucible, Mo, Ta boats	—	Wets Mo
SnO_2	1127[e]	—	Res W boat	—	Antistatic coatings and transparent heating elements
Ti	1727	—	Res W boat, EB	Deposits contain traces of W	—
TiO_2	1850	2200	EB	—	Multilayer and laser coatings
TiO	1750	1700–2000	EB, Res W	—	Antireflective films, decorative coatings, beam splitters
W	3382	—	Res W wire, EB	Sputters well	—
U	1132	—	EB, Res W wire or boat	Sputters well	—
V	1697	—	Res Mo, W boats	Evaporates just at molten state	—
Zn	419	—	Res alumina crucible	Use external heater	—
ZnS	1900	1000–1100	Res Mo, Ta or C crucible	—	Beam splitter, very hygroscopic
ZnSe	—	600–900	Res Mo, Ta	—	IR filters, color filters
Zr	2127	—	Res W	Deposits contain traces of W	—
ZrO_2	2700	2500	EB	—	Antireflective films, multilayer coatings

[a] From Klabunde [8]. Data for table originally from Feldman *et al.* [17], Baer [18], Adams [19], the Sylvania Emissive Products Catalog, and the Vacuum Evaporation Sources Catalog (published by R. D. Mathis Company).

[b] Arranged alphabetically according to name of compound or element (approximately).

[c] Res refers to resistive heating. Thus Res W would mean to vaporize the substance directly from a W wire, filament, or crucible. Res alumina crucible would mean an integral $W\text{–}Al_2O_3$-coated crucible. However, if under Remarks it states to use an external heater would mean an alumina crucible placed in a W wire heater. EB means electron-beam vaporization works best.

[d] Vapor-deposited metal films are generally used for decorative, protective, or electronics applications.

[e] Decomposes.

This last process, cluster formation, is the precursor of film formation and will be examined in more detail in Chapter 2.

In the absence of some added reactant, condensation of free atoms and particles does not lead to new chemical compounds but instead to polymers/films. With the availability of excellent high-vacuum technology and high-temperature sources, industry has made great use of this process. Table 1.3 lists some of the materials normally vaporized industrially for film preparation [8].

Actually Table 1.3 is only a sampling. The possibilities for further production of films are staggering. Remember that with the availability of vacuum technology, high-temperature sources, electron beams, lasers, sputtering apparatus, and induction heating, experimental difficulties will be largely overcome. Our only limitations will be on synthesis of new vaporizable solids and applications of the resultant films.

In conclusion, the use of free atoms and particles for film formation could be exploited considerably more. The film-forming process (the polymerization of the reactive atom or particle) has a low activation energy and can be carried out at low temperature. However, controlling the polymerization to yield special films, perhaps expitaxial films, presents some problems. Thus, it is worthwhile to look in some detail at the embryonic stages of clustering of free atoms and particles. Chapter 2 deals with these topics.

REFERENCES

1. K. J. Klabunde, "Chemistry of Free Atoms and Particles," pp. 2, 5, Academic Press, New York, 1980.
2. P. L. Timms, *Adv. Inorg. Chem. Radiochem.* **14**, 121 (1972).
3. S. Dushman, *in* "Scientific Foundations of Vacuum Technique" (J. M. Lafferty, ed.), p. 691, Wiley, New York, 1962; S. Dushman, "Vacuum Technique," p. 745, Wiley, New York, 1949.
4. H. Y. Wu and P. G. Wahlbeck, *J. Chem. Phys.* **56**, 4534 (1972); P. W. Gilles, K. D. Carlson, H. F. Franzen, and P. G. Wahlbeck, *J. Chem. Phys.* **46**, 2461 (1967).
5. K. J. Klabunde, M. P. Kramer, A. Senning, and E. K. Moltzen, *J. Am. Chem. Soc.* **106**, 263 (1984); E. K. Moltzen, A. Senning, M. P. Kramer, and K. J. Klabunde, *J. Org. Chem.* **49**, 3854 (1984); A. Krebs, A. Gunter, A. Senning, E. K. Moltzen, K. J. Klabunde, and M. P. Kramer, *Angew. Chem.* **96**, 691 (1984).
6. W. Reichelt, *Angew. Chem., Int. Ed. Engl.* **14**, 218 (1975).
7. K. J. Klabunde, *Chem. Tech.* **6**, 624 (1975).
8. K. J. Klabunde, *in* "Reactive Intermediates," Vol. 1 (R. Abramovitch, ed.), pp. 42, 46, 47, 49, 65, Plenum, New York, 1979.
9. P. L. Timms, *in* "Cryochemistry" (M. Moskovits and G. Ozin, eds.), p. 61, Wiley (Interscience), New York, 1976.
10. E. A. Korner von Gustorf, O. Jaenicke, D. O. Wolfbeis, and C. R. Eady, *Angew. Chem., Int. Ed. Engl.* **14**, 278 (1975).

11. D. E. Powers, S. G. Hansen, M. E. Guesic, A. C. Puiu, J. B. Hopkins, T. G. Dietz, M. A. Duncan, P. R. Langridge-Smith, and R. E. Smalley, *J. Phys. Chem.* **86**, 2556 (1982).
12. J. L. Dumas, *Rev. Phys. Appl.* **5**, 795 (1970); M. Burden and P. A. Walley, *Vacuum* **19**, 397 (1969).
13. J. L. Margrave, *High Temp.–High Pressures* **2**, 583–586 (1970).
14. M. Moskovits and G. Ozin, *in* "Cryochemistry" (M. Moskovits and G. Ozin, eds.), p. 261, Wiley (Interscience), New York, 1976.
15. R. H. Hauge, S. E. Grandsen, and J. L. Margrave, *J. Chem. Soc., Dalton Trans.* 745 (1979).
16. Y. Imizu and K. J. Klabunde, *Inorg. Chem.* **23**, 3602 (1984).
17. J. Feldman, M. Friz, and F. Stetter, *Res. Dev.* **27**, 49 (1976).
18. C. A. Baer, *Res. Dev.* **25**, 51 (1974).
19. F. Adams, *Res. Dev.* **26**, 45 (1975).

2 CLUSTERING OF FREE ATOMS AND PARTICLES: POLYMERIZATION AND THE BEGINNING OF FILM GROWTH

George C. Nieman

Department of Chemistry
Monmouth College
Monmouth, Illinois

Kenneth J. Klabunde

Department of Chemistry
Kansas State University
Manhattan, Kansas

I. HISTORICAL CONSIDERATIONS

A. Evaporative Coating

One of the important observations of early chemists was that metals could be vaporized in a vacuum. No practical applications were made of this phenomena until Soddy [1] proposed the vaporization of calcium as a method of reducing the residual pressure in a sealed tube. This was the first example of clean-up or gettering, and after considerable study and development it was adopted by the newly emerging electronics industry. A number of metals exhibit this gettering action, which occurs by both chemical reactions of the residual gas with the metal atoms and by adsorption of the gas on the continuously forming metal film [2].

Vacuum evaporation of a metal to form a reflective coating on glass was first reported in 1912 [3]. With improvements in vacuum technology, evaporative coating technology underwent rapid development and by 1940 was a standard method of optical coating [4] and largely displaced chemically deposited silver films. For example an aluminized mirror could be produced and the Al film protected by further evaporative coating of SiO_2, SiO, or beryl on the top. The easily controlled deposition rates and film thicknesses have led to the use of multilayer coatings of different materials for lasers, interference filters, reflectance and antireflectance coatings, camera lenses, silicon chips, and a host of other materials.

In 1952 Auwärter introduced reactive evaporation as a clever extension of evaporative coating [5]. In this procedure a lower boiling substance is evaporated and upon condensation to form a film the molecules/film are changed to another form. For example, evaporation of SiO to form a film of SiO in the presence of oxygen (discharge activated) allows the growing film to be that of SiO_2

$$\mathrm{SiO + O \longrightarrow SiO_2\ film.}$$

Further developments on the laboratory and industrial scale have been reported by Heitmann [6] and Zollinger [7]. Another approach in reactive evaporation is to allow the sample to chemically react with the crucible to yield a new high-temperature species [5]. For example SiO_2 heated in a metal (M) crucible will yield MO and SiO vapor. Reactive evaporation has been used in film preparations of Mo, Ta, W, C, ThO_2, BeO, ZnO_2, Al_2O_3, MgO, TiO_2, SiO_2, and SiO [5].

Evaporative coating by straight evaporation/deposition and by reactive evaporation has become a major industry. Reichelt [8] discusses the production and condensation of metal vapors in large quantities, which sometimes approach a rate of kilograms per hour, by using electron-beam evaporation technology. He points out that these practical applications require a precise knowledge of evaporation rates at various temperatures, which can be calculated according to the equation of Hertz

$$G = \alpha ps\sqrt{M/(2\pi RT)},$$

where G is the evaporation rate in grams per square centimeters per second, ps the saturation vapor pressure, and α the Langmuir coefficient (approximately unity for most metals evaporated from the melt). Reichelt [8] also discusses some of the practical problems encountered in electron-beam evaporations including crucible materials, power consumption and efficiency, and problems with parasitic electrons and x rays. A comparison of electron-beam evaporation and sputtering is made. Currently sputtering is less power efficient (5–10%) than e-beams (35%) and so does not allow such large quantities to be evaporated per unit time (sputtering 500 g/h; e-beam evaporation >30 kg/h).

B. Chemical Vapor Deposition Compared with Evaporative Coating

In recent years chemical vapor deposition (CVD) has become important commercially. In this method volatile metal compounds [for example, SiH_4 or $Ga(CH_3)_3$] are passed over hot surfaces and the compounds are decomposed leaving a deposited film

$$\underset{\text{gas}}{SiH_4} \xrightarrow[\text{substrate}]{\text{hot}} \underset{\text{solid}}{Si} + 2H_2,$$

$$\underset{\text{gas}}{2GeI_2} \longrightarrow \underset{\text{solid}}{Ge} + \underset{\text{gas}}{GeI_4}$$

$$\underset{\text{gas}}{Ni(CO)_4} \longrightarrow \underset{\text{solid}}{Ni} + 4CO.$$

Sometimes combinations of gases are used

$$\underset{\text{gas}}{SiCl_4 + 2H_2} \longrightarrow \underset{\text{solid}}{Si} + \underset{\text{gas}}{4HCl},$$

$$\underset{\text{gas}}{TiCl_4} + \underset{\text{gas}}{2H_2O} \longrightarrow \underset{\text{solid}}{TiO_2} + \underset{\text{gas}}{4HCl},$$

$$\underset{\text{gas}}{Ga(CH_3)_3} + \underset{\text{gas}}{AsH_3} \longrightarrow \underset{\text{solid}}{GaAs} + 3CH_4.$$

Pyrolytic CVD reactions take place mainly at atmospheric pressure or in slight vacuum. Sometimes a carrier gas, such as hydrogen, is used. Film growth rates on unusually shaped objects occur at more than 100 Å/s at 1000°C [5]. This compares favorably with evaporative coating processes when strict tolerances are needed. Although the growth rate for evaporative coating is an order of magnitude slower, it has the important advantage of control over the substrate temperature. In fact, many substrates must be kept at low temperatures especially for optical and electronic applications. A disadvantage of evaporative coating compared with CVD is, of course, the need for good vacuum (10^{-3}–10^{-7} mbar).

C. Initial Steps in Film Growth

A comparison of mechanisms of film growth for CVD versus evaporative coating brings us to the main topic of this chapter. In the case of CVD it is still not clear whether free metal atoms are involved [9, 10]. In most cases it appears that the sequence is (1) surface adsorption, (2) pyrolytic loss of ligands (R), and (3) atom deposition. Migration may take place during steps (2) or (3):

$$PH_3 + GaR_3 + \sim\sim \longrightarrow \underset{\sim\sim\sim}{GaR_3\ PH_3} \xrightarrow[\text{migration}]{-R\cdot \text{ and } -H\cdot}$$

$$R_2Ga{-}PH_2 \xrightarrow[-2R\cdot]{-2H\cdot} \longrightarrow Ga{-}P.$$

It is possible that the presence of one or more deposited metal atoms aids subsequent deposition of more metal atoms. For example, GaP formed on the surface was found to catalyze the unimolecular decomposition of $Ga(CH_3)_3$ and PH_3 [10]. It is believed that $Ga(CH_3)_3$ and AsH_3 behave similarly;

$$\underset{\sim\sim}{GaP} + Ga(CH_3)_3 \xrightarrow{-\cdot CH_3} \begin{matrix} Ga(CH_3)_2 \\ \underset{\sim\sim\sim}{Ga{-}P} \end{matrix} \xrightarrow{-2\cdot CH_3} \begin{matrix} Ga \\ \setminus \\ \underset{\sim\sim\sim}{Ga{-}P}. \end{matrix}$$

At these early stages the CVD process resembles the microcluster growth process that occurs in evaporative coating:

$$\sim\sim + \text{Ga atom} \longrightarrow \underset{\sim\sim}{Ga} \xrightarrow{\text{Ga atom}} \xrightarrow{\text{Ga atom}} \begin{matrix} Ga \\ \setminus \\ \underset{\sim\sim\sim}{Ga{-}Ga}. \end{matrix}$$

At this stage of microcluster growth many questions arise. Do the clusters grow in two dimensions or three? Do many microclusters grow simultaneously on the surface? Do the microclusters migrate? Is the process reversible? At what size do the microclusters become metallic? These questions and the embryonic growth of microclusters will be considered in the next section. Theoretical and experimental studies will be included.

II. MICROCLUSTERS—THE BEGINNING OF FILM GROWTH BY THE CLUSTERING OF FREE ATOMS

A. Introduction

The general topic of microclusters has received much attention in the literature [11–16]. Much of this interest centers around the fact that these particles bridge the gap between the chemists' detailed knowledge of small molecular systems and the physicists' understanding of the collective behavior of matter. For our purposes we will define microclusters to be collections of atoms (or molecules) containing at least three but not more than about a million individual atoms (or molecules) that have a collective existence. For the most part we will be concerned with relatively small, homonuclear, metallic clusters, i.e., clusters containing fewer than 100 identical metal atoms.

The study of microclusters spans many time periods and many fields. In the mid-19th century, Michael Faraday studied gold colloids and his preparative methods are still used today [17]. Richard Zsigmondy received the 1925 Nobel Prize in Chemistry for his study of colloidal solutions [18]. Today, the use of microclusters as models for heterogeneous catalysts is among the most active areas of research [19]. Many industrially important catalysts consist of active metal particles in the microcluster size range, most often on some inactive support. Much more will be said about this topic in this and other chapters in this book. Among other numerous applications of nucleation and cluster growth theory are studies of thin films and coatings, radiation damage, latent image development in photographic films, colloids, weather and cloud formation, particulates and smoke, and the theory of phase transitions [20].

The study of clusters and their chemistry and physics can be classified by either their composition or their environment. Thus, there are clusters composed of rare-gas atoms, alkali-metal atoms, transition-metal atoms, small molecules, or ionic species. The nature of the bonding is quite different in each case. Likewise, there are clusters in molecular beams, in cold rare-gas matrices, on solid supports, as colloids in solution, or as cluster compounds. Here the interactions with the environment are quite different.

In this section we shall consider general comments about surface/volume ratios and thermodynamics followed by a section on theoretical studies of cluster growth. We shall then investigate the experimental evidence from studies of molecular beams and matrix isolation experiments.

B. Surface/Volume Ratio and Thermodynamics

One of the interesting characteristics of clusters that makes their study difficult is that clusters have a large fraction of their atoms on the surface. In order to get a perspective for the small size of clusters and the relationship between the number of atoms and the linear, surface, and volume dimensions, let us consider a simplified model. While the details are obviously incorrect, the general conclusions are good to well within a factor of two, especially for the larger clusters.

Consider that clusters are spherical bodies of radius R that are collections of N identical spherical atoms of radius r. Further, let us neglect the details of the packing and the packing fraction (which is 0.87 for the closest packed structures). With these assumptions we have $V = \frac{4}{3}\pi R^3 = Nv = \frac{4}{3}N\pi r^3$, where V and v are the volumes of single clusters and atoms, respectively. At this level of approximation $N = (R/r)^3$ or $R = N^{1/3}r$.

By using these same approximations, we can also estimate the number of surface atoms N_s by projecting the cross-sectional area of atoms onto the outer sphere. Thus, $A = 4\pi R^2 = N_s \pi r^2$ and $N_s = 4(R/r)^2 = 4N^{2/3}$. The fraction of surface atoms is then given by $N_s/N = 4N^{-1/3}$. Example calculations are summarized for several numbers of atoms assuming an atomic radius of 0.1 nm (1 Å) in Table 2.1.

In spite of the rather crude approximations used in deriving the numbers in Table 2.1, the values provide a reasonably accurate picture. For example, consider the 7th Mackay icosahedra (discussed later) which contain a total of 923 atoms of which 362 are on the surface [14]. This nearly spherical structure has a measured N_s/N value of 0.39 as compared to an estimated N_s/N value of 0.41. Thus, even for clusters of 1000 atoms approximately 40% of them are surface atoms and "surface effects" are bound to be large.

In the realm of microclusters several standard thermodynamic variables, such as surface tension, lose their meaning. Nevertheless, it is instructive to look at this macroscopic view, if only briefly [21]. The fundamental difficulty is that classical thermodynamics would seem to preclude the formation of

TABLE 2.1

Estimates of Size Parameters for Microclusters[a]

Number of atoms N	Radius R (nm)	Fraction of surface atoms N_s/N	Vapor pressure enhancement P/P_0
10^2	0.464	0.86	9.6
10^3	1.00	0.40	2.9
10^4	2.15	0.19	1.6
10^5	4.64	0.086	1.3
10^6	10.0	0.040	1.1

[a] See the text for a description of the model used.

nuclei and their subsequent growth. They have too high a chemical potential to ever exist!

Consider a liquid–gas interface or surface where the atoms experience an attractive force towards the interior of the liquid without a counter balancing force from the gas phase. This imbalance leads to the concept of surface tension γ, which exerts a force in opposition to attempts to increase the surface area. Now, if the surface is curved with a radius of curvature r, we can derive the Young–Laplace equation for the pressure differential across the interface

$$P_{\mathrm{in}} - P_{\mathrm{ex}} = 2\gamma/r,$$

where P_{in} and P_{ex} are, respectively, the internal (toward the center of curvature) and external pressures. Thus, since $\gamma > 0$, the internal pressure of a liquid drop is higher than the external pressure, i.e., the internal atoms are compressed. It is interesting to note that the inter-atomic distances (bond lengths) measured for many clusters decrease with decreasing cluster size.

Over 100 years ago Kelvin treated the problem of the vapor pressure of small drops. The Kelvin equation

$$\ln(P/P_0) = 2M\gamma/(RT\rho r)$$

gives the natural logarithm of the ratio of the equilibrium vapor pressure P of a drop to that of the flat liquid P_0 in terms of the molecular weight M, the surface tension γ, the gas constant R, the absolute temperature T, the density ρ, and the drop radius r. The last column of Table 2.1 lists the vapor pressure enhancement for water at 25°C for each size drop as an illustration of the size of this effect. While these numbers should only be considered as examples of what to expect, we can see that a 100-molecule water droplet might be expected to have a vapor pressure about 10 times the equilibrium vapor pressure of a flat surface. Thus, the droplet would evaporate quickly, that is, if it ever formed in the first place. Similar arguments apply to interfaces between other phases. Clearly applying bulk properties, such as surface tension, to microclusters leads us into trouble quickly.

The correct way to calculate the surface free energy and entropy of such microcluster systems is still a matter of investigation [22, 23]. Recent calculations on liquid metals indicate that the surface atoms are significantly more ordered than those below the surface [24]. Even the size at which the concepts of surface tension and surface free energy are no longer valid ideas is a matter of debate. Many present calculations are limited to calculating the energy effects at 0 K, and thereby they avoid some of these difficulties.

C. Theoretical Studies

There are several approaches to the theory of cluster growth, which we will look at in turn. We can consider a macroscopic kinetics approach and solve the rate equations for growth of either free clusters or on extended surfaces. We may also perform Monte Carlo or molecular dynamics calculations. Finally we may perform *ab initio* or parametrized calculations of cluster structure.

1. Kinetics

The various detailed reaction mechanisms proposed to explain the kinetics of cluster and surface growth share several features in common which are outlined below [25–27]. The usual kinetics approach is to consider the following sort of mechanism:

$$A_n + A_i + M \longrightarrow A_{n+i} + M, \quad K_{g,n}, \tag{1}$$

$$A_n \longrightarrow A_{n-1} + A_1, \quad K_{e,n}. \tag{2}$$

Equation (1) represents the growth of a cluster of size n by aggregation of a cluster of size i. It is usually assumed that $i = 1$, i.e., the growth is by accretion of monomers. The involvement of a third body M is necessary to remove the energy of condensation, otherwise there is always sufficient energy to evaporate an atom from the newly formed cluster. This is especially important for very small clusters. For many cluster sizes there may well be a time delay between this energy removal collision and the formation collision because the excess energy of condensation is rapidly distributed among the internal modes of the cluster. Equation (2) represents the spontaneous evaporation of monomers from the surface of a cluster. Energetics would indicate that monomers are the only particles that easily leave a homonuclear cluster.

Making the usual approximation that $i = 1$ (growth by monomers), we have for the rate of change of n-size clusters

$$\begin{aligned} d[A_n]/dt = {} & K_{g,n-1}[A_{n-1}][A_1][M] + K_{e,n+1}[A_{n+1}] \\ & -K_{g,n}[A_n][A_1][M] - K_{e,n}[A_n]. \end{aligned} \tag{3}$$

This set of coupled differential equations (one for each n) must now be solved. We may wish to apply the steady-state approximation or direct integration, but the solution is made difficult by the coupling among the differential equations. For larger clusters, in which case it is possible to ignore the third-body nature of the growth terms, we have growth by means of binary processes and decay by means of unimolecular processes. Hoare has recently treated this sort of quadratic transport problem in detail [28].

Numerous schemes have been used to estimate the various rate constants in Eq. (3). The rate constants for growth K_g are often estimated by consideration of collision probabilities derived from the kinetic theory of gases with unitary sticking coefficients [29]. In the case of cluster growth on extended surfaces, account must be taken for migration over the surface in addition to gas-phase impingement, since the former is often the faster and more important process [27]. The rate constants for evaporation K_e are often estimated from the growth constant and bulk free energy (vapor pressure) measurement assuming equilibrium conditions [25]. On extended surfaces the evaporation models attempt to take into account the number of neighboring like atoms.

These various models usually give rise to the notion of a critical-size cluster. Clusters below this critical size decay faster than they grow. We are dependent upon statistical fluctuations to get past this critical size barrier. In the case of surface adsorption we often introduce the idea of a roughening transition, below which growth is slow [30]. A common result of most of these models is that the population of clusters decreases exponentially with number of atoms for free clusters. Most experimental studies of cluster-size distributions confirm these predictions.

Some recent measurements on molecular (CO_2) clusters indicate that growth by coagulation of larger clusters rather than by monomer aggregation dominates for clusters larger than about 100 individual units [31]. This leads to a Gaussian distribution in the logarithm of the cluster size. The monomer population remains high, but intermediate-sized clusters disappear. This result indicates that evaporation of monomers dominates cluster cooling and that the exchange of monomers provides a means of energy (velocity) exchange among clusters. Growth by coagulation is energetically favorable since the energy released upon coagulation of an i-sized cluster on a j-sized one is significantly less than the energy released by the aggregation of i individual particles.

The fate of the heat of condensation has been studied in theoretical models by Soler and García [32]. The internal and translational temperatures are not necessarily in thermal equilibrium. Small clusters may well be "liquid" because of this extra energy and the fact that the melting point decreases with cluster size. Also, the lack of third-body, energy-removing collisions in the case of interstellar grains has led to speculation about some interesting chemistry in space [33].

2. Electronic Structure

While the kinetics approach attempts to give size distributions, other computer-based approaches attempt to describe the detailed structure of individual clusters. The usual approach taken by chemists is to construct the cluster from individual atoms and to minimize the energy as a function of shape. In this way the evolution from small clusters toward the bulk is unraveled [13, 34]. In contrast, the physicists' approach is to start with a band-type description of the bulk phase and to investigate how this breaks down as the size gets smaller [16, 35].

In the latter case a primary goal is often the calculation of the density of states function of metals, especially near the Fermi level. Early in this century considerable progress was made in describing the color of colloids, such as gold sols, by means of Mie's theory and the idea of plasma resonances [16]. More recently, plasmons have been invoked in an attempt to explain the surface-enhanced Raman effect. The calculation of the electronic properties of small particles becomes complicated because the spacing between adjacent levels may become large compared to kT, and classical continuum models break down. Thus, using Frohlich's model of a free electron gas in a cube with 10-nm sides, we have a level spacing corresponding to about 40 K at the Fermi energy. While such a simple model greatly overemphasizes the spacing, it is clear that quantum mechanical effects are very important when considering the electronic band structure of small metal clusters. The magnetic and superconducting properties of small clusters are also of considerable interest.

For many metals the location of the d and s bands and their degree of overlap is used as a guide to the accuracy of the calculations. In the bulk phase of most transition metals the s band is contained within the d band. Calculations using the $X\alpha$–SW approach show this sort of behavior with copper clusters as small as 8 atoms. However, Hartree–Fock calculations on the same copper clusters show completely distinct s and d bands for 13-atom clusters, as do other methods [36]. Thus, the $X\alpha$ results may be an artifact of the method.

Because of computational limitations, *ab initio* calculations for metals have for the most part been limited to very small "clusters", i.e., dimers and trimers. The most extensive calculations have been done on lithium. The assumption that the core electrons can be treated by using a pseudopotential allows the extension of these calculations to the other low-atomic-number metals [37]. Recently calculations using a pseudopotential core have been done on several transition and coinage metal clusters (Ni, Pd, Cu, Ag) of up to approximately 10 atoms [13]. While these calculations often assume a fixed geometry (and thus do not give structural information), they do allow us to estimate various electronic energies.

Calculation of the binding energy indicates a significant increase in binding energy per atom with cluster size for metals. Calculations have been carried out on metals such as lithium and copper for 13-atom or smaller clusters and there is a nearly linear relationship between number of atoms and binding energy. Although the absolute energies are very dependent on the basis set and calculation procedures used, the dimer would appear to have a binding energy of about one-quarter the bulk value and the 13-atom cluster has roughly two-thirds the bulk value [36].

The calculation of ionization potentials provide a good test of theoretical models since they can be measured by various photoemission techniques. In general, there is a decrease by roughly a factor of two from the ionization potential of the atom to the work function of the bulk metal. The decrease is not monotonic but depends very much upon the cluster geometry. There is also an odd–even alternation with the odd-atom clusters having a lower ionization potential, presumably because they are odd electron systems as compared to closed-shell structures for the even electron systems [38]. More will be said about the experimental measurements of bonding energies and ionization potentials in the experimental section.

3. Geometrical Structure

In addition to calculation of the electronic properties of metal clusters, there have been numerous calculations of the geometrical shape of clusters. These calculations have spanned the complete range of bond types including van der Waals clusters of rare gases, ionic clusters of salts, and metal clusters. Somewhat surprisingly, the rare gases and the metals often are predicted to have similar shapes perhaps reflecting the nondirectionality of the binding forces. The more directional ionic forces, caused by the presence of both positive and negative local charges, lead to more differentiated structures. A common goal of many of these calculations is to try to predict the bulk three-dimensional crystal structure. In general, this goal has not been met as the small clusters often have quite different structures from the bulk and the transition to bulk geometries occurs only gradually and for rather large clusters (>500 atoms or particles). As an example of the difficulty, most calculations for rare gases (and metals) predict clusters with fivefold symmetry while it is commonly known that no extended three-dimensional structure may have fivefold symmetry. Let us look at these calculations organized by cluster type.

One of the more complete sets of structural calculations for rare gases has been performed by Hoare [14]. He considers various pair potentials, especially the Lennard–Jones, and minimizes the energy as the coordinates are allowed to change. Such molecular dynamics calculations clearly show a lower energy for tetrahedral groupings of atoms as compared to octahedral structures. The 13-atom case is most interesting because it represents the smallest structure that can have an internal atom, i.e., one that is not on the surface. Two of the most important 13-atom structures, the cuboctahedron and the icosahedron, are shown in Fig. 2.1. The cuboctahedron is derived from a face-centered cubic (fcc, closest packed) structure and may be pictured

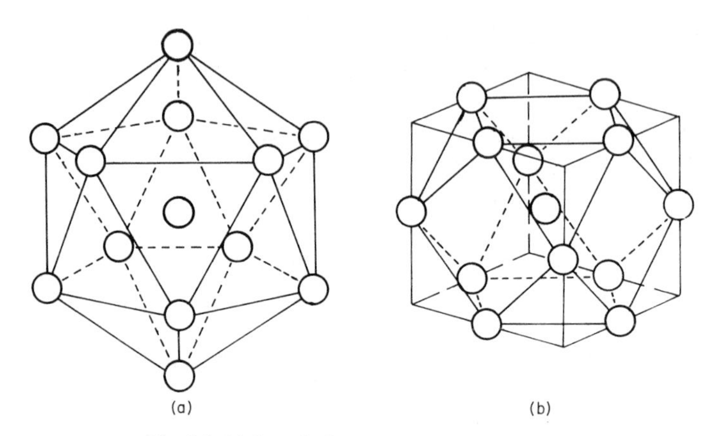

Fig. 2.1 (a) Icosahedron and (b) cuboctahedron.

as a central atom in a cube surrounded by 12 equivalent atoms at the centers of each edge. This figure has 8 triangular faces and 6 square faces. The icosahedron consists of a central atom surrounded by layers of 5 atoms each above and below. Each of these layers has in turn a central atom capping the figure. All 20 faces are triangular and all 12 vertices have fivefold symmetry. This leads to a more closely packed surface for the icosahedron than for the cuboctahedron. The icosahedron can be constructed from twenty tetrahedral figures packed so that they each share three faces with only minimal distortion (the dihedral angle of a tetrahedron is 70.53° as compared to 72° for the pentagonal angles). The icosahedral structure is dynamically the most stable 13-atom cluster. Inclusion of small three-center forces does not change this conclusion [39].

There exist some 988 distinct minimal 13-atom structures in these dynamic calculations. Thus we ought to anticipate much amorphous and/or fluctional character to these small clusters. We might even wish to ask whether they are solid or liquid. For larger clusters there is experimental evidence that the melting point decreases significantly as the cluster gets smaller.

Various experiments involving the condensation of gases show the presence of "magic numbers," i.e., clusters with certain numbers of atoms are significantly more prevalent than others [40, 41]. Often these magic numbers correspond in size to the nearly spherical Mackay icosahedra (1, 13, 55, 147, 309, 561, 932, . . .). Continuous deformations of these structures can lead to fcc cuboctrahedral structures. These deformations involve the transformation of sites of tetrahedral symmetry to sites of octahedral symmetry.

In addition to the problem represented by the fact that icosahedra (tetrahedra) cannot form extended three-dimensional structures, the Mackay icosahedra do not grow smoothly from one into another by simply placing new atoms in the centers of each triangular face. Such a growth pattern by the formation of caps might be expected to be energy efficient and does lead to other magic numbers that are experimentally seen. For the larger clusters a high-symmetry geometry was assumed and the evolution, if any, of the cluster with time was followed.

While these calculations using pair potentials can be extended to clusters of several hundred atoms, they are only appropriate to systems, such as the rare gases, in which non-nearest-neighbor interactions can be neglected. This is clearly not the case for metal clusters in which valence electrons move readily from atom to atom and electronic band structures are important. Because of computational difficulties most *ab initio* calculations start with an assumed geometry and look for electronic properties. Early calculations on lithium clusters seem to favor linear chains over planar or three-dimensional structures [13, 38]. However, inclusion of 2p orbitals seems to destroy this preference for linear structures [37].

There have been some attempts to use CNDO and extended Hückel calculations to address the geometry issue. Again linear structures appear to be more stable for many metals, such as silver. The relative stability of icosahedral versus cuboctahedral structures would appear to depend on the d-orbital occupation. Thus, palladium with a $d^{9.2}s^{0.8}$ configuration prefers an icosahedral structure whereas silver with a $d^{10}s^{1}$ configuration is predicted to be cuboctahedral [42]. Qualitatively these results may be understood in terms of the more closely packed surface of the icosahedral structure. This leads to a larger d-orbital bandwidth and to a higher energy for a completely filled d band. As with the pair potential calculations the energy does not vary strongly with geometry and many structures are calculated to be nearly isoenergetic.

Before leaving the topic of calculated geometry we should mention some work with ionic clusters, such as sodium chloride [43]. Again linear chains are predicted to be most stable for many smaller clusters. There are also many stable three-dimensional structures involving stacked rings of 6 ions $[(NaCl)_3]$. Ionization of these alkali halides usually results in the loss of a neutral halogen atom as well as loss of the electron since the halogen atom is only weakly bound. The stabilities of various $M_nX_{n-1}^{+}$ clusters have also been calculated.

D. Experimental Results

The experimental study of microclusters can be conveniently broken down into three categories: molecular beams, matrix isolation, and clusters on surfaces. There is an increasing interaction with the environment in going from molecular beams, which should reflect the properties of the free particles, to clusters on surfaces, which may show very strong interactions between the cluster and the surface (support).

1. Molecular Beams

The mass analysis of an effusive beam of metal particles from a Knudsen cell allows the calculation of equilibrium constants for the formation of dimers, trimers, etc. [44]. Performing similar studies over a wide range of temperatures then allows us to obtain the complete set of thermodynamic quantities: free energy, enthalpy, and entropy of formation. Unfortunately, the multimer concentrations are typically so small that this technique is limited to an analysis of the thermodynamics of only dimers except in favorable cases. Nevertheless, these studies provide a convenient check for spectroscopic measurements of bond energies.

In recent years there have been numerous studies using supersonic beams [45, 46]. In this case, a high-pressure gas expands adiabatically through a small nozzle and cools greatly. Now either the supersaturated gas itself or another material seeded into the gas condenses forming clusters. The low temperature of the seeded beams not only produces condensation but greatly simplifies the electronic spectra allowing the detailed study of even weakly bound molecules such as van der Waals complexes. However, even with the simplification caused by the low temperature, only small systems such as dimers and trimers have been studied in detail. Our interest is with larger clusters.

Most pure-gas experiments have investigated the structure of clusters of rare gases or small molecules, such as CO_2 or SF_6 [41, 47]. Sattler and co-workers have used a mass spectrometer to determine cluster size. For small supersaturations and clusters smaller than 100 atoms, they observe a gradual decrease in cluster probability with increasing size. In certain cases, especially xenon, the decrease is not monotonic, but rather certain sizes are over-represented in the distribution. Such magic numbers often correlate with the Mackay icosahedra or these figures capped with several atoms. The magic number 13 is the most often found case. The appearance of such magic numbers is taken as evidence for a preference for icosahedral over closest packed structures for small clusters of rare gases.

Higher levels of supersaturation lead to growth by coagulation of large clusters as opposed to monomer growth. This gives a Gaussian distribution of cluster size as indicated earlier. This effect comes into play especially for molecular gases such as CO_2 and N_2O for which cluster sizes of over 1000 are easily produced and measured [40].

There are several potential difficulties associated with the use of a time-of-flight mass spectrometer to measure cluster size distributions. First and foremost is the fact that the cluster must be ionized in order to see it. In the case of weakly bound clusters this can lead to severe fragmentation. Thus, we see the mass spectrum of the fragments and not that of the nascent beam. The photoionization efficiencies of small argon clusters have been studied in detail and they show both ionization and fragmentation to be cluster size and wavelength dependent [48]. Even in strongly bound clusters we see fragmentation due to the multiple ionization of medium-size clusters. These Coulomb explosions occur when the repulsion between two charges exceeds the binding forces [49]. Finally, the detection efficiency on an electron multiplier decreases with increasing cluster size unless stripping grids are used to ensure that the detector sees only a constant size particle coming from the mass spectrometer [31].

While a mass spectrometer may give a count of the number of particles in a cluster, it does not give any direct information about the bonding and shape of the cluster. Such shape information can be obtained from electron diffraction studies. Unfortunately, the electron diffraction techniques must simultaneously deal with diffraction patterns from a distribution of cluster sizes. One of the more complete diffraction studies of the rare-gas clusters is that of Kim and Stein [47]. They find that for clusters of more than 1000 atoms the diffraction patterns resemble closely those predicted to come from cuboctahedral clusters, i.e., the bulk phase. These patterns switched over to those expected from icosahedral packing as the cluster size became smaller, approaching 50–100 atoms.

An important difficulty associated with electron diffraction experiments concerns the temperature of the clusters. Fast cluster growth in nozzle expansions leaves much of the heat of condensation in the internal modes of the cluster. The distributions of internal and translational energy are not necessarily in equilibrium [32]. Both the internal and translational temperatures increase with cluster size, the latter showing a rather sharp change in the 10–15-molecule range for H_2O [50].

In a seeded beam a condensable material is mixed with a large excess of a rare gas that does not condense under the experimental expansion conditions (usually helium). This technique is used with metals and other materials that have too low a vapor pressure to be expanded directly. Low-boiling metals, such as the alkalis and alkaline earths, can be heated in a simple Knudsen cell [51]. Stronger heating has been used for the more refractory transition metals [52, 53]. Most recently laser evaporation has been used to produce beams of many metals. This technique, which uses a pulsed, focused laser to evaporate metal from a rod, is well suited to the use of a time-of-flight mass spectrometer for cluster analysis. Analysis by laser-induced fluorescence or laser ionization also allows many spectroscopic studies to be conducted on small clusters. In this way bond lengths and other parameters have been determined for dimers of Cr, Cu, and Sn to name just a few [54–56].

For larger clusters much spectroscopic detail is lost, but measurements of the photoionization thresholds provides information concerning cluster ionization potentials. Sodium clusters show a relatively smooth decrease in ionization potential from 5.1 eV for the atom to 3.5 eV for the 14-atom cluster [57]. This is still significantly above the 2.3-eV work function of bulk sodium. For the smaller clusters the odd sizes have a lower ionization potential than the neighboring even-sized clusters because of effects due to open- versus closed-shell configurations. Kaldor and co-workers have measured the ionization potential of iron clusters up to 25 atoms [58]. In this experiment the ionization potentials were bracketed by the use of various ionizing lasers. There is a decrease in the ionization potential from 7.870 eV for an iron atom to the 4.7-eV work function of the bulk metal, but the trend is by no means monotonic [59]. Thus the ionization potential of Fe_2 is about 5.9 eV, while those of Fe_3 and Fe_4 are above 6.42 eV. Clusters in the range of 9–12 atoms have ionization potentials below 5.58 eV while those in the range of 13–18 atoms are above 5.58 eV. The ionization potential of the 25-atom cluster still exceeds the bulk value by 0.3 eV.

Chemical reactions of metal clusters in a beam are just beginning to be studied. Sodium clusters (Na_n) react with chlorine atoms by the abstraction of a sodium atom forming NaCl and electronically excited Na_{n-1} [60]. Riley and co-workers have measured the rate constants for reaction of H_2 with iron clusters [61]. They find that the hydrogen always appears to add as a dimer and that the reaction rate is a very strong function of cluster size. Clusters below 8 atoms are slow to react as are clusters of 15 to 18 atoms. There appears to be a strong correlation between low ionization potential and fast reaction rate, but more work needs to be done to confirm this.

Some preliminary isotopic exchange reactions involving H_2 and D_2 on iron clusters have been carried out, which seem to show a preference for H_2 adsorption that is size dependent [61]. An opposite preference for deuterium was seen in mixed H_2O–D_2O clusters, in which case differences in the evaporation rates of H_2O and D_2O can explain the observations [62]. These same authors see breaks in the size distribution of mixed HNO_3–H_2O clusters at ratios corresponding to concentrations of bulk nitric acid that are known to decompose. This may indicate that certain bulk chemical properties may exist in clusters of only a few particles. The observation of iron deficiencies much like those seen in bulk FeO, for small clusters (e.g., Fe_9O_{10}) should also be noted in this connection [63].

Before leaving the discussion of gas-phase studies mention should be made of some work by Frurip and Bauer [64]. These workers studied light scattering from metal atom clusters formed by thermal decomposition of metal carbonyls in a shock tube. They measured cluster growth rates and various thermal parameters. Thermal parameters for hydrogen-bonded systems (mostly dimers) have also been measured by thermal conductivity studies [65].

2. Matrix Isolation

In the technique of matrix isolation, the atom or molecule (or precursor) to be studied is mixed with a large excess of an inert gas upon condensation on a cold surface. The inert gas is often argon in excess by a factor of about 1000 and the temperature is in the neighborhood of 10 K. A paper by Ozin summarizes much of this work [15]. This technique has the advantages of freezing rather reactive materials, concentrating them by collecting over time, and holding them for a more leisurely study. Often spectral lines are very sharp, although they can be broadened by inhomogeneous lattice effects. In some cases laser excitation can give sharp line emission from only a limited number of sites. Matrix relaxation also leads to line broadening [66].

A principal difficulty with matrix isolation studies is that they are often limited to small clusters of indeterminate size. Large dilutions favor monomer deposition. Comparison of the growth of spectral features as a function of dilution ratio with statistical models allows the correlation of new spectra with various-sized clusters. However, there is significant migration of atoms within the growing surface before the condensing gas becomes rigid. The degree of this migration can be enhanced by increasing the temperature, either during or after the deposition or by using a lower melting gas. Silver atoms undergo a photomigration when they are excited in an argon matrix with atom absorptions being bleached and dimer absorptions growing [67].

While many spectroscopic studies have been on dimers, the most extensive polymer studies have been with Ag, Na, and Cu clusters [67–69]. As might be expected much of the interest in silver relates to the photographic process in which it appears that a 4-atom silver cluster on a silver halide surface leads to reduction by developer, whereas a 3-atom cluster does not [70]. The electron spin resonance (ESR) spectra of sodium in argon deposits confirm that the trimer is covalently bonded and not an equilateral triangle [71]. Ultraviolet photoelectron spectroscopy (UPS) of Cu clusters indicates that the d band is separate from the s band, unlike in the bulk or in the $X\alpha$ calculations mentioned earlier [72].

Even under matrix isolation conditions some chemical reactions occur. A large number of reactions with atoms (especially the alkalis) have been observed. Reactions between CO and Cu or Ni clusters have been studied by infrared spectroscopy [73]. The CO stretching frequencies on clusters of one to four nickel (or copper) atoms rapidly approach the values found for CO chemisorbed on the polycrystalline bulk metal. Thus, the CO-to-metal bonding appears to be local as expected.

In mixed deposits of metal, methane, and argon, of some 18 metals tried only Al appears to react at 10 K without photolysis [74]. The reactivity of Al appears to be unique and probably due to its radical-like ^2P state. Theoretical studies lend support to the idea that atoms with partially filled p orbitals would be most reactive in C–H insertion processes [75].

The question as to whether clusters or atoms are more reactive in carbon–halogen bond breaking processes was recently considered by Imizu and Klabunde [76]. These workers found that in an argon matrix at 10 K Mg_2 and Mg_3 react with CH_3Br (presumably to form CH_3Mg_2Br and CH_3Mg_3Br), but Mg atoms do not. This higher reactivity was attributed to the lower ionization potential of these small clusters and to the fact that in the free state Mg_2 and Mg_3 are very weakly bound, but in the product should be strongly bound. Similar results have been found for Ca, Ca_2, and Ca_3 [77]. Theoretical results of Jaisen and Dykstra convincingly support these experimental results [78].

Kinetic studies of metal-atom aggregation in cold matrices have received some attention. A statistical frozen matrix approach [79, 80] (calculate probability that M and M are neighbors and react to give M_2) and a highly mobile metal-atom approach (diffusion is rapid in quasi-liquid layer) have been used. It was found that the diffusion mechanism was supported best by the experimental results. The eventual M_2 concentration was found to be proportional to the square of the M/substrate ratio. Concentrations of higher metal aggregates vary as some higher power of the M/substrate ratio.

Moskovits and Hulse [81] have carried this analysis further and have shown that the statistical approach underestimates the formation of clusters in the matrix. Both dilute and high-concentration matrices are dealt with experimentally and mathematically. The best model was found to be one that simulates freeze-out by assuming that the reactions stop abruptly after a certain time. They also point out that the trend in product distribution as a function of metal concentration is adequately described by this model and greatly aids spectroscopic assignment of bands to metal atoms and clusters [81].

3. Clusters on Surfaces

A film produced by deposition of atoms or particles on a surface forms in several stages: (1) nucleation, (2) cluster growth, (3) coalescence, (4) further thickening, and (5) recrystallization (perhaps) [26, 82]. We will be concerned with steps (1)–(3). A review by Mason is particularly pertinent [83].

Clusters on surfaces, or supported clusters, lend themselves to a variety of x-ray and electron spectroscopies, such as extended x-ray absorption fine-structure (EXAFS), x-ray and ultraviolet photoelectron spectroscopy (XPS and UPS), and transmission electron microscopy (TEM) to name a few. Most studies of model systems have used sodium chloride or carbon substrates and ultrahigh-vacuum techniques.

Neidermayer [26] as well as Weeks and Gilmer [27] have modeled the kinetics of cluster growth. Nucleation occurs at specific sites, often associated with lattice defects on the substrate. Cluster growth commences as atoms impinge near each other; in a series of experiments it was found that gold atoms will be captured if they land within 6.5 Å of a growing cluster [84].

As cluster growth continues the energetics regarding two-dimensional or three-dimensional growth must be considered. Calculations indicate that if the heat of vaporization of the metal (Λ_0) is greater than three times the heat of desorption of the metal atom from the surface (E_{des}), three-dimensional clusters should form. However, if the heat of metal vaporization is less than three times the energy of desorption minus the energy of diffusion (E_{diff}), two-dimensional cluster growth is favored [26]. Depending on the surface, E_{des} and E_{diff} will vary as

$$\Lambda_0 \gtrsim 3E_{des} \qquad \text{(three-dimensional favored)},$$

$$\Lambda_0 \lesssim 3(E_{des} - E_{diff}) \qquad \text{(two-dimensional favored)}.$$

These theoretical considerations predict that metal clusters growing on most clean metal surfaces and semiconductors would grow initially in two dimensions, and this is found experimentally [26].

Small clusters on surfaces have some unusual properties. Their geometrical shapes usually do not resemble the bulk element. For example, using moire interference patterns [85] and TEM, the smallest colloids of gold are shown to have pentagonal symmetry [86]. The presence of multiply twinned tetrahedra such as icosahedra, again suggests that icosahedral structures are preferred over the bulk structure during the initial growth phase. A short review of gold colloid work has been given by Kohlschütter [87].

If perfect crystallites are not formed in the initial stages of growth, there will be some strain energy in the cluster. As metal thickness increases, the desorption energy of a metal atom on the surface changes. For Na on a tungsten surface, E_{des} is initially 2.5 ± 0.2 eV, but after four layers are deposited it becomes 1.06 eV, which corresponds closely to the bulk heat of vaporization of Na. Since the Na–Na bonding energy is so high initially, lattice relaxation via two-dimensional dislocations is difficult, so more strain energy is created in the growing Na cluster [26].

The importance of these unusual growth characteristics has stimulated theoretical work on the kinetics of surface cluster growth. Neidermayer has treated cluster growth as a polymer growth problem [26]. The growth process becomes governed by an equilibrium between impingement and desorption, which is indicated by a constant value of monomer concentration over a considerable time period. So at a certain substrate temperature it should be possible to establish equilibria with very small cluster sizes. However, at low substrate temperatures no such equilibria can be established, and cluster and nucleus concentrations rise very steeply. Generally, the cluster growth for Au under low-temperature conditions (80 K) on a clean surface involves an induction period of 10^{-5} s, twin formation until 10^{-3} s, constant growth to 10^{-1} s, and then coalescence to a film [26].

Small clusters on surfaces can also be affected electronically, and so supported clusters have been extensively investigated [83]. Of particular interest is the behavior of the d electrons. X-ray photoelectron spectroscopy indicates that the d-electron binding energy decreases and the d-electron bandwidth increases with increasing cluster size. These effects can be explained in terms of changes in the d–s(p) orbital hybridization. Thus the $3d^8 4s^2$ configuration of atomic nickel becomes $3d^9 4s^1$ in bulk nickel. Similar increases in d-orbital occupation with increasing cluster size are expected for other metals. Because of their more localized nature, d electrons are repelled more by the core electrons and the electron binding energy is expected to decrease with increasing d-orbital occupancy (cluster size). From a simple molecular orbital picture, the larger the number of like neighbors, the larger should be the d-electron bandwidth. Because of overlap integral effects, this increasing width is asymmetric with the weakly bonding, "antibonding" orbitals shifting more, i.e., the binding energy decreases with increasing size (number of neighbors). An ultraviolet photoelectron spectroscopic study of silver and iodine-covered silver clusters has led to a similar interpretation of cluster-size effects on the d-electron bandwidth and ionization potential [88].

Gold clusters on weakly interacting substrates, such as carbon or alkali halides, have been most extensively studied [83]. In this case, changes in d-electron binding energy as a function of cluster size are exactly paralleled by similar changes observed in gold–cadmium alloys as a function of gold concentration. The only important parameter appears to be the average number of like nearest neighbors. Dilute gold alloys behave like small clusters. Thus, there is a linear increase in d-electron binding energy with decreasing concentration. When the substrate has localized p or d orbitals that overlap the d orbitals of the cluster, there is a strong interaction that usually leads to a decrease in the d-electron binding energy. Many studies of chemical reaction on supported and unsupported clusters are reviewed by Davis and Klabunde [12].

III. REACTIVE PARTICLES THAT POLYMERIZE TO FILMS

Throughout this book numerous references to reactive, short-lived particles polymerizing to films will be found. This is especially true of the chapter on discharge processes (Chapter 3). The intent of this section is simply to summarize the methods and give a few unusual examples.

A. Evaporation

Table 1.2 shows numerous substances that can be evaporated yielding interesting reactive particles in the gas phase. These species polymerize upon condensation. Some examples are TiO, $NiCl_2$, ZnS, MgF_2, and SiO.

The first stages of polymerization (clustering) have not been studied in much detail. The matrix isolation work of Weltner [89] and Margrave and co-workers [90] should be noted. A large number of MX_2 molecules were isolated and studied spectroscopically. They found that $MgCl_2$ and $MgBr_2$ are linear molecules, which agrees with the electron diffraction data of Klemperer and co-workers [91]. Further work has shown the presence of dimers, which are halide bridged species [92] (see [11], p. 42 and throughout for further discussion):

$$2\,MgCl_2 \longrightarrow Cl{-}Mg\overset{Cl}{\underset{Cl}{\lessgtr\gtrless}}Mg{-}Cl.$$

It must be realized that the vaporization of metal halides, oxides, and sulfides often leads to dimer, trimer, or tetramer species in the gas phase, and decomposition is frequent. For example, HgCl vaporizes as both Hg and $HgCl_2$ [93, 94]. Likewise, CdS vaporizes as both Cd and S_2, and it is interesting to note that during the production of CdS photocells by evaporation onto a cool plate, the distance of the vaporization source affects the properties of the deposit. Near the vaporization source the deposit is rich in S while further away it is rich in Cd [95–98]. Furthermore, the vaporization of CdS is greatly increased under the influence of UV light. It is proposed that charge transfer is an important step in CdS vaporization, and UV light of greater than band gap energy can affect this charge transfer step [96]. Initial stages of polymerization of reactive particles need much more study. New methods of characterization of such "telomers" are needed as well.

B. Reactive Evaporation *

This method was summarized earlier in this chapter and necessitates a chemical reaction either prior to evaporation or upon condensation. For example, TiO_2 films can be produced by condensation of TiO in the presence of electronically excited oxygen atoms

$$\underset{\text{forming film}}{TiO} + O \longrightarrow \underset{\text{film}}{TiO_2}$$

C. High-Temperature and Low-Temperature Chemical Reactions

When gaseous SiF_4 is passed over silicon metal at 1250°C, gaseous SiF_2 is formed. The chemistry of this interesting species has been reviewed [99]. When SiF_2 is allowed to condense on a cold surface $F_2Si{-}SiF_2$ first forms followed by polymerization to yield a white solid $(SiF_2)_n$. This material is plastic-like and tough but pyrophoric in air and reactive with water. Pyrolysis at about 200°C yields a mixture of perfluorosilanes from SiF_4 to $Si_{16}F_{34}$. Thus, the polymer film appears to be the silicon analog of teflon $(CF_2)_n$, but unfortunately is much more susceptible to decomposition and degradation.

Similarly, when BF_3 gas is passed over granular boron at 2000°C, high yields of gaseous BF can be generated [100]. This species polymerizes readily and pyrophoric $(BF)_n$ can be prepared. When BF gas is codeposited on a cold surface with SiF_4 gas, only a $(BF)_n$ polymer resulted [101]. However, cogeneration of SiF_2 with BF yields adducts of the two reactive species, and presumably copolymers could be prepared (see [11] pp. 171 and 198 and throughout for a discussion of the chemistry of these species).

Sometimes the codeposition of reactive particles with unsaturated organic molecules can lead to copolymers. For example SiO codeposited with alkynes or alkenes leads to new silicone polymers [102, 103]:

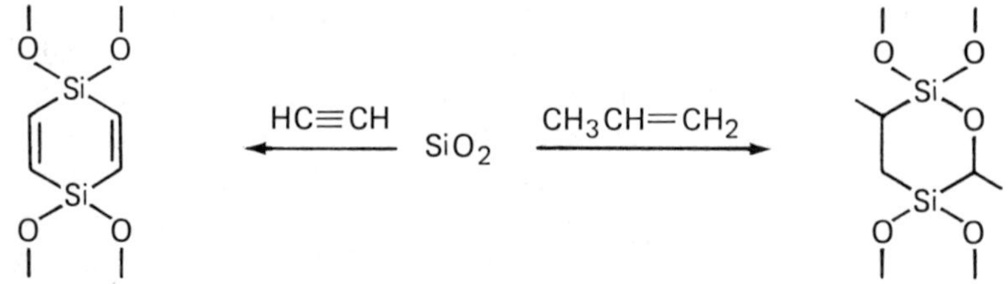

$$\text{[cyclic Si–O polymer]} \xleftarrow{HC{\equiv}CH} SiO_2 \xrightarrow{CH_3CH{=}CH_2} \text{[cyclic Si–O polymer]}$$

Similar results have been obtained with $SiCl_2$ and SiF_2.

The possibilities for preparing new reactive particles that polymerize to thin films or copolymerize with organics are enormous. High-temperature

* See Auwärter [5].

reactions followed by low-temperature polymerizations and copolymerizations will lead to many new materials in the next few years.

D. Discharge Methods

As a prelude to the next chapter, in which discharge methods are discussed in detail, a few examples of reactive particles prepared by discharge methods are shown here.

If a reactive particle can be generated in a discharge region and lives long enough to be passed out of the discharge area, its chemistry and polymerization characteristics can be studied. For example, SeF has been prepared by passing the effluent of a microwave discharge on CF_4 into a stream of COSe as

$$CF_4 \xrightarrow{\text{discharge}} [F\cdot] \xrightarrow{O=C=Se} SeF$$

An ESR spectrum of SeF was obtained in a gas phase flow tube [104, 105].

The generation of CS has been carried out by a variety of methods, including several discharge procedures [106–108]. Synthetic amounts of CS can be prepared from the action of a high-voltage discharge on CS_2.

Upon condensation of CS a white solid is formed (monomer). Upon slight warming from −196 to −160°C violent polymerization takes place. Heat and light are given off and a dark brown polymer is formed that is very resistant to oxidation or dissolution. A thin film of this tough $(CS)_n$ polymer can be prepared on essentially any surface, even unusually shaped ones, if the temperature of the substrate is held about 25°C. The film does not conduct electricity and is very difficult to remove from glass surfaces.

The characteristics of this polymer suggest that it is highly cross-linked and does not contain many unpaired electrons. The exothermicity of the polymerization is so great that tremendous amounts of heat are generated along with electronically excited intermediates; perhaps as shown:

$$CS + CS \longrightarrow [SC\text{–}CS]^* \xrightarrow{-h\nu} [SC\text{–}CS]$$

[SC–CS]* ↓CS ↓CS ↓CS; [SC–CS] ↓CS ↓CS ↓CS

```
        C
        |
        S
        |
(CS)n ← S–C–C–S ←
            | |
            S C–S
            |
            C
```

This process is poorly understood as is the case for most reactive particle polymerization processes.

REFERENCES

1. F. Soddy, *Proc. R. Soc. London, A* **78**, 429 (1907).
2. S. Dushman, "Scientific Foundations of Vacuum Technique," J. Wiley, New York, 1949.
3. P. Pringsheim and R. Pohl, *Dtsch. Phys. Ges., Verh.* **14**, 506 (1912).
4. J. Strong, "Procedures in Experimental Physics," Chapter 4, Prentice-Hall, New York, 1938.
5. M. Auwärter, *New Synth. Methods* **3**, 43 (1975); also reported in *Angew. Chem., Int. Ed. Engl.* **14**, 207 (1975).
6. W. Heitmann, *Appl. Opt.* **10**, 2414 (1971).
7. E. Zollinger, DOS 2305359 (1973), Balzers Vacuum Technology, West Germany.
8. W. Reichelt, *New Synth. Methods* **3**, 79 (1975); also reported in *Angew. Chem., Int. Ed. Engl.* **14**, 218 (1975).
9. H. M. Manasevit and W. I. Simpson, *J. Electrochem. Soc.* **116**, 1725 (1969); H. M. Manasevit, *J. Electrochem. Soc.* **118**, 647 (1971).
10. D. J. Schlyer and M. A. Ring, *J. Electrochem. Soc.* **124**, 569 (1977).
11. K. J. Klabunde, "Chemistry of Free Atoms and Particles," Academic Press, New York, 1980.
12. S. C. Davis and K. J. Klabunde, *Chem. Rev.* **82**, 153 (1982).
13. R. C. Baetzold and J. F. Hamilton, *Prog. Solid State Chem.* **15**, 1 (1983); D. W. Goodman, *Acc. Chem. Res.* **17**, 194 (1984); E. L. Muetterties and M. J. Krause, *Angew. Chem., Int. Ed. Engl.* **22**, 135 (1983); F. A. Cotton, *ASC Symp. Ser.* **211**, 209 (1983); R. E. McCarley, *ACS Symp. Ser.* **211**, 273 (1983); E. L. Muetterties, R. R. Burch, and A. M. Stolzenberg, *Ann. Rev. Phys. Chem.* **33**, 89 (1982).
14. M. R. Hoare, *Adv. Chem. Phys.* **40**, 49 (1979).
15. G. A. Ozin and S. A. Mitchell, *Angew. Chem., Int. Ed. Engl.* **22**, 674 (1983).
16. J. A. Perenboom and P. Wyder, *Phys. Rep.* **78**, 173 (1981).
17. M. Faraday, *Philos. Trans. R. Soc. London* **147**, 145 (1857).
18. Encyclopaedia Britannica, 15th ed. (1980).
19. E. Shustorovich, R. C. Baetzold, and E. L. Muetterties, *J. Phys. Chem.* **87**, 1100 (1983).
20. *Surface Sci.* **106** (1981).
21. W. J. Moore, *in* "Physical Chemistry," 4th ed., Chapter 11, Prentice-Hall, Englewood Cliffs, New Jersey, 1972.
22. D. J. Frurip, M. Blander, and C. Chatillon, *ACS Symp. Ser.* **179**, 207 (1982).
23. J. P. Borel, *Surf. Sci.* **106**, 1 (1981).
24. M. P. D'Evelyn and S. A. Rice, *J. Chem. Phys.* **78**, 5225 (1983).
25. S. H. Bauer and D. J. Frurip, *J. Phys. Chem.* **81**, 1015 (1977).
26. R. Neidermayer, *Angew. Chem., Int. Ed. Engl.* **14**, 212 (1975).
27. J. D. Weeks and G. H. Gilmer, *Adv. Chem. Phys.* **40**, 157 (1979).
28. M. R. Hoare, *Adv. Chem. Phys.* **56**, 1 (1984).
29. J. L. Katz and M. D. Donohue, *Adv. Chem. Phys.* **40**, 137 (1979).
30. H. Müller-Krumbhaar, *Festkörperprobleme* **19**, 1 (1979).
31. J. M. Soler, N. García, O. Echt, K. Sattler, and E. Recknagel, *Phys. Rev. Lett.* **49**, 1857 (1982).
32. J. M. Soler and N. García, *Phys. Rev. A* **27**, 3300 (1983); *Phys. Rev. A* **27**, 3307 (1983).
33. M. Allen and G. W. Robinson, *Astrophys. J.* **195**, 81 (1975).

34. R. P. Messmer, *Surf. Sci.* **106**, 225 (1981).
35. G. B. Bachelet, F. Bassani, M. Bourg, and A. Julg, *J. Phys. C* **16**, 4305 (1983).
36. J. Demuynck, M. M. Rohmer, A. Strich, and A. Veillard, *J. Chem. Phys.* **75**, 3443 (1981).
37. G. Pacchioni, D. Plavŝić, and J. Koutecký, *Ber. Bunsenes. Phys. Chem.* **87**, 503 (1983).
38. H. Stoll and H. Preuss, *Int. J. Quantum Chem.* **9**, 775 (1975).
39. T. Halicioglu and P. J. White, *Surf. Sci.* **106**, 45 (1980).
40. K. Sattler, *Festkörperprobleme* **23**, 1 (1983).
41. O. Echt, K. Sattler, and E. Recknagel, *Phys. Rev. Lett.* **47**, 1121 (1981).
42. R. C. Baetzold, *J. Phys. Chem.* **80**, 1504 (1976).
43. T. P. Martin, *Phys. Rep.* **95**, 168 (1983).
44. K. A. Gingerich, *ACS Symp. Ser.* **179**, 109 (1982).
45. O. F. Hagena, *Surf. Sci.* **106**, 101 (1981).
46. R. E. Smalley, L. Wharton, and D. H. Levy, *Acc. Chem. Res.* **10**, 139 (1977).
47. S. S. Kim and G. D. Stein, *J. Colloid Interface Sci.* **87**, 180 (1982).
48. P. M. Dehmer and S. T. Pratt, *J. Chem. Phys.* **76**, 843 (1982).
49. K. Sattler, J. Mühlbach, O. Echt, P. Pfau, and E. Recknagel, *Phys. Rev. Lett.* **47**, 160 (1981).
50. D. Dreyfuss and H. Y. Wachman, *J. Chem. Phys.* **76**, 2031 (1982).
51. E. Schumacher, W. H. Gerber, H. P. Härri, M. Hofmann, and E. Scholl, *ACS Symp. Ser.* **179**, 83 (1982).
52. R. S. Bowles, J. J. Kolstad, J. M. Calo, and R. P. Andres, *Surf. Sci.* **106**, 117 (1981).
53. S. J. Riley, E. K. Parks, C-R. Mao, L. G. Pobo, and S. Wexler, *J. Phys. Chem.* **86**, 3911 (1982).
54. S. J. Riley, E. K. Parks, L. G. Pobo, and S. Wexler, *J. Chem. Phys.* **79**, 2577 (1983).
55. D. E. Powers, S. G. Hansen, M. E. Geusic, D. L. Michalopoulos, and R. E. Smalley, *J. Chem. Phys.* **78**, 2866 (1983).
56. V. E. Bondybey, M. Heaven, and T. A. Miller, *J. Chem. Phys.* **78**, 3593 (1983).
57. A. Herrmann, E. Schumacher, and L. Woste, *J. Chem. Phys.* **68**, 2327 (1978).
58. E. A. Rohlfing, D. M. Cox, and A. Kaldor, *Chem. Phys. Lett.* **99**, 161 (1983).
59. R. C. Weast, ed., "CRC Handbook of Chemistry and Physics," Chem. Rubber Publ. Co., Cleveland, Ohio, 1982–1983.
60. W. H. Crumley, J. L. Gole, and D. A. Dixon, *J. Chem. Phys.* **76**, 6439 (1982).
61. S. C. Richtsmeier, E. K. Parks, K. Liu, L. G. Pobo, and S. J. Riley, *J. Chem. Phys.* **82**, 3659 (1985).
62. A. W. Castleman, Jr., B. D. Kay, V. Hermann, P. M. Holland, and T. D. Märk, *Surf. Sci.* **106**, 179 (1981).
63. S. J. Riley, E. K. Parks, G. C. Nieman, L. G. Pobo, and S. Wexler, *J. Chem. Phys.* **80**, 1360 (1984).
64. D. J. Frurip and S. H. Bauer, *J. Phys. Chem.* **81**, 1007 (1977); H. J. Freund and S. H. Bauer, *J. Phys. Chem.* **81**, 994 (1977).
65. L. A. Curtiss, D. J. Frurip, and M. Blander, *J. Phys. Chem.* **86**, 1120 (1982).
66. G. A. Ozin and S. A. Mitchell, *ACS Symp. Ser.* **211**, 303 (1983).
67. G. A. Ozin and H. Huber, *Inorg. Chem.* **17**, 155 (1978); C. Steinbrüchel and D. M. Gruen, *Surf. Sci.* **106**, 160 (1981).
68. G. A. Ozin and H. Huber, *Inorg. Chem.* **18**, 1402 (1979); M. Hofmann, S. Leutwyler, and W. Schulze, *Chem. Phys.* **40**, 145 (1979).
69. M. Moskovits and J. E. Hulse, *J. Chem. Phys.* **67**, 4271 (1977).
70. R. C. Baetzoid, *ACS Symp. Ser.* **200**, 59 (1982).
71. G. A. Thompson and D. M. Lindsay, *J. Chem. Phys.* **74**, 959 (1981).
72. K. Jacobi, D. Schmeisser, and D. M. Kolb, *J. Chem. Phys.* **75**, 5300 (1981).
73. M. Moskovits and J. E. Hulse, *J. Phys. Chem.* **81**, 2004 (1977).

74. K. J. Klabunde and Y. Tanaka, *J. Am. Chem. Soc.* **105**, 3544 (1983).
75. C. B. Lebrilla and W. F. Maier, *Chem. Phys. Lett.* **105**, 183 (1984).
76. Y. Imizu and K. J. Klabunde, *Inorg. Chem.* **23**, 3602 (1984).
77. K. J. Klabunde and A. Whetten, unpublished results; A. Whetten, M.S. thesis, Kansas State University, Manhatten, Kansas, 1984.
78. P. G. Jaisen and C. E. Dykstra, *J. Am. Chem. Soc.* **105**, 2089 (1983).
79. E. P. Kundig, M. Moskovits, and G. A. Ozin, *New Synth. Methods* **3**, 151 (1975); also see *Angew. Chem., Int. Ed. Engl.* **14**, 292 (1975).
80. M. Moskovits and G. A. Ozin, *in* "Cryochemistry" (M. Moskovits and G. A. Ozin, eds.), p. 395, Wiley (Interscience), New York, 1976.
81. M. Moskovits and J. E. Hulse, *J. Chem. Soc., Faraday, Trans. 2* **73**, 471 (1977).
82. J. R. Anderson, ed., "Chemisorption and Reactions on Metallic Films," Vols. 1 and 2, Academic Press, New York, 1971.
83. M. G. Mason, *Phys. Rev. B* **27**, 748 (1983).
84. J. F. Hamilton, D. R. Preuss, and G. R. Apai, *Surf. Sci.* **106**, 146 (1981).
85. J. Woltersdorf, A. S. Nepijko, and E. Pippel, *Surf. Sci.* **106**, 64 (1981).
86. N. Uyeda, M. Nishino, and E. Suito, *J. Colloid Interface Sci.* **43**, 264 (1973).
87. H. W. Kohlschütter, *New Synth. Methods* **3**, (1975); also see *Angew. Chem., Int. Ed. Engl.* **14**, 193 (1975).
88. R. C. Baetzold, *Surf. Sci.* **106**, 243 (1981).
89. A. Weltner, Jr., *Adv. High Temp. Chem.* **2**, 85 (1969).
90. S. P. Randall, F. T. Greene, and J. L. Margrave, *J. Phys. Chem.* **63**, 758 (1959).
91. A. Buechler, J. L. Stauffer, and W. Klemperer, *J. Am. Chem. Soc.* **86**, 4544 (1964).
92. D. L. Cocke, C. A. Chana, and K. A. Gingerich, *Appl. Spectrosc.* **27**, 260 (1973).
93. A. Smith and A. W. C. Menzies, *J. Am. Chem. Soc.* **32**, 1541 (1911).
94. A. Smith, *Z. Elektrochem.* **22**, 33 (1916).
95. F. Y. Pikus and G. N. Talnova, *Fiz. Tverd. Tela (Leningrad)* **12**, 1355 (1970).
96. G. A. Somorjai, *Surf. Sci.* **2**, 298 (1964).
97. G. A. Somorjai and D. W. Jepsen, *J. Chem. Phys.* **41**, 1389, 1394 (1964).
98. L. Gombay and M. Zollei, *Acta. Phys. Chem.* **2**, 28 (1956).
99. J. L. Margrave and P. W. Wilson, *Acc. Chem. Res.* **4**, 145 (1971).
100. P. L. Timms, *J. Am. Chem. Soc.* **89**, 1629 (1967).
101. R. W. Kirk and P. L. Timms, *J. Am. Chem. Soc.* **91**, 6315 (1969).
102. P. L. Timms, *Acc. Chem. Res.* **6**, 118 (1973).
103. E. T. Schaschel, D. N. Gray, and P. L. Timms, *J. Organomet. Chem.* **35**, 69 (1972).
104. A. Carrington, G. N. Currie, T. A. Miller, and D. H. Levy, *J. Chem. Phys.* **50**, 2726 (1969).
105. J. M. Brown, C. R. Byfleet, B. J. Howard, and D. K. Russell, *Mol. Phys.* **23**, 457 (1972).
106. K. J. Klabunde, M. P. Kramer, A. Senning, and E. K. Moltzen, *J. Am. Chem. Soc.* **106**, 263 (1984).
107. E. K. Moltzen, A. Senning, M. P. Kramer, and K. J. Klabunde, *J. Org. Chem.* **49**, 3854 (1984).
108. G. Gatton and W. Behrendt, *Top. Sulfur Chem.* **2**, 1 (1977); C. W. White, M.S. thesis, University of North Dakota, Grand Forks, North Dakota, 1974.

3 ANALYSIS OF GLOW DISCHARGES FOR UNDERSTANDING THE PROCESS OF FILM FORMATION

M. Venugopalan

Department of Chemistry
Western Illinois University
Macomb, Illinois

R. Avni

Division of Chemistry
Nuclear Research Center Negev
Beer-Sheva, Israel

I. INTRODUCTION

The transitory existence of atoms and free radicals has long been established in spectroscopy. Line spectra, of course, have their origin in free atoms, and investigations of band spectra have quite definitely shown the existence of free radicals. Generally, high temperature is needed to produce these particles, so the chemistry investigated is necessarily that of energetic species in the gas phase. Their high reactivity results in their rapid disappearance by reaction with themselves or with other substances that may be present. Often, condensation or *deposition* from the vapor phase results on a surface (sometimes also known as a *substrate*), which relieves the species of their high energies. If the coating thus formed is less than about 1×10^{-6} m, it is called a *thin film*. Clearly, the deposition process involves a phase transformation which can be understood from thermodynamics and kinetics studies.

Because of important industrial applications, thin-film technology has grown rapidly during the past two decades or so. Originally, thin films were produced by pyrolytic chemical vapor deposition (CVD) and deposition of inorganic or metal films from solution by plating, anodization, etc. Subsequently, physical methods were developed for deposition of films. One such method* makes use of glow discharges, which produce a unique state of matter called *plasma*. In this chapter we shall attempt a short review of the process of film formation as understood from analyses of glow discharge plasmas rather than of the films they produce. The latter topic, which has substantially contributed to the practical aspects of thin film technology, will be covered in other chapters in this book. The reader is cautioned that descriptions of the plasma will be made in the context of the glow discharge process only. For a rigorous discussion of plasmas and plasma chemistry the reader is referred to a book edited by Venugopalan [1].

A. The Plasma State

A plasma is a collection of charged and neutral particles resulting from the partial ionization of the atoms or molecules of a gas generally by an external electric field. In the type of plasmas discussed in this chapter, the degree of ionization is typically only 10^{-4}, so the gas consists mostly of neutrals. The Coulomb interaction of the charged particles with each other and with whatever electric (and magnetic) fields are externally applied is both strong and long-range so that collective behavior constitutes the prime charac-

* In ion-beam deposition substrates are not immersed in a glow discharge but are in a relatively good vacuum.

teristic of a plasma. The neutral particles, on the contrary, interact with each other and with the charged particles via the short-range forces that come into play only during close encounters. The essential mechanisms in the plasma are excitation and relaxation and ionization and recombination. We may characterize the effects of all these interactions in terms of macroscopic variables such as pressure, density, and temperature.

The plasma itself is virtually electric field free, i.e., equipotential. It is this potential that is called the plasma potential V_p. The electron density n_e and ion density n_i are equal (on average) so that the plasma is quasi-neutral; their number, which is much less than the density of neutrals, is often known as the plasma density. Typical glow discharge plasmas have densities of 10^9–10^{12} cm^{-3}. There is a distribution of electron energies $f(E)$ that is at best described as non-Maxwellian. The average electron energies are around 2–10 eV, which corresponds to electron temperatures T_e of 10^4–10^5 K. However, the ions have only temperatures T_i somewhat above ambient, say 500 K, so that there is a *nonequilibrium* nature of the plasma that leads to deposition of a film at much lower temperatures than are possible with pryolytic CVD. This is the main reason for the sustained interest in the glow discharge as a means for film deposition.

Another reason we may cite is the possibility of using temperature-sensitive substrates. However, substrates immersed in a plasma acquire a negative floating potential with respect to the plasma. Consequently, the condensing species on the substrates would be subjected to ion bombardment, which can affect the heterogeneous reactions occurring on the substrate surface and, thus, influence the properties of the growing film. The ion impact may set up a series of collisions between atoms of the substrate, possibly leading to the ejection or sputtering of one of these atoms. While *sputtering* is in itself a process used for the deposition of thin films, in glow discharges it may result in undesirable structural rearrangements of the deposited film. Nevertheless, by the application of glow discharge plasmas films of novel materials and commercial value have been produced. Furthermore, plasma-deposited films are conformal in nature and produce a lower pinhole density compared to line-of-sight deposition techniques such as evaporation and sputtering.

B. Phenomena and Parameters Relevant to Film Formation

In the preceding chapter the theory of thin-film formation was discussed. In general, the substrate will have a chemical nature different from that of the film material so that we must consider a third phase in which gaseous species are adsorbed on the substrate but have not yet reacted with other adsorbed species. Figure 3.1 shows a substrate indicating the adsorbed phase and the

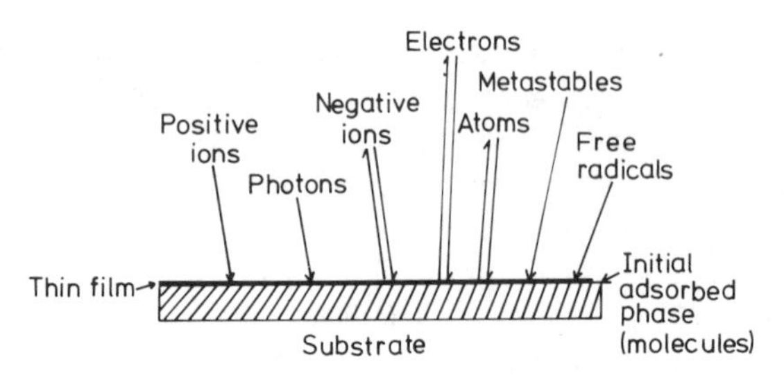

Fig. 3.1. Particles encountered by the substrate in glow discharge deposition.

likely particles bombarding it when immersed in a glow discharge. Not only the nature and temperature of the substrate but also the adsorption sites on its surface are important in determining the nature of the deposited film. During the film growth, the substrate and growing film will be subject to many types of bombardment. A description of this will be attempted after the scenario of the glow discharge is presented in the next section.

Species impinging on the substrate may be adsorbed on the surface and then migrate, evaporate, or collide and combine eventually to form clusters in a process known as nucleation (Fig. 3.2). This is the onset of a condensation process in which nuclei may contain tens or hundreds of atoms and typically have densities of 10^{10} cm^{-2}. The buildup and/or growth of several nuclei may result in contact between nearest neighbors and the so-called formation of islands of film material. Each island will contain one or a few crystallites. On a polycrystalline substrate, the orientation of each island will be random, so that coalescence of the islands may produce a

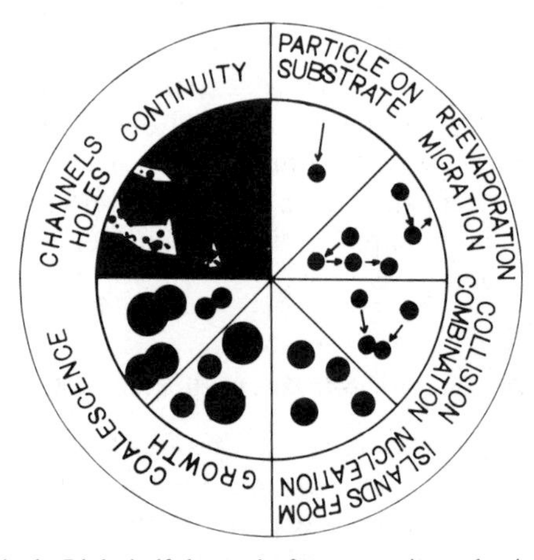

Fig. 3.2. Thin film clock. Right half shows the four stages in nucleation on substrate, and left half shows the four stages in growth of film on substrate.

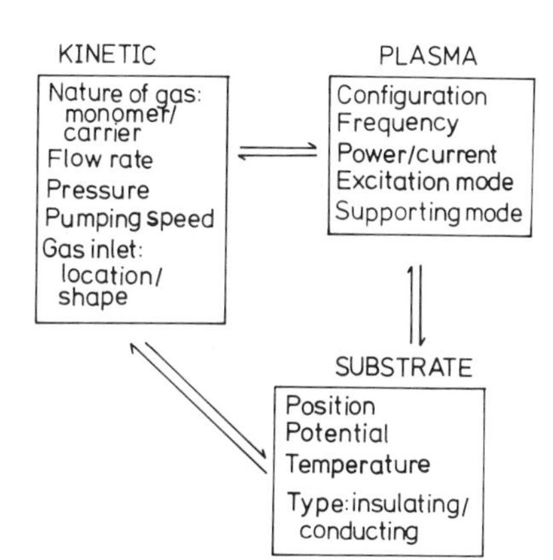

Fig. 3.3. Parameter interaction in plasma deposition of films.

polycrystalline film.* Channels or holes, if any, during the coalescence stage fill via secondary nucleation to give a continuous film (Fig. 3.2). As in crystal growth, the mobile surface species tend to seek out low-energy positions. Mobility is enhanced by increased surface temperature, but locating favorable lattice sites is time consuming so that a high substrate temperature and low deposition rate lead to large grains, low density of crystal defects, and large film thickness for continuity.

Phenomena such as these are described in terms of the *sticking coefficient*, which is the ratio of the amount of species condensed on the substrate to the amount that has impinged. So far we have not delineated the plasma species that may undergo these phenomena. Of course, ions and electrons are present in any plasma. Recombination of positive ions and electrons gives rise to highly excited states that may dissociate into neutral fragments but more probably dissociate into radicals. A considerable body of analytical evidence has shown that a wide variety of free radicals are formed in glow discharges. Experiments and/or theory from the field of surface science offer nothing more than intuitive guidelines about the interaction of radicals with surfaces. It is, however, known that radicals frequently chemisorb on surfaces that appear inert to the parent molecule. In a discussion of this topic Winters [3] has suggested that radicals have reasonably large sticking coefficients but that they are not unity.

Figure 3.3 illustrates the parameter problem relevant to film formation. It is obvious that the results of film deposition are subject to large number of variables that are interdependent. For this reason experimental results from one deposition system to another are often not easily reproducible. The key

* In the case of a single-crystal substrate, the island orientations may be determined by the substrate structure so that growth and coalescence leads to a single-crystalline film. This is known as the phenomenon of *epitaxy*, described by Bauer and Poppa [2].

insight that is needed is the identity, abundance, and energy of all species incident on the substrate surface. This depends on the electrical element (plasma) of the system, which in turn depends on a number of gas kinetic parameters. The complexity of polyatomic molecular gas discharges often used for thin films would appear to preclude quantitative information about species concentrations being derived from n_e and $f(E)$ even if these can be determined, since much of the cross-section data for ionization/dissociation or ion–molecule and radical–molecule reactions does not exist presently. Thus, mechanistic understanding in relation to the plasma becomes difficult. However, process control should be feasible by the use of film thickness/composition monitors, mass spectrometry, and optical spectroscopy for detecting gas-phase intermediates and for monitoring their concentration.

II. GLOW DISCHARGES

A. Architecture of Glow Discharges

Consider a glass tube containing a gas at low pressure, say 1 torr. Into each end of the tube is inserted an electrode. Between the electrodes a direct current (dc) power supply is connected. When a voltage is first applied, a very small current flows in random bursts (Fig. 3.4). As the voltage is increased, the current increases steadily. As the current increases, the gas begins to glow visibly and the potential across the tube drops to a constant value. We say that a glow discharge has been established in the tube. The early work in gas discharges centered around the study of the visual behavior of the glow discharge and led to a classification of the various regions illustrated in Fig.

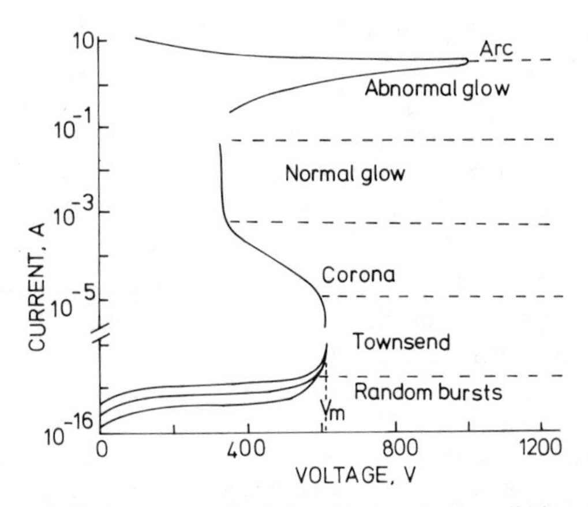

Fig. 3.4. Voltage–current relationship in a dc glow discharge.

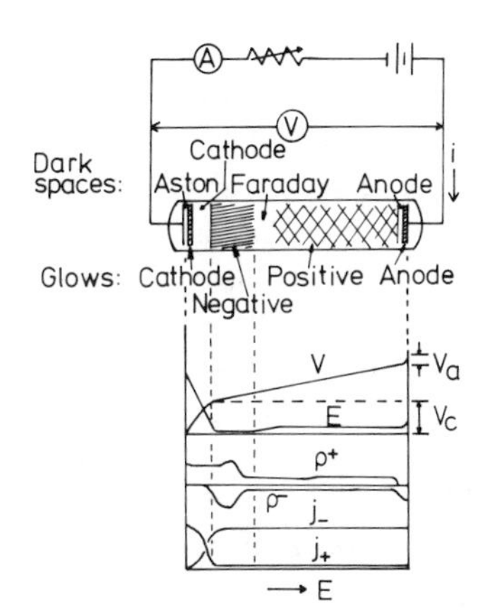

Fig. 3.5. Spatial distribution of luminous zones and electrical parameters in a dc glow discharge tube. Here V is the potential, E the electric field, ρ the space charge density, and j the current density.

3.5. There are excellent descriptions of these in many textbooks such as Cobine [4], von Engel [5], and Nasser [6]. Before we analyse the spatial characteristics let us examine the voltage–current characteristics in some detail.

Ordinarily, there exists in a gas a small number of ions and electrons due to ionization by cosmic radiation. The initial application of a dc potential causes these charges to move and cause random bursts of nearly constant current. The cathode emits electrons by a mechanism called *field emission* in which the electric field near the cathode surface extracts electrons directly from the metal. As the voltage is increased, the charged particles acquire more energy and undergo collisions with the electrodes and with neutral gas atoms, thus producing more ions. If the voltage between the electrodes is high enough, the ions striking the cathode can release electrons. This secondary electron emission causes the current to increase at a constant voltage known as the threshold or breakdown potential V_m.* When the number of electrons produced is just sufficient to produce enough ions to regenerate the same number of electrons, a self-sustaining discharge is obtained. It is at this point the gas begins to "glow," the voltage drops, and the current rises abruptly (Fig. 3.4).

The breakdown voltage V_m is crucial to the formation of the glow discharge. Its magnitude is governed, among others, by the interelectrode separation and the gas pressure, which determines the mean free path of secondary electrons. If the pressure and/or separation is too large, ions generated in the gas are slowed by inelastic collisions, and they strike the cathode with insufficient energy to produce secondary electrons. On the

* The term *Townsend discharge*, after the physicist who studied this phenomenon, is used to delineate this stage in the formation of a glow discharge.

other hand, if the gas pressure is too low or the electrode separation too small, the secondaries cannot undergo a sufficient number of ionizing collisions before they are collected at the anode.

The secondary electron emission ratio of most materials is of the order of 0.1 so that several ions must bombard a given area of the cathode to produce another secondary electron. Initially, the bombardment is concentrated near the edges of the cathode or at other irregularities on the surface. As more power is supplied, the bombardment covers the whole cathode surface and a constant current is achieved. Further increases in power produce both increased voltage and current in a region known as the abnormal glow (Fig. 3.4). This mode is usually used in virtually all glow discharge processes including film deposition. If thermionic emission occurs as a result of the cathode heating to a high temperature, the discharge will transcend into an electrical arc characterized by a low-voltage, high-current profile (Fig. 3.4).

Within a glow discharge there exist distributions of potential, field, space charge, and current density as shown in Fig. 3.5. Visually these are seen as regions of varied luminosity. The application of a voltage primarily affects the region near the cathode, where it facilitates electron emission. Adjacent to the cathode, there is a narrow luminous region known as the cathode glow. The light emitted is characteristic of both the cathode material and the incident ion. Thus, it is the region in which the positive ions formed at the cathode, and the incoming discharge ions are neutralized by a variety of processes.

Secondary electrons are repelled at high velocity from the cathode and make collisions with neutrals at a distance away from the cathode corresponding to their mean free path. Because the electrons lose their energy by collisions, nearly all of the applied voltage appears across this dark space. Since the mobility of ions is very much less than that of electrons, the predominant species in the dark space are ions. The high net positive space charge present in the cathode dark space forms a *sheath* and causes the sudden increase in potential between the cathode and the leading edge of the negative glow. This part of the potential is referred to as the cathode fall and is typically 100–400 V in magnitude.

Acceleration of secondary electrons from the cathode results in ionizing collisions in the *negative glow* region. It so happens that most plasma deposition systems are essentially negative-glow-type discharges. The general consensus regarding the electron energy distribution in the negative glow is well summarized by Chapman [7] as consisting of:

(1) primary electrons with typical densities of 10^6 or so at 1 torr pressure, which enter from the cathode sheath with nearly the full dark space potential and decay in energy primarily by inelastic collisions;

(2) secondary electrons ($n_e = 10^7$ or so) of considerably lower energy, which are the product of ionizing collisions or primaries that have lost much of their energy; and

(3) ultimate electrons ($n_e > 10^9$), which have energies in the 0.1–3-eV range depending on the pressure and local electric fields.

The last group* is the net result of the energy exchange processes.

The Faraday dark space and positive column† (Fig. 3.5) are nearly field-free regions and are characterized by nearly equivalent concentrations of ions and electrons. For these reasons the positive column most nearly resembles a plasma. Unfortunately, in many glow discharge systems the interelectrode separation needs to be small so that the anode is located in the negative glow; therefore, the positive column and the Faraday dark space do not exist. The plasma potential V_p is now essentially the potential of the negative glow, which is the most positive potential anywhere in the system. This potential must be more positive than the next highest potential on any large surface in the discharge by an amount that is at least as great as the first ionization potential of the gas.

Now that an overview of the architecture of dc glow discharges has been presented, we can consider a small electrically isolated substrate placed in the glow. Initially it will be bombarded by electrons and ions, but since electrons have greater velocities than ions, the substrate immediately starts to build a negative charge and hence negative potential with respect to the plasma. A negatively charged substrate will repel electrons and attract ions. Thus, the electron flux decreases, but the substrate continues to charge negatively until the electron flux is reduced by repulsion just enough to balance the ion flux. The floating potential V_f is the potential at which equal numbers of electrons and ions arrive at a surface that is not externally biased or grounded.‡ The floating potential is related to the electron temperature by the approximate relation given by Chen [8]:

$$V_f = (1/2e)kT_e \ln[\pi m_e/2m_i], \qquad (1)$$

where k is the Boltzmann constant, e the unit electron charge, m_e the electron mass, and m_i the ion mass. In the case of a discharge tube having insulating walls (Fig. 3.5), these walls also require zero steady-state net flux, so that *wall potential* and floating potential are related terms. Since V_f is such as to repel

* In radio-frequency (rf) discharges, to be discussed in Section II.B, the electric fields in the plasma volume may raise the average electron energy so that secondary and ultimate electron groups are not distinguishable.

† In a self-maintained discharge, the emission is not unlimited so that the plasma column far from the cathode retains a positive charge. Hence the usage of the term "positive column."

‡ In this case it is the same as the substrate potential V_s.

electrons, $V_f < V_p$. In the absence of a reference, only $V_p - V_f$ is meaningful; this difference determines the maximum energy with which ions collide with the electrically floating substrate.

B. Types and Supporting Modes

In the preceding section we described the classical type of glow discharge that is operated by dc potentials. When an alternating current (ac) power supply is used (Fig. 3.5), each electrode alternately acts as cathode and anode. On each half-cycle a dc-type discharge is established once the breakdown voltage is surpassed. At radio frequencies (5–30 MHz) the electrons oscillating in the glow space have sufficient energies to cause ionizing collisions, thus reducing the dependence of the discharge on secondary electrons and lowering the breakdown voltage. Because rf voltages can be coupled through any kind of impedance it is possible to use reactors without internal electrodes.

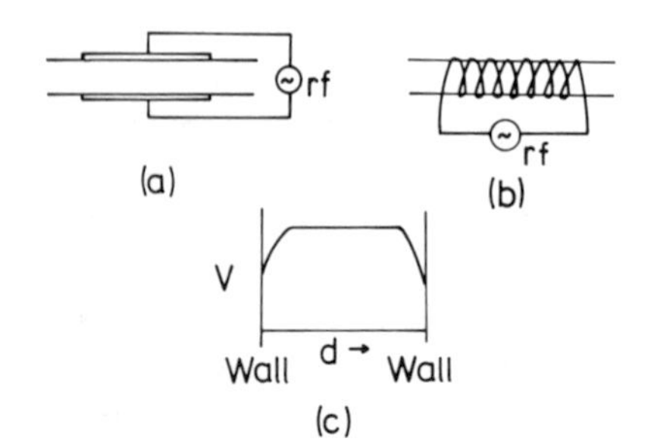

Fig. 3.6. Simplified diagram of (a) capacitively and (b) inductively coupled tunnel reactors; (c) potential distribution across the diameter of the tube.

Two most common techniques for coupling rf power into a glass tube, often referred to as a tubular or tunnel reactor, are shown in Fig. 3.6. In the capacitively coupled reactor rf power is coupled through the capacitance of the dielectric wall of the tube; in the so-called inductively coupled reactor, the rf power is also coupled capacitively through the wall, except that coupling is not uniform in that air gaps may be present. Flat-bed reactors (Fig. 3.7), on the other hand, have a configuration in which the substrate rests on a grounded or floating table with the powered rf electrode above.* Note that a series capacitor is usually inserted between the rf generator and the powered electrode.

For a tubular reactor the distribution of potential in space taken across the diameter of the tube is symmetrical. In the flat-bed reactor the largest potential difference is found between the plasma and the capacitively coupled

* A variation of the reactor in which the substrate rests on the powered electrode was used by Holland and Ojha [9] for deposition of carbon films.

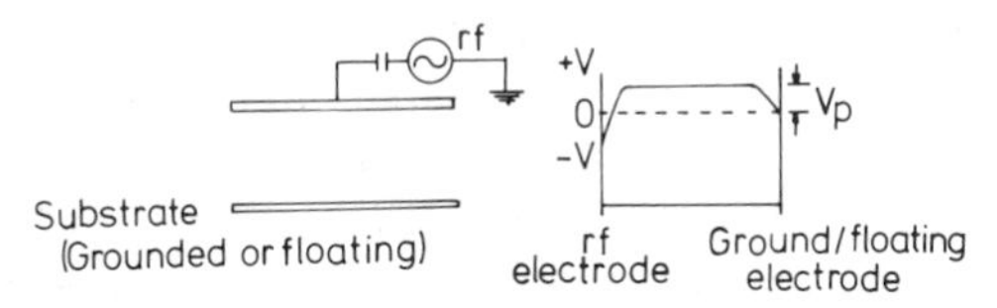

Fig. 3.7. Cross section of the electrode arrangement for a flat-bed reactor and its non-symmetrical potential distribution.

electrode, but significant potentials can develop between the plasma and grounded or floating surfaces in contact with the discharge. Koenig and Maissel [10] have shown that the ratio of these potentials is area dependent. The difference between the plasma potential and that of the powered capacitively coupled surface largely determines how much sputtering of that surface will occur. Ground potential in flat-bed reactors is always negative with respect to the plasma, so the difference between the plasma potential and ground determines the amount of ion bombardment (sputtering) that occurs on grounded surfaces in contact with the discharge. Vossen [11] has determined that the plasma potential and floating substrate and wall potentials in a parallel-plate deposition system and an "inductively coupled" tunnel reactor can vary substantially, depending on the discharge conditions.

Because of the difference in mobility between electrons and ions, the static i–V characteristics of the discharge resemble that of a leaky rectifier as shown by Butler and Kino [12] in Fig. 3.8. If an rf signal is applied to this type of load, a very large electron current will flow during one-half of the cycle while a relatively small ion current flows on the other half of the cycle. Since no charge can be transferred through the dielectric wall (or capacitor) between the rf source and the discharge, the electron flow to the surface must

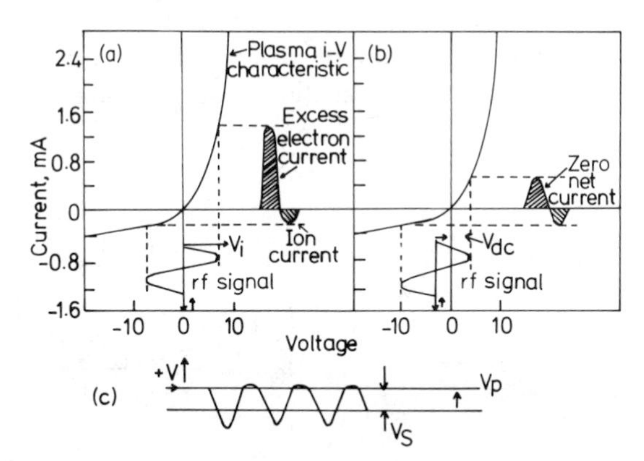

Fig. 3.8. Development of a pulsating negative sheath on a capacitively coupled surface in an rf glow discharge. (From Butler and Kino [12].) (a) Initial and (b) steady-state application of rf voltage that results in (c) stable rf signal. (From Anderson *et al.* [13].)

decrease. With respect to the applied rf signal, this means that the voltage self-biases negatively as shown in Fig. 3.8b, until the stable condition, where equal numbers of electrons and ions reach the surface on each half-cycle, is attained. This pulsating negative signal sketched in Fig. 3.8c by Anderson *et al.* [13] has an average value V_s known as the sheath potential; eV_s is the average energy with which ions enter the surface sheath on their way to colliding with the surface. Ions can enter the sheath at any time, and so the distribution of ion energies ranges from 0 to almost $2eV_s$. It is the negative self-bias that makes the inside walls of tunnel reactors and internal powered electrodes of flat-bed reactors sputtering targets.

There are numerous variants of the basic designs of rf deposition systems described earlier. We may mention here a system described by Rosler and Engle [14], which combines the higher throughput of wafers in tunnel reactors with the uniformity of film thickness and properties in parallel-plate reactors. Sokolowski [15] and Sokolowski *et al.* [16] used a high-energy pulsed plasma rather than a continuous discharge for avoiding excessive substrate heating. Veprek [17] has described a reactor in which a material is transported from a solid phase by plasma-dissociated gas, such as hydrogen, and decomposed by thermal and plasma activation in another part of the reactor.

In general, most investigators in the field of plasma film deposition have used dc and rf discharges (internally or externally excited). Only a few experiments have been reported that use microwave discharges. We cite here the work of Avni *et al.* [18] and Shiloh *et al.* [19]. Typical microwave generators are magnetron or klystron devices radiating at 2.56 GHz. The energy is transferred via waveguides, which are metal tubes usually of rectangular or circular cross section, or resonant cavities (Evenson, Broida, etc.), which are generally cylindrical in shape (Fig. 3.9). Since the glass discharge tube is usually placed along the axis of the cylindrical cavity, the mode of oscillation of the electromagnetic radiation should be chosen so that

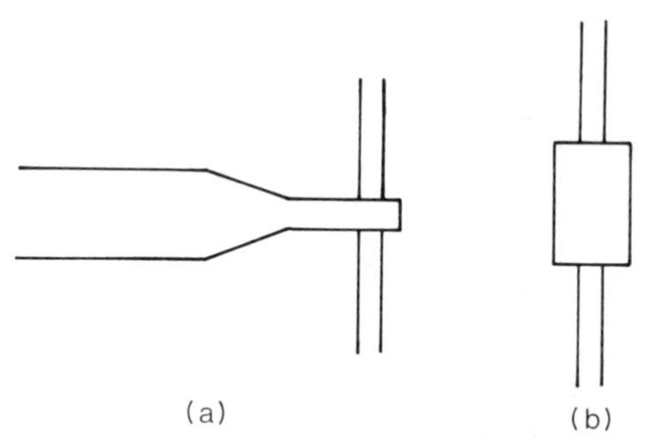

Fig. 3.9. Microwave discharges use (a) waveguides or (b) resonant cavities.

the electric field lines near the center are parallel to the cavity's axis. This condition is fulfilled in the TM_{01} mode, for which the radius of the cylinder should be selected equal to 0.383λ, where λ is the wavelength of the microwave radiation.

There are basic differences in the fundamental mechanism operating at microwave frequencies. In an alternating field, the free electrons oscillate 90° out of phase with the field and cannot derive power from the field (on the average). To acquire energy their collisions with neutral molecules must alter their phase relationship with the field. If we assume that the excitation frequency f is greater than the elastic collision frequency, the maximum displacement x of an electron due to the high-frequency field is given by

$$x = eE/2\pi^2 m_e f^2, \tag{2}$$

where E is the field strength. In a typical microwave excitation, $E = 30$ V/cm, so that for $f = 2.45$ GHz, substitution in Eq. (2) gives $x < 10^{-3}$ cm. The corresponding maximum electron energy acquired during the cycle is eEx, about 0.02 eV. Thus, the electrons slowly accumulate the energy necessary to undergo inelastic, ionizing collisions and to sustain the discharge.

As noted in Section I, only a few percentages of the gas atoms in a glow discharge are ionized. Various configurations to enhance ionization are described by Holland [20]. For increasing the ionization efficiency magnetic fields have been interposed. A magnetic field normal to the cathode surface constrains secondary electrons to follow a helical, rather than straight-line path. The effect is to give electrons a longer path length for a fixed mean free path, thus increasing the probability of ionizing collisions before the electron reaches the anode. The magnetic field pinches the discharge in toward the center of the cathode, resulting in nonuniformity of film thickness. On the other hand, a magnetic field parallel to the cathode surface can restrain the primary electron motion to the vicinity of the cathode and thereby increase the ionization efficiency. However, the $E \times B$ motion causes the discharge to be swept to one side. This difficulty has been overcome by using cylindrical cathodes that allow the $E \times B$ drift currents to close on themselves. This is the general magnetron geometry. Remarkable performance is achieved when end losses are eliminated. The first operation in the magnetron mode for sputter deposition was achieved by Penning [21]. Since 1969, considerable development in this field of work has taken place and the reader is referred to articles by Thornton and Penfold [22], Fraser [23], and Waits [24]. Morosoff *et al.* [25] have reported magnetically enhanced deposition rates in rf glow discharges in ethylene.

In addition to magnetic fields, ionization enhancement may be accomplished by using hot filament cathodes and hollow cathodes. What is distinctive about the hot filament discharge is that electron emission from the cathode is primarily by thermionic emission (rather than by ion impact), which can be increased almost indefinitely in the presence of an anode that withdraws the electron space charge surrounding the filament (usually made of tungsten). In the hollow cathode discharge [5] a cylindrical cathode is usually used and the ionization is increased by faster electrons from one dark space, which enter the other.

C. Processes Occurring in Discharges

There are several pathways by which various kinds of particles are produced and react in a glow discharge. Nasser [6] has given an excellent discussion of the ionization and deionization processes. Kaufman [26] has reviewed the basic physical processes leading to the production of atoms and simple radicals. Because of the complexity involved in discharge phenomena associated with polyatomic molecules of the sort used for deposition of films, say SiH_4, little or no discussion is to be found in the published literature. Glow discharges in water vapor, carbon dioxide, ammonia, and simple hydrocarbons are good examples of the chemical complexity of polyatomic discharge systems [27, 28].

Figure 3.10 summarizes the many processes occurring in glow discharges. The most important ionization processes are the ionization of gas molecules by electron collision and by absorption of radiation, also known as *photoionization* [1]. The ultraviolet/visible photons in glow discharges have

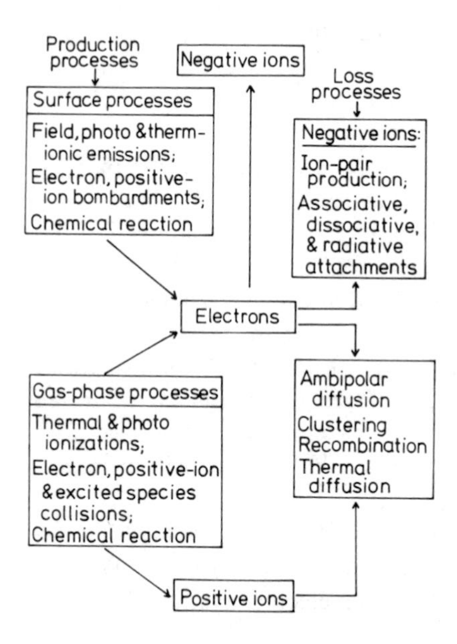

Fig. 3.10. Schematic diagram showing the major possible processes for production and loss of the various kinds of charged particles.

energies in the 3–40-eV range [5]. The formation of negative ions occurs when free electrons are available to attach themselves to neutral atoms or molecules, usually one or two electrons deficient in the outer shells. The widely used rf electrodeless discharges (Fig. 3.6) are almost similar to the positive column of dc discharges (Fig. 3.5), which has a small and constant axial voltage drop and only a small imbalance of charge carriers, because although electrons initially diffuse to the tube wall faster than ions, the resultant radial field prevents further charge separation and forces electrons and ions to diffuse equally fast. This process is called *ambipolar diffusion* and contributes significantly to the disappearance of charged particles.

Because of ambipolar diffusion the currents of electrons and positive ions reaching the wall must be equal, i.e.,

$$\begin{aligned} i_e &= -D_e \nabla n - n\mu_e E_s \\ &= -D_i \nabla n + n\mu_i E_s \\ &= i_i, \end{aligned} \tag{3}$$

where $n = n_e = n_i$ is the plasma density, ∇n the density gradient, D the diffusion coefficient, μ the mobility, and E_s the field caused by the (small) space charge. By eliminating E_s in Eq. (3), it is seen that

$$\begin{aligned} i &= -[(D_e \mu_i + D_i \mu_e)/(\mu_i + \mu_e)]\nabla n \\ &= D_a \nabla n. \end{aligned} \tag{4}$$

This is the defining equation for the ambipolar diffusion coefficient D_a.

The disappearance of charged species by ambipolar diffusion in the absence of a source term is given by the diffusion equation

$$\delta n/\delta t = D_a \nabla^2 n. \tag{5}$$

By solving the equation for the case of an infinite cylinder, it is seen that the diffusion-controlled loss under steady-state conditions with a spatially well distributed source term is closely approximated by a first-order rate constant

$$k/\mathrm{s}^{-1} = 5.78 D_a/r_0^2. \tag{6}$$

In a plasma deposition reactor using a tube 1 cm in diameter, Eq. (6) will yield $k = 1 \times 10^5\ \mathrm{s}^{-1}$, so that electron impact ionization is the major source term for charged species in the discharge. When negative ions are also present, as for example in glow discharges of electronegative gases such as SF_6, their principal effect is to accelerate the ambipolar diffusion of the electrons.

If the Penning mechanism, which involves direct ionization by collision with metastables, were to occur, the ionization energy of the metastable must exceed the ionization potential of the other reactant. Typical energies of metastables are in the 0–20-eV range and are to be compared with the energies of glow discharge electrons (0–20 eV) and ions (0–2 eV).

Several radiative two-body and three-body recombination mechanisms are recognized for the disappearance of the charged particles. Of these, the fast reactions are the dissociative recombination of electrons and positive molecular ions

$$AB^+ + e^- \longrightarrow A + B, \tag{7}$$

where the products may be electronically excited atoms or free radicals. The generally observed range for the rate constant of this type of reaction is 1–5 × 10^{-7} cm^3/molecule s. Such products are also formed by the dissociation decay of molecules excited by electron impact:

$$e^- + AB \longrightarrow e^- + (AB)^* \longrightarrow e^- + A + B. \tag{8}$$

For this event to occur the energy of the electrons must exceed the dissociation energy of the molecule; the transition must occur within the limits set by the Franck–Condon principle. The energies available in many glow discharges (0–20 eV) favor dissociation by electron impact via excited states.

Electron impact excitation of vibrational energy is a highly specific process, the cross section of which can differ by two orders of magnitude from one molecule to another. The dissociation will usually come about via the repulsive upper state that dissociates upon its first pseudovibration, i.e., in about 10^{-14} s (Fig. 3.11a). If the upper state is bound but the molecule is on the repulsive part of its potential energy curve at a point above its dissociation energy (Fig. 3.11b), dissociation will occur on its first vibration, also in about 10^{-13} s. Excitation to a stable excited state followed by a shift to an unstable excited state (Fig. 3.11c) gives rise to a condition known as predissociation. Depending on the degree of mixing of the states, this predissociation process is much slower than the other two processes. The overall cross section and rate constant for dissociation will normally equal

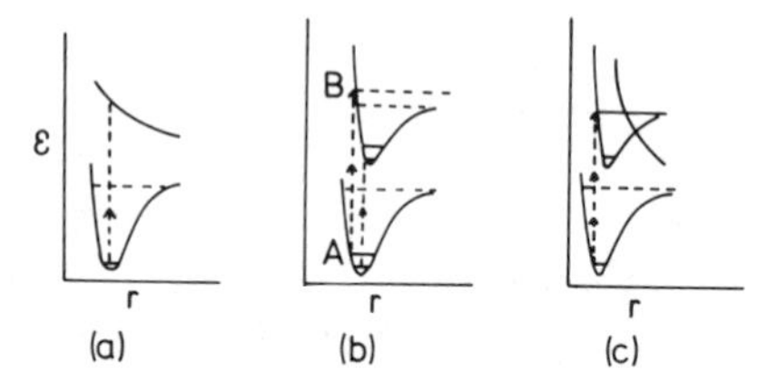

Fig. 3.11. (a) Excitation to unstable excited state with immediate dissociation; (b) excitation to stable excited state *B* at an energy level above the dissociation energy leading to dissociation; and (c) excitation to stable excited state followed by a shift to unstable excited state (predissociation).

that of the primary excitation step. That is to say, there will be near unit probability for the subsequent dissociation, because collisional or radiative lifetimes of the excited states cannot be shorter than about 10^{-7}–10^{-8} s.

Dissociative attachment reactions such as

$$e^- + AB \longrightarrow A^- + B \tag{9}$$

have rate constants in the range of 10^{-12}–10^{-11} cm^3/molecule s and are considered less important sources of atoms and free radicals. Furthermore, many associative detachment reactions such as

$$A^- + B \longrightarrow AB + e^- \tag{10}$$

have rate constants of 1–5 × 10^{-10} cm^3/molecule s so that negative ions are unlikely to be of importance in rapidly pumped steady-state glow discharges. Two-body ion–ion recombinations have rate constants in the same range as Reaction (7) and must be included particularly in glow discharge systems of electron-attaching gases.

Because of strong ion–dipole and ion-induced dipole interactions, charge transfer and ion–molecule reactions are important, i.e.,

$$A^{\pm} + BC \longrightarrow A + BC^{\pm}, \tag{11}$$

$$A^{\pm} + BC \longrightarrow AB^{\pm} + C. \tag{12}$$

If the reactions are exothermic they have little or no activation energy. Frequently, their rate constants are of the order of 10^{-9} cm^3/molecule s and will, therefore, go to completion in a small fraction of the residence time of the gas in the glow discharge.

From the standpoint of surface chemistry it is necessary to consider catalytic effects of which little is known. As Venugopalan and Veprek [29] pointed out, the lack of understanding of adsorption of discharge products by reactor walls and the catalysis of reactions by these walls is a major problem that is associated with all plasma deposition systems. Ion-impact-induced fragmentation of physisorbed molecular layers followed by chemisorption of the fragments and subsequent reaction may lead to deposition. The activation of the gas phase enhances reactions such as

$$A(s) + (m/n)B_n(g) \longrightarrow AB_m(s) \tag{13}$$

whenever the dissociative chemisorption of the species B_n on the AB_m surface is the rate-determining step. When high species fluxes impinge on the surface, the rate-determining step becomes the bulk diffusion. The non-deposition (evaporation, etching, sputtering, etc.) process

$$aA(s) + (b/n)B_n(g) \longrightarrow A_aB_b(g) \tag{14}$$

can be catalyzed either via formation of reactive radicals, e.g., atoms, in the plasma phase or by bombardment of the surface with ions or electrons possessing sufficient energy. A number of these processes have been reviewed by Winters [3].

The review of results to be presented in Section IV will consider many of these processes in important glow discharge deposition systems.

D. Problems under Film Deposition Conditions

From the standpoint of reactor design and physical effects of glow discharges, there are several other processes that are essentially the same as plasma deposition. For example, glow discharge polymerization and etching uses equipment similar to that used for plasma deposition. In plasma polymerization, the starting gas is a monomer, which, when decomposed by the glow discharge, forms species that polymerize into an organic film. In plasma etching the gases used are selected for their ability to produce reactive species upon decomposition, which react chemically with a surface to be etched to form gaseous components that are easily pumped away. With fluorocarbons both polymerization and etching may occur, as Kay *et al.* [30] have demonstrated, depending on slight perturbations in the system. The kinetic and mechanistic aspects of these processes have been reviewed by a number of authors, among them we cite Bell [31], Carmi *et al.* [32], Coburn [33], Flamm and Donnelly [34], and Vinogradov *et al.* [35].

There are as many reactor designs and experimental conditions to consider as there are published studies on plasma deposition. Nevertheless, for understanding the complexities involved in film formation it is sufficient to take a close look at the deposition region. This is usually the substrate area that is subjected not only to plasma radiation but also to fluxes of energetic and reactive charged and neutral particles in various excited states. Figure 3.12 shows the luminous region around a graphite substrate that is floated, grounded, or biased at -100 V in a 16% C_3H_6 + Ar rf induction plasma under otherwise the same experimental conditions. It is obvious that the width of the luminous region, which we shall call the plasma layer (PL), is the largest for the biased substrate and the smallest for the floated substrate. The PL wraps around the grounded and the negatively biased substrate (even in the direction opposing the gas flow) while on the floated substrate the PL is formed only on the surface facing the direction of the gas flow. Furthermore, the PL luminosity is often very much higher than that of the plasma bulk (PB), suggesting that there is a surface neutralization/recombination layer (NL) interposed between the substrate bulk (SB) and the PL. These regions are delineated in Fig. 3.13.

Fig. 3.12. Luminous region around a graphite substrate immersed in a 16% C_3H_6 + Ar rf plasma at 5 torr and 500 W. The substrate is (a) floated, (b) grounded, and (c) biased -100 V. (From R. Avni, unpublished work, 1984.)

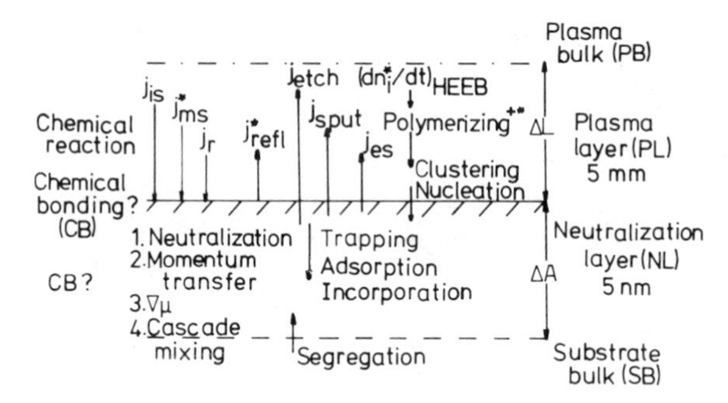

Fig. 3.13. Fluxes of particles and possible processes in the plasma and neutralization layers between the substrate and plasma bulk.

Since electrons are repelled by the potential difference $V_p - V_f$ (Section II.A), it follows that the substrate will acquire a positive space charge sheath around it. According to Bohm [36], between this sheath and the PB there is a quasi-neutral transition region of low electric field, the effect of which is to increase the velocity of the positive ions to enable them to penetrate the sheath and bombard the substrate. Hagstrum [37] has shown that the probability of an ion being neutralized upon collision with a surface is greater than 99% for a number of ions over a range of kinetic energies. Therefore, it is reasonable to conclude that the ion flux j_{is} reaching the substrate surface in a deposition experiment will be neutralized and adsorbed in the NL. Khait *et al.* [38] describe the neutralization of positive ions as short-lived hot spots (SLHS) with a lifetime of the order of 10^{-14} s. Fluxes of excited neutrals such as metastables j_{ms} and radicals j_{rs} will also reach the surface, release their energy, and become adsorbed or trapped in the NL.

Following Hagstrum [39], the possible mechanisms for an ion or excited particle to neutralize at substrate (metal) surfaces involve deexcitation (1) by emission of radiation, (2) by an Auger process* followed by emission of secondary electrons, and (3) by a resonance process whereby an electron is transferred from the conductance band of the metal to an equivalent energy level in the incoming ion or a similar transition where the electron goes from the ion to the metal. Secondary electron emission from the substrate surface back into the plasma bulk takes place by a potential ejection that depends on the ionization potential of the incident ion and/or by a kinetic ejection, which

* The *Auger effect* is the emission of a second electron after high-energy radiation or particle has expelled another. Its mechanism is that the first electron leaves a hole in a low-lying orbital into which a higher-energy electron drops. The energy from this release may result either in the generation of radiation (which gives x-ray fluorescence) or in the ejection of another electron, the "secondary" electron of the Auger process.

depends on the energy of the particle projectile. The loss of energy of the excited particle to the lattice upon impact is a momentum transfer process that sets the atoms in motion by a process known as a "collision cascade." Although a binary collision approximation qualitatively describes the path and energy loss of the incident particle, the consequences of the collision cascade are many and varied: sputtering, back-scattering, lattice damage, and the formation of chemical potential gradients $\nabla\mu$ (Fig. 3.13). This last consequence may enhance diffusion and segregation of atoms from the substrate bulk to the neutralization layer and thus modify the surface chemistry. Numerous studies by Gruen *et al.* [40] have demonstrated gibbsian-type segregation on metallic surfaces.

The yield (potential and kinetic) of secondary electrons (γ_i, γ_m, and γ_r) is inversely proportional to the mass but only slightly dependent on the energy of the incoming projectile. It varies, however, depending on the condition of the surface and is generally lower for nonmetals (higher work function) than for metals. For 600-eV Ar^+ ions Chapman [7] gives

$$\gamma_i = a \times 10^{-2} \tag{15}$$

with $a < 10$ for nonmetals and $a \geqslant 10$ for metals. According to Kaminski [41], $\gamma_m \simeq \gamma_i$ for metals, again with Ar as the incoming projectile.

The secondary electrons traveling the sheath along z (normal to the substrate surface) gain energy from the high electric field E_s and have an energy distribution function $F_{es}(\varepsilon_{es})$ that is anisotropic along z with a maximum around 100 eV. Figure 3.14 shows the values calculated by Khait *et al.* [38] for the field and particle energies in the sheath as a function of gas pressure in a system with a grounded graphite substrate. This high-energy electron beam (HEEB), with average energy $\bar{\varepsilon}_{es}$ and flux j_{es}, collides with the

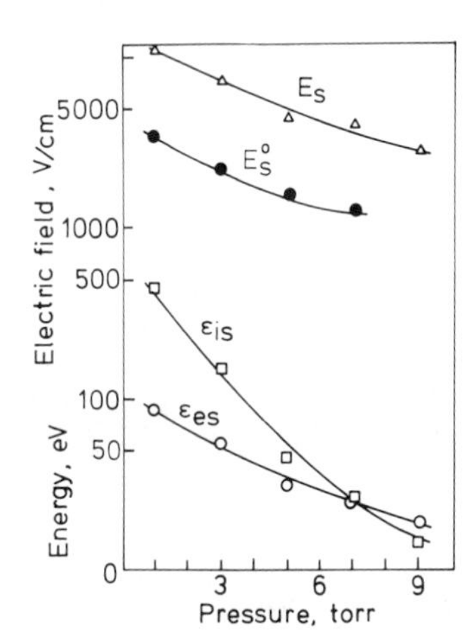

Fig. 3.14. Calculated electric field with (E_s^0, ●) and without (E_s, △) collisions in the plasma sheath, and the ion (ε_{is}, □) and electron (ε_{es}, ○) energies in this region as a function of gas pressure. (From Khait *et al.* [38].)

particles populating the plasma bulk and loses its energy along z. The inelastic collisions with atoms and molecules will result in excitation and ionization and a decrease in HEEB energy to the average value for PB electrons $\bar{\varepsilon}_{PB}$. Thus, a boundary condition can be set along z between PL and PB, i.e.,

$$\bar{\varepsilon}_{es(z=\Delta L)} = \int_{z=0}^{z} \varepsilon_{es} F_{es}(\varepsilon_{es}, \varepsilon_{is}, z, P, W)\, d\varepsilon_{es} \rightarrow \bar{\varepsilon}_{e(PB)}, \tag{16}$$

where ΔL is the thickness of PL, P the gas pressure, and W the net power input.

In what follows we shall consider the evidence from work of Khait *et al.* [38, 42] to show that the contribution of HEEB electrons to excitation and ionization in PL is significantly greater than that of PB electrons in PL. For the ionization process, by analogy with known phenomena produced by the PB electrons, we write

$$dn_i/dt = sj_{es}q_{es}[1 + (a_i n^2 k_i \Delta L/j_{es}q_{es})], \tag{17}$$

where s is the cross section of PL, q_{es} the efficiency of ionization by HEEB electrons, a_i the degree of ionization ($n_e = n_i = a_i n$) in PB, k_i the rate coefficient of ionization in PB, and n the plasma density. Figure 3.15 is a plot of the ratio $(dn_i/dt)_{PL}/(dn_i/dt)_{PB}$ as a function of gas pressure. It is obvious that the contribution by secondary electrons to the ionization in PB is very much greater than the earlier estimates ($\simeq 10\%$) made by Chapman [7] and Winters [3].

The fraction A of the electron distribution function of HEEB performing particle xcitation to a higher quantum level with a cross section $\sigma * (\bar{\varepsilon}_{es})$ is given by

$$A_{(HEEB, z=\Delta L)} = [2\gamma \bar{u}_{(PB)}\sigma * \varepsilon_{e(PB)}/\bar{u}_e \sigma * \bar{\varepsilon}_{es}]B_{(PB)}, \tag{18}$$

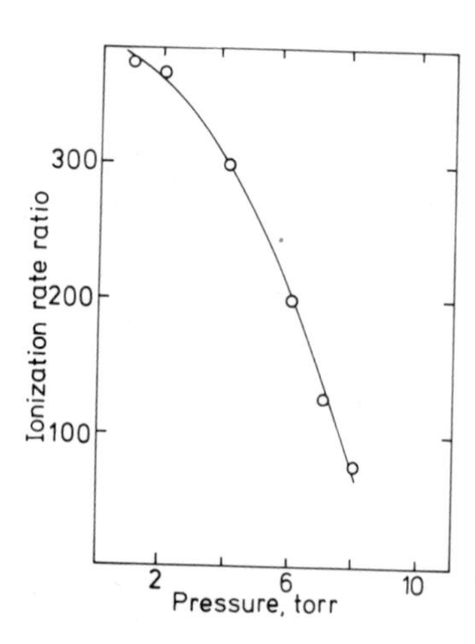

Fig. 3.15. Ionization rate ratio $(dn_i/dt)_{PL}/(dn_i/dt)_{PB}$, as a function of gas pressure.

where B is the fraction of PB electrons causing excitation in PL, $\gamma \ll 1$ the secondary electron coefficient and $\bar{u}_e$ and $\bar{u}_{PB}$ the average velocities of HEEB and PB electrons, respectively, in PL. Figure 3.16 shows the spectral line/band intensities ratio between PL and PB, i.e., I_{PL}/I_{PB} as a function of gas pressure. These measurements were made in a microwave-cavity-excited plasma using graphite and silicon substrates, grounded or biased with -100 V, in flowing pure Ar or mixtures of Ar with 5 v/o of N_2 or H_2. The increase in the ratio of Ar, N_2, CN, and integral intensity (as well as for CH, and C_2 not shown) with increasing gas pressure suggests that decreasing the mean free path rather than ΔL causes more collisions and results in a higher population of the excited states for the various species. The correlation between the ratio of SiH and H_α, and of CN and N_2^+, on the other hand, indicates the interaction of excited positive ions and radicals with substrate surface and/or the sputtered away particles Si and C, respectively.

By neglecting elastic collisions, Khait *et al.* [42] approximated the inelastic collisions causing excitation and ionization to the total number α of collisions with gas particles. In each α collision the HEEB electrons undergo scattering and their motion can be described by a Brownian motion with a drift velocity v_α that decreases rapidly along z toward PB. Accordingly, the thickness ΔL of PL is estimated as

$$\begin{aligned}\Delta L &= \alpha^{1/2}\langle\lambda_{\text{HEEB}}\rangle \\ &= [\bar{\alpha}_{\text{in}}\langle\sigma(\bar{\varepsilon}_{\text{es}(z)})/\bar{\sigma}_{\text{in}}(\bar{\varepsilon}_{\text{es}(z)})\rangle]^{1/2}kT/P\langle\sigma\rangle_{\Delta L},\end{aligned} \tag{19}$$

where σ is the total collision cross section, $\bar{\sigma}_{\text{in}}$ the average value of the cross section for inelastic collisions, $\bar{\alpha}_{\text{in}}$ the number of inelastic collisions, and

$$\lambda_{\text{HEEB}} = 1/n\sigma = kT/P\sigma \tag{20}$$

is the mean free path of HEEB electron in PL at plasma gas temperature T and total pressure P. The same authors [38] considered the electron

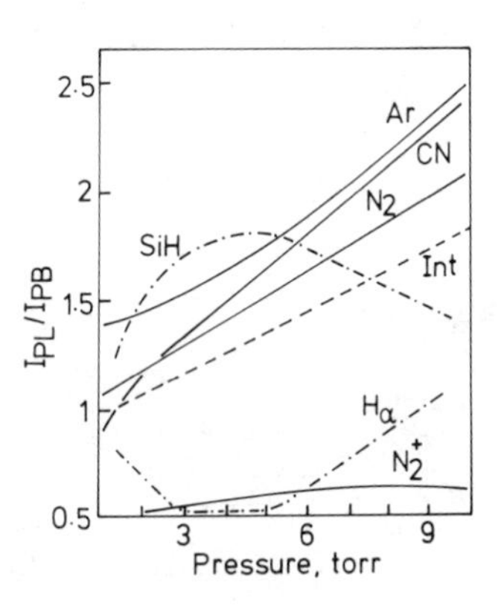

Fig. 3.16. Spectral intensity ratio I_{PL}/I_{PB} versus gas pressure in a microwave discharge at 100 W.

trajectory b due to elastic and inelastic collisions of the HEEB electrons in PL and evaluated

$$\Delta L = b/n\sigma \tag{21}$$

having defined

$$b = ([(\varepsilon_{es} - \bar{\varepsilon}_{es(z=\Delta L)})/\delta\bar{\varepsilon}_{in}]\langle\sigma/\sigma_{ion} + \sigma_{ex}\rangle)^{1/2}, \tag{22}$$

where $\delta\bar{\varepsilon}_{in} = \langle\delta\varepsilon_{ion} + \delta\varepsilon_{ex}\rangle_{\Delta L}$ is the energy lost by the average HEEB electrons in inelastic collisions for ionization (ion) and excitation (ex) in PL, while σ_{ion} and σ_{ex} are the inelastic collision cross sections resulting in ionization and excitation, respectively. By combining Eqs. (19), (21), and (22), we finally get the following relationship for the thickness of PL

$$\Delta L = ([(\varepsilon_{es} - \bar{\varepsilon}_{es(z=\Delta L)})/\delta\bar{\varepsilon}_{in}]\langle\sigma/\sigma_{ion} + \sigma_{ex}\rangle)^{1/2}kT/P\langle\sigma\rangle_{\Delta L}. \tag{23}$$

Note that the thickness of PL is dependent on P, T, the nature of the gas particles, and the power input into the plasma. Figure 3.17 shows that values of ΔL measured by optical emission of the excited gas and sputtered particles in PL are in reasonable agreement with the calculated values.

Besides secondary electrons there are other particles released from NL into PL, such as reflected particles j_{refl}, sputtered atoms j_{sputt}, and chemically etched molecules j_{etch} (Fig. 3.13). As previously noted, the sputtering and etching phenomena influence film deposition. Sputtering experiments can provide information on the thickness of NL. In reviewing the work of a number of authors Winters [3] concluded that the thickness of NL is about 30 Å. Naturally, this will depend on the mass, energy, and angle of impact of the incident ion as well as on the surface conditions, such as crystal lattice and surface coverage, and type of substrate (polycrystalline, amorphous,

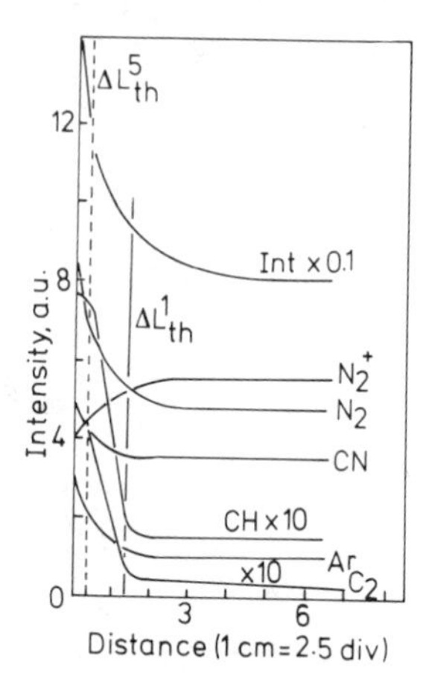

Fig. 3.17. Spectral intensity as a function of the distance from substrate surface in a microwave discharge at 1 torr and 100 W.

metallic, or semiconductor). If we sum all ions arriving from PB and formed in PL by interaction with HEEB electrons we get for the incident flux of ions

$$j_{is} = \frac{1}{4} \sum a_i n u_i . \tag{24}$$

Their maximum energy ε_{is} depends on $V_p - V_s$ or $V_p - V_S$, where V_s and V_S are the sheath and substrate potentials, respectively, and are a function of

$$\varepsilon_{is} \propto f[(kT_e/m_i)^{1/2}; P, W, E_s, F_{es}(\varepsilon_{es}), V_s, V_S]. \tag{25}$$

In glow discharges, PL usually tends to follow the contour of the surface, and, therefore, most ions approach the surface at normal incidence.

The sputtering yield, which is defined as the number of substrate (target) atoms ejected per incident ion, has been determined for numerous metals and a variety of ions with different energies. The interested reader is referred to tabulations by Laegried and Wehner [43], Kaminski [41], Vossen and Cuomo [44], Chapman [7], Roth [45], and Thornton [46]. We have collated some data in Table 3.1 for those elements that may be encountered in one glow discharge deposition system or another. These data suggest that sputtered particles with a flux 0.1–3 times that of incident ions ($n_i = 10^{11}$ cm^{-3}) will enter PL, where they will undergo excitation and/or ionization by interacting with PB and HEEB electrons. Therefore, some of the sputtered particles will return to the surface as ions and enhance its chemistry, for example, as in ion nitriding, boridizing, and carburizing processes.

Some of the sputtered particles will contribute to cluster formation in PL (Fig. 3.13). Table 3.2 gives data for the degree of ionization and the normalized amount of polymerized species in both PB and PL (around a grounded graphite substrate). In comparison with PB there is a higher concentration of positive ions (n_i) and excited polymerized (n_{poly}) molecules in PL because of the interaction of sputtered particles with HEEB electrons. Along the z axis, toward the substrate surface, where a large number of collisions with HEEB electrons take place, n_i and n_{poly} increase to such magnitudes that the supersaturation condition described by Lothe and Pound [49] for the formation of clusters is established. Phenomenologically, we may consider two processes along the z axis from PL toward the substrate surface:

(1) The ionization rate due to HEEB electron interaction is higher near the sheath than at the boundary between PL and PB. Near the sheath the HEEB electrons reach their maximum energy (Fig. 3.17). The ions thus formed with a long-range dipole moment attract and attach polymerized molecules such that polymer clusters are formed. Evidence for prenucleation near the surface has been found in experiments using focused UV lasers by Ehrlich *et al.* [50], Brueck and Ehrlich [51], and Ehrlich and Tsao [52].

TABLE 3.1

Sputtering Yields for Argon Ion Bombardment at Various Energies[a]

	Threshold (eV)	Ar^+/eV				
		60	100	200	300	600
Ag	15	0.22	0.63	1.58	2.20	3.40
Al	13		0.11	0.35	0.65	1.24
Au	20		0.32	1.07	1.65	2.43
Be	15		0.074	0.18	0.29	0.80
Cr	22		0.30	0.67	0.87	1.30
Cu	17	0.10	0.48	1.10	1.59	2.30
Fe	20	0.064	0.20	0.53	0.76	1.26
Ge	25		0.22	0.50	0.74	
Mo	24	0.027	0.13	0.40	0.58	0.93
Nb	25	0.017	0.068	0.25	0.40	0.65
Ni	21	0.067	0.28	0.66	0.95	1.52
Pd	20		0.42	1.00	1.41	2.39
Pt	25	0.032	0.20	0.63	0.95	1.56
Re	35		0.10	0.37	0.56	0.91
Si			0.07	0.18	0.31	0.53
Ta	26	0.01	0.10	0.28	0.41	0.62
Ti	20		0.081	0.22	0.33	0.58
V	23	0.03	0.11	0.31	0.41	0.70
W	33	0.008	0.068	0.29	0.40	0.62
Zr	22	0.027	0.12	0.28	0.41	0.75

[a] The yield S is given as $S/(1 + \gamma)$, where γ is the secondary electron emission coefficient. Data collated from Laegreid and Wehner [43] and Stuart and Wehner [47].

TABLE 3.2

Ionization Degree and Polymerization Degree in A 16 v/o C_3H_6 + Ar rf Plasma Bulk and in Plasma Layer (Graphite Substrate)[a,b]

	Plasma bulk	Plasma layer
n_i/n	3×10^{-5}	5×10^{-3}
n_{poly}/n	2×10^{-1}	$>8 \times 10^{-1}$

[a] From Inspektor *et al.* [48].

[b] Measurements made in position F with respect to Fig. 39, i.e., a position immediately beyond the plasma; n is the concentration of C_3H_6 in the gas mixture; pressure is 1–6 torr; power is 150 W at 13.56 MHz.

(2) Polymerized species in the PL are in a three-coordinate system, but they lose one coordinate upon reaching the surface and must remain on the solid in a two-coordinate system. As de Boer [53] pointed out, this increases their concentration as well as the probability for clustering and nucleation.

Thus, we see that many of the particles coming towards the surface react with the substrate, perhaps by chemical bonding, and form diffusion-controlled layers or deposits of films in NL. Ion-enhanced dry chemical etching takes place in the opposite direction, particularly in systems containing hydrogen and halogens. The etching process returns species deposited on the surface back to PL at rates that depend upon etch parameters described by Flamm and Donnelly [34] and Coburn [33]. Both processes proceed simultaneously; depending on their rates, deposition or etching will result. Apparently, under film formation conditions deposition prevails because of a higher rate of deposition than etching. The fluxes of ions, metastables, and radicals are continuous, as are the sputtering and etching losses during deposition. Consequently, we may expect the material balance in the film to change with deposition time.

In summary, the "dialogue" between the plasma of a glow discharge and the substrate immersed in it involves the following processes (Fig. 3.13):

(1) impact of positive ions, excited radicals and metastable states with the grounded or negatively biased substrate surface;

(2) neutralization, momentum transfer, trapping, adsorption, and incorporation of these species in the surface resulting with the formation of NL in which, due to a gradient of chemical potential and cascade mixing, diffusion and segregation of particles from NL to SB and vice versa occurs;

(3) emission of secondary electrons (HEEB), sputtered particles, and chemically etched species from NL back to the plasma forming the PL; and

(4) enhancement of excitation and ionization processes in PL by HEEB electrons. As a result cluster formation and nucleation take place leading to film deposition.

III. DIAGNOSTICS: THEORY AND EXPERIMENT

The history of homogeneous reactions in the PB is transmitted along to the heterogeneous plasma–surface interactions controlling the film deposition processes. This continuity approach is schematically shown in Fig. 3.18. The monomeric carrier gases or vapors fed to the plasmas are partly excited (k_{ex}) and ionized (k_i) by energetic electrons; interaction with fresh monomer molecules results in its dissociation (k_D) and fragmentation (k_F) reaction rates. The homogeneous reactions with fresh monomer molecules

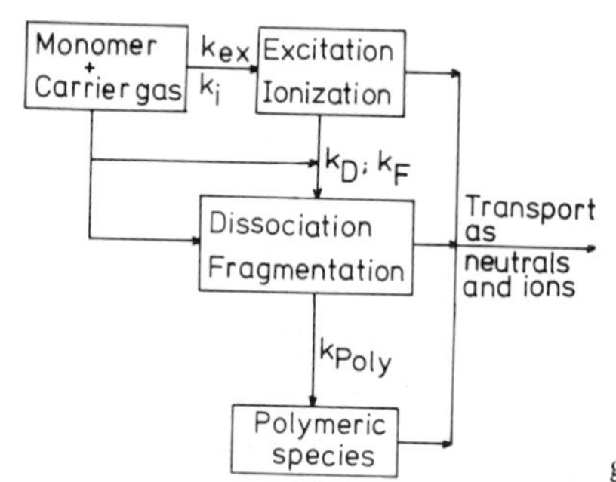

Fig. 3.18. Schematic representation of homogeneous reactions in plasmas.

continues with the formation of new polymeric species with k_{Poly} reaction rates. Transported by convection, diffusion, electric fields, and the flow velocity in the plasma system, these new particles reach the substrate surface region. The kinetic mechanisms and the rate determining step of monomer dissociation and formation of polymeric species are the link between the PB homogeneous reactions and the plasma–surface heterogeneous reactions.

Methods for determining the temperature, density, and composition of plasmas are the subject of an important part of experimental plasma physics that has come to be called "diagnostics." Many physical phenomena occurring in plasmas can be utilized to provide diagnostics. For example, the intensity and spectral composition of radiation emitted by a plasma depends on the electron energy, its distribution function, and the plasma density. With proper care, the electron density, its temperature, and energy distribution in a plasma can be determined simultaneously by an electrostatic (Langmuir) probe. Methods such as mass spectrometry and electron paramagnetic resonance can provide information on atomic and molecular, ionic, and free radical species in the plasma. The interpretation of these measurements involves theories with limiting assumptions. The literature is abundant with general as well as specialized treatises on plasma diagnostics. (See, for example, Huddlestone and Leonard [54], Lochte-Holtgreven [55], Podgornyi [56], Venugopalan [1], and Eubank and Sindoni [57].)

The confident characterization of a given plasma generally requires that several measurement methods give a consistent picture. Unfortunately, we find, particularly in the early publications, only rarely the application of even one diagnostic tool in film deposition studies. Many recent works, however, use a combination of methods such as probes and mass spectrometry and occasionally include emission spectroscopy. Even then it is difficult to assess whether proper care has been exercised in the measurement and interpretation. We must know the fluxes and energies of ions and electrons incident usually on the substrate surface. Because the energy of positive ions incident on surfaces is determined by $V_p - V_s$, it is necessary to know these potentials as well. Even if spatially resolved measurements are made from as close a position as the substrate surface to the other end of the glow the fluxes thus obtained will include for a given species not only those responsible for film deposition but also those that are etched and sputtered in deposition processes. In those cases in which deposition has been reported outside the glow region, it is also necessary to perform diagnostics of the afterglow and sometimes even the foreglow regions. While such cases have resulted occasionally in good-quality films, clearly due to the absence of positive ion bombardment, these are more often the result of poor apparatus geometry or configuration and parameter control. The following discussion of theory and experiment is directed at some of these problems as well.

A. Spectroscopic Methods

Spectroscopic plasma diagnostics can give information about the concentration of various species and their temperatures without disturbing the plasma. The simplest technique is the direct analysis of the plasma emission using a spectrograph with adequate resolution. Line radiation in the visible and IR spectra is useful for identifying excited species. Relative line intensities give the temporal and spatial variations of species in the PB and PL. Under some conditions it is possible to relate absolute line intensity measurements to the density of energetic electrons. Absorption measurements using a suitable radiation source can be used to determine the population of nonradiating metastable levels or first-excited states that radiate in the vacuum UV. Both emission and absorption spectroscopies are nonintrusive *in situ* techniques for investigating and monitoring the complex processes occurring in the glow discharge PB and PL. In the following section we briefly survey the theory and experiment involved in these methods of diagnostics. We shall also review the more recent methods of laser-induced fluorescence spectroscopy and anti-Stokes Raman scattering spectroscopy.

1. Emission Spectroscopy

The radiation emitted from a glow discharge may exhibit a variety of spectra as indicated in Table 3.3, which is taken from Cabannes and Chapelle [58]. Its analysis by spectroscopic measurements of continuum intensities, line intensities, line profiles, etc. is the basis of the diagnostic technique of emission spectroscopy. Although a wide range of frequencies is emitted, it is the optical region that has been extensively studied. The optical radiation is due to radiation arising from the capture of free electrons into various excited states of the atoms and ions and from the acceleration of electrons (bremsstrahlung) during collisions with ions or atoms. Its interpretation is

TABLE 3.3

Spectra Observed in Glow Discharge[a]

Particle	Degree of freedom	Type of spectrum	Spectral region
Atom or ion	Electronic excitation	Line	UV–VIS–IR
	Ionization	Continuum	UV–VIS–IR
	Translation	Line profiles	
Electrons	Recombination	Continuum	UV–VIS
	Free–free transitions	Continuum	IR
Molecules	Rotation	Line	Far IR
	Vibration–rotation	Band	IR
	Electronic excitation	Band systems	UV–VIS–IR

[a] From Cabannes and Chapelle [58]. Copyright 1971 John Wiley & Sons, Inc.

based on the assumption of an electron distribution determined exclusively by particle collision processes.

The glow discharges used in film deposition operate at gas pressures in the range of 10^{-2} to 10 torr, meaning that local thermal equilibrium (LTE) conditions do not prevail in the plasma. The temperature values of the electrons (T_e) are much higher than its values for excitation (T_{ex}), ionization (T_i), or translation (T_{gas}). Under such conditions the absolute or relative spectral line intensities for the evaluation of T_{ex} and T_i should be treated with great care. The same care has to be applied in measuring the spectral line profile for Stark, Doppler, or pressure broadenings (n_e, T_e, and T_{gas} values, respectively). At low pressures, where plasma density $n_e \geqslant 10^{12}$ cm^{-3}, the normal procedures of Stark broadenings for H or He line spectra, which are valid for $n_e \geqslant 10^{15}$ cm^{-3}, cannot be used (see [55, 59]). Instead, spectral lines with high quantum number may be used, for example, $H_n > 10$ in the hydrogen Balmer series as indicated by Bergsted *et al.* [60].

The experimental techniques have been described by a number of authors. Griem [59], Turner [61], Kimmitt *et al.* [62], and Lochte-Holtgreven and Richter [55] are excellent sources of information. Greene and Sequeda-Osorio [63] and Greene [64] describe experimental setups suitable for glow discharge sputter deposition systems; Harshbarger *et al.* [65] describe geometries convenient for plasma etching; and Matsuda *et al.* [66], Kampas and Griffith [67], Taniguchi *et al.* [68], Inspektor *et al.* [48], Perrin and Delafosse [69], Knights *et al.* [70], and many others describe apparatus for glow discharge decomposition systems that deposit films. Apparatus for a typical glow discharge optical spectroscopy (GDOS) work is shown in Fig. 3.19. A series of slits placed along the optical axis between the discharge reactor and the spectrometer are used to focus the spectrometer on a narrow

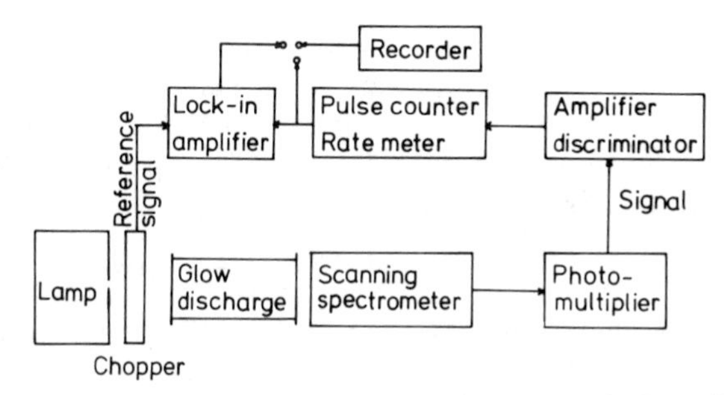

Fig. 3.19. Schematic diagram of a glow discharge optical and absorption spectroscopy system.

region of the glow along the reactor axis. The optical windows should be placed as far from the discharge as possible to minimize loss of transmission by film deposition. With adequate design it is quite possible to obtain intensity versus position profiles in the reactor. In the bell-jar systems described originally by Bradley and Hammes [71] and modified by several others (for example, Williams and Hayes [72]), optics extending through the base plate may be used to achieve the same results. Either dc or ac optical detection schemes may be employed. In case where maximum sensitivity is desired, pulse counting techniques are preferable to minimize random electronic noise from the photomultiplier tube.

The emission intensity at any position x in the discharge for a given transition is governed by the excitation efficiency or probability, which depends on n_e and $f(E)$, and is generally unknown for the complex species encountered in many deposition systems. Therefore, for quantitative diagnostic information it is necessary to calibrate the signal intensity with standard samples under known and comparable discharge conditions. While this can be done for the stable molecular species without much difficulty, it is impossible with reactive intermediates such as free radicals. Consequently, in most instances the optical emission from deposition plasmas is monitored for the presence or absence of lines to signify the spatial distribution of species of interest for several plasma parameters. For this last case the following approach described by d'Agostino *et al.* [73, 74] is typical.

If the direct electron impact from the ground to the emitting state is by far the most important excitation process we can write

$$\mathrm{e} + \mathrm{X} \xrightarrow{k_e} \mathrm{X}^* + \mathrm{e}, \tag{26}$$

$$\mathrm{X}^* \xrightarrow{k_R} \mathrm{X} + h\nu, \tag{27}$$

$$\mathrm{X}^* + \mathrm{M} \xrightarrow{k_Q} \mathrm{X} + \mathrm{M}, \tag{28}$$

where X* is the emitting state and X any lower state including the ground state. The ks represent the appropriate rate constants for the direct electron impact excitation process (k_e), the radiative process ($k_R = 1/\tau$, where τ is the natural lifetime), and for collisional deactivation (k_Q). At steady state, the intensity of optical emission is given by

$$I_{x^*} \propto k_e n_e[X]/(1 + [M]k_Q/k_R) = C\eta_x[X]. \quad (29)$$

The equality is obtained by assuming $[M]k_Q/k_R$, which is generally less than unity, to be a constant at constant pressure. The excitation efficiency $\eta_x = k_e n_e$ is clearly a function of the density and energy distribution function of the electrons, which are all related to the discharge parameters.

2. Inert Gas "Discharge" Actinometry

The possibility of using an actinometric technique for plasma diagnostics was suggested by Kaufman [75]. Since then increasing attention has been paid to a technique that has come to be known as inert gas (discharge) actinometry.

Coburn and Chen [76] added a small amount of Ar to a reactive plasma and monitored the Ar emissions concurrently with those of the reactive particle (X):

$$I_{Ar} = k_{Ar} n_{Ar} \eta_{Ar}, \quad (30)$$

$$I_x = k_x n_x \eta_x, \quad (31)$$

where k_{Ar} and k_x are proportionality constants. Since n_{Ar} is known, the excitation efficiency η_{Ar} of any of its levels is determined simply by I_{Ar}/n_{Ar}. If the excited state responsible for the Ar ($\simeq$13.5 eV) optical emission matches closely in energy with the level responsible for an emission line from a reactive species (F, $\simeq$14.5 eV), then the same group of electrons will be responsible for the excitation of both levels. The excitation efficiencies of these levels of Ar and the reactive particle will then have similar dependence on glow discharge parameters and we may assume $n_x = k\eta_{Ar}$ as discharge parameters are varied. Thus, the relative changes in the reactive particle density can be determined by combining its emission intensity with the excitation efficiency of Ar:

$$n_x/n_{Ar} = KI_x/I_{Ar}, \quad (32)$$

where K is a constant independent of the discharge parameters. Coburn and Chen [76] tested the technique indirectly in a $CF_4 + O_2$ system by establishing a correlation between F atom emission intensities, normalized to Ar emission intensities and F atom concentrations measured downstream of the discharge by chemical titration.

Similar work was subsequently reported by d'Agostino *et al.* [73, 74, 77] who used both Ar and N_2 as inert gas actinometers in $CF_4 + O_2$ and $SF_6 + O_2$ glow discharges; Tiller *et al.* [78] used Ar in CCl_4 discharges; and Ibbotson *et al.* [79] used Ar in Br_2 discharges. Ibbotson *et al.* [79] in a direct test found that normalized Br atom emission, from a Br_2 plasma with an Ar actinometer, correlates with Br atom concentrations determined by *in situ* Br_2 optical absorption. However, Donnelly and Flamm [79] have obtained time-resolved emission spectra that suggest that Cl emission, normalized to Ar, from a Cl_2 discharge may not be a reliable measure of the instantaneous Cl atom density. Further work, which will establish directly and unambiguously the regions of validity, is called for before actinometry is widely employed for diagnostic purposes.

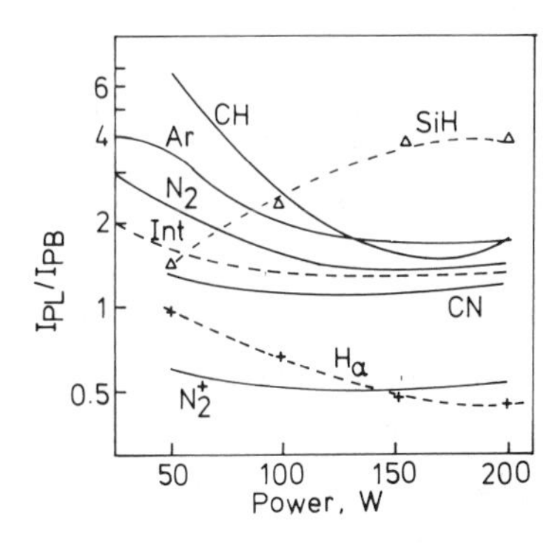

Fig. 3.20. Spectral intensity ratio I_{PI}/I_{PB} versus microwave power in a discharge at 1 torr.

Emission spectroscopy was used to measure the relative intensity ratio of the same spectral line and band in the PL and in the PB (I_{PL}/I_{PB}) as shown in Fig. 3.16. The I_{PL}/I_{PB} ratio allows the correlation between the projectile particles to and the sputtered particles from the substrate surface as function of rf power input as shown in Fig. 3.20 for H and SiH on a single Si crystal and for N_2^+ and CN on a graphite substrate. The relative spectral intensity was used for scanning the emission spatial distribution in the PL. Selwyn and Kay [80] reported the use of optical fibers for spatially resolving the emission of fluorocarbons in rf plasmas.

3. Absorption Spectroscopy

The major drawback of emission spectroscopy is that information is obtained only about species that are already excited. This may often represent only a very small proportion of the total number of the species in question. Absorption spectroscopy would be advantageous here, but higher resolution is then required because with low resolution sharp lines are readily lost in the background. Another difficulty is the possible absence of active species (transients) in concentrations sufficient to give rise to detectable absorption in easily accessible regions of the optical spectrum. Nevertheless, with high resolution the *relative* populations in the different rotational and vibrational levels can be deduced. It should be noted that the intensities of interest may often be modified by absorption due to other parts of the species, or by overtones and combination bands.

In a typical experimental setup a lamp or laser beam that emits the appropriate frequency is used as the radiation source (Fig. 3.18). In sputter–deposition systems a hollow-cathode lamp containing the element of interest is desirable. The light beam is modulated and the absorbance of the discharge is determined by using a phase-sensitive lock-in detector synchronously tuned to the chopper frequency. Tuned ac detection, of course, subtracts the discharge emission. Depending on the setup it may be necessary to apply corrections for self-absorption in the lamp and the possibility that the spectral bandwidth of the spectrometer is larger than both the emission linewidth from the lamp and the absorption linewidth.

The absorption of radiation in a plasma can be described by the Lambert–Beers law

$$I_\nu(l) = I_\nu(0)\exp(-\kappa_\nu l), \tag{33}$$

where $I_\nu(0)$ and $I_\nu(l)$ are the specific intensities of the radiation entering and leaving the plasma, l is the length of the absorbing path, and the absorption coefficient κ_ν is given by

$$\kappa_\nu = \sum_k n_k \sigma_{\nu k}. \tag{34}$$

In Eq. (34) the summation is over all absorbing species and states, n_k are the number densities and $\sigma_{\nu k}$ the absorption cross sections. If the incident radiation has a frequency $\nu = (E_m - E_n)/h$, then line absorption according to equation

$$X_n + h\nu \xrightarrow{\sigma_{nm}} X_m \tag{35}$$

will be the dominant process if a strong transition $n \rightarrow m$ is chosen. Note that photoionization from the ground state is excluded in this case. For resonance lines and the strongest nonresonance transitions, photoionization from some excited states is negligible and a measurement of $\kappa(\nu)$ will yield the population n_n of the lower state. A discussion of detailed theory and experiment is given by Cabannes and Chapelle [58]. Examples of the diagnostic application of absorption spectroscopy to glow discharges are to be found in the papers by Greene [64], Klinger and Greene [81], Knights *et al.* [70], and many others.

Use of absorption spectroscopy in the visible and IR regions is frequently hindered by the plasma emission. Therefore, species effusing from the plasma, through a small orifice, have been trapped and isolated in a solid matrix of an inert gas at cryogenic temperatures (4–20 K) and subsequently identified by absorption spectroscopy. This is the technique of *matrix isolation spectroscopy* described in the classical books of Bass and Broida [82] and Minkoff [83]. Its application for elucidating the sputtering process has been described by Gruen *et al.* [84]. In the apparatus used by Veprek *et al.* [85] for glow discharge deposition, the molecular beam effuses from the discharge tube through a small orifice and is trapped on a liquid-helium-cooled, optically polished copper block simultaneously with a suitably chosen matrix gas such as Ar, Kr, Xe, or N_2. After the desired matrix has been deposited, the copper block, together with the reservoir for the liquefied gases, is rotated through 180° and the absorption spectrum of the matrix is scanned through a KBr window. A large number of papers demonstrate the applicability of this technique for detecting and measuring the concentrations of free radicals and atoms as well as ions that are present in glow discharges without the problems resulting from plasma radiation and electron impact fragmentation, as in mass spectrometry (Section III.C).

4. Fluorescence Spectroscopy

A particle excited by the absorption of radiation may undergo radiative transitions in which light of the same or a different frequency is emitted. When such transitions occur between states of the same multiplicity the particle is said to emit *fluorescence* radiation, the intensity of which is proportional to the particle density. In principle the density of molecules, atoms, and ions in the ground state as well as in a metastable or unstable excited state can be determined. Since fluorescence rise times are considerably shorter than a microsecond, it is necessary to use laser pulse widths of the order of nanoseconds. Therefore, the technique has come to be known as *laser-induced fluorescence* (LIF). The importance of this technique as a plasma diagnostic tool was recently emphasized by Miller [86] in a review article.

With a continuous wave (CW) laser it is possible to continuously excite and detect fluorescence, but at a level much lower than the pulsed laser and often below the noise level resulting from the background emission. A tunable pulsed dye laser has a wider wavelength coverage (2500–8500 Å), and the number of photons available during a pulse is considerably greater than that available from a CW laser. By saturating a given transition with the pulsed laser, all the excited molecules can be made to emit in a time period comparable to the excited state lifetime. This laser-induced signal may be considerably greater, over that given time period, than even a relatively bright continuous background emission from the plasma. The high sensitivity of this technique was demonstrated by Bondybey and Miller [87]. The technique also provides excellent spatial and temporal resolution.

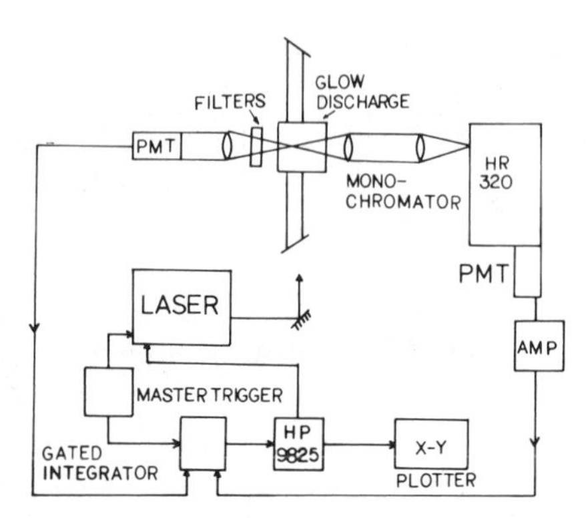

Fig. 3.21. Schematic diagram of laser fluorescence experiment. (From Miller [86].)

Figure 3.21 is a schematic diagram of the setup suggested by Miller and Bondybey [88] for an LIF experiment. The plasma reactor is equipped with long arms containing light baffles and Brewster-angle windows to minimize laser light scattering. The laser beam passes through the plasma along this axis and fluorescence is detected in a plane perpendicular to this axis by a photomultiplier. Often a diagnostic arrangement is configured with a pulsed dye laser and a gated detection system, as shown in Fig. 3.21, to reject most of the continuous background emission from the reactor. An excitation spectrum is recorded by sweeping the laser frequency and monitoring the total fluorescence. In some experiments monochromator-based optical detection systems have been used to monitor the continuous emission from the reactor and/or to resolve the LIF wavelength resulting from a feature of the excitation spectrum obtained by tuning the laser.

Initially LIF was used mainly in the field of plasma wall interaction in magnetic confinement experiments and laser fusion research. A number of authors, among them Stern [89], Burakov *et al.* [90], Burakov [91], Bogen and Hintz [92], and Elbern *et al.* [93], demonstrated that besides neutral H and D atoms, several metals can be detected with high sensitivity. Elbern [94] has reported measurements on the relaxation and diffusion of Fe atoms in an argon glow discharge. Subsequently, Pellin *et al.* [95] and Wright *et al.* [96] used both CW and pulsed lasers to measure the velocity distribution of sputtered Zr atoms.

A number of species of the type encountered in plasma chemistry have also been observed by Cook *et al.* [97], Katayama *et al.* [98], Miller and Bondybey [99], Bondybey and Miller [100], Miller *et al.* [101], and Sears *et al.* [102]. These include metastable states such as $N_2(A^3\Sigma_u^+)$ and $CO(a^3\Pi)$, whose presence would be very difficult to infer from emission spectroscopy or mass spectrometry. Molecular ions (N_2^+, O_2^+, CO^+, CO_2^+), including moderately large organic cations such as $C_6F_6^+$, $C_6H_3F_3^+$, and their analog, have also been detected, but, in general, the determination of absolute concentrations has been difficult. Free radicals such as CH, CH_2, CN, and OH were observed in combustion experiments rather than glow discharge tubes. A review of this work is given by Eckbreth *et al.* [103]. With regard to glow discharge deposition the LIF technique has yet to be applied. Of related interest is the work of Brueck and Pang [104], who made spatially resolved measurements of CF_2 concentrations and vibrational and rotational temperatures in CF_4, CF_4/O_2, and CF_4/H_2 glow discharges of the type used for etching of Si and SiO_2.

Walkup *et al.* [105] have combined LIF with laser optogalvanic (LOG) spectroscopy to study a glow discharge in N_2. The LOG signals were measured by monitoring the voltage across the discharge through a dc blocking capacitor and processing this voltage with a boxcar averager. The spatial distribution of N_2^+ (which is closely related to the distribution of electric fields) was obtained by LIF and LOG methods. The measurements, which included kinetic (translational) and internal (rotational and vibrational) energy distributions of N_2^+, suggested the importance of charge exchange collisions, particularly in the cathode sheath region. However, the kinetic energy of 2.2 ± 0.7 eV for N_2^+ in a 1-torr N_2 discharge measured by Doppler broadening is really too low. Nevertheless, with further development LOG spectroscopy will provide a uniquely sensitive probe of positive ions in the substrate region of glow discharge deposition systems.

5. Coherent Anti-Stokes Raman Spectroscopy

Several systems have been observed to scatter radiation. The inelastic scattered radiation occurs at lower (Stokes lines) or higher (anti-Stokes lines) frequencies than the frequency of the incident radiation, corresponding to the rotational or vibrational transition frequencies (or both) of the sample. This property is the basis of a technique that was initially demonstrated by Rado [106] and has come to be known as coherent anti-Stokes Raman spectroscopy (CARS). With respect to glow discharges and plasmas in general, CARS is better suited than fluorescence because of its excellent background light rejection capability. Besides the shorter-wavelength side spectral position of the signal, the other advantages are the high signal power, signal coherency, and good spatial resolution. For a review, see Nibler [107].

A schematic diagram of a CARS setup used by Hata *et al.* [108] is shown in Fig. 3.22. Typically, two laser beams of frequency ν_1 and ν_2 ($\nu_1 > \nu_2$) are colinearly guided into a glow discharge to produce a $2\nu_1 - \nu_2$ signal beam through nonlinear (third-order) susceptibility $\chi^{(3)}$ of the sample. The absolute value of $|\chi^{(3)}|$ is a function of ν_1 and ν_2 and resonantly increases as $(\nu_1 - \nu_2)$ approaches the Raman active transition frequency ν_R of the species under

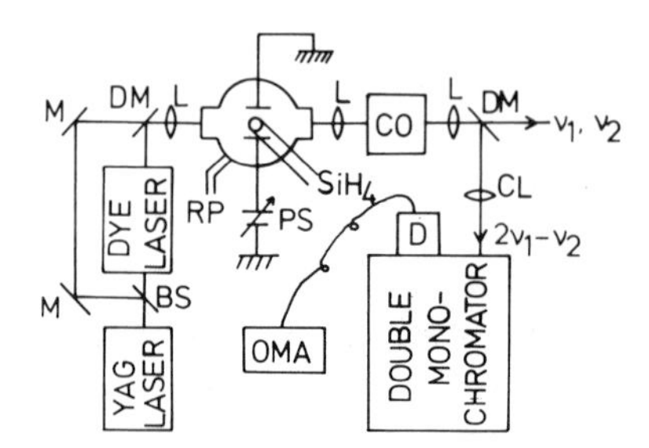

Fig. 3.22. Block diagram of a CARS system: BS, beam splitter; M, mirror; DM, dichroic mirror; L, lens; RP, rotary vacuum pump; PS, dc power supply; CL, cylindrical lens; D, multichannel detector. (From Hata *et al.* [108(a)].)

investigation in the discharge. Therefore, the CARS power signal, which is proportional to $|\chi^{(3)}|^2$, can be treated as a function of ν_2 when ν_1 is fixed and will exhibit some structure near the frequency $\nu_2 = \nu_1 - \nu_R$. Alternatively, ν_1 may be mixed with a broad ν_2 source and a broad range of the CARS spectrum may be detected simultaneously by using an optical multichannel detector attached to a monochromator exit.

Figure 3.23 shows the CARS power signal as a function of $(\nu_1 - \nu_2)$ for a 5-mA dc glow discharge in silane for (a) discharge off and (b) discharge on conditions. Here CO is the reference signal at 2143 cm^{-1}. The peak intensity of the CARS signal is a measure of the number density of the species being investigated. Therefore,

$$N^2 = N_o^2[I(SiH_4)/I(CO)]/[I_o(SiH_4)/I_o(CO)], \quad (36)$$

where N and I represent the number density and peak intensity of the CARS line, respectively, and the subscript o denotes the discharge-off condition, i.e., the plasma-free gas flow. Since N_o is known from pressure measurements, N is readily evaluated.

Originally, CARS was applied by Nibler *et al.* [109] for the measurement of rotational and vibrational temperatures of a gas such as D_2 in a glow discharge. On the assumption of a Boltzmann equilibrium among only the lowest vibrational levels, Shaub *et al.* [110] made direct measurements of nonequilibrium vibrational-level populations of electrically excited N_2 and

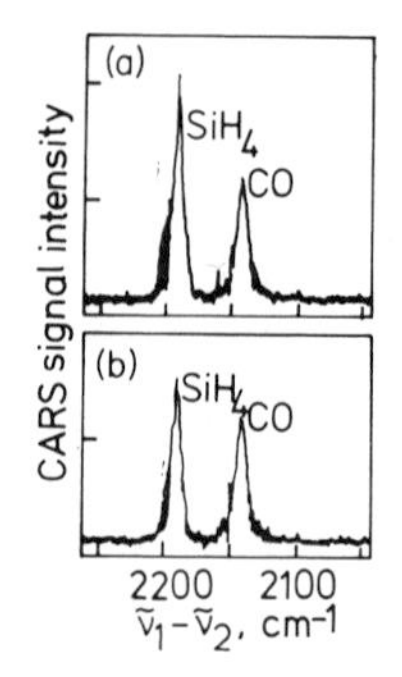

Fig. 3.23. CARS power signal as a function of wave-number difference $\tilde{\nu}_1 - \tilde{\nu}_2$ of the two pumping lights for (a) discharge-off condition and (b) discharge-on condition (From Hata *et al.* [108(a)].)

found them to be much hotter than discharged N_2. Since then some publications have appeared, among them Péalat *et al.* [111], who showed that the population of vibrational states $v = 0, 1$, and 2 in a low-pressure H_2 discharge has a non-Boltzmann distribution, and Hata *et al.* [108], who measured the rotational temperatures and densities of species such as H_2 and SiH_4 in a silane glow discharge. The latter work is of particular interest since it demonstrated the existence of the SiH_2 radical (2030 cm^{-1}) in its ground electronic state in concentrations of 10^{14} cm^{-3}.

B. Probe Analysis

The electric probe method has found wide application in the classical physics of gas discharges. With proper care, the local plasma density n, electron temperature kT_e, and plasma potential V_p can be determined simultaneously by inserting a metallic electrode, usually a wire, into the plasma (Fig. 3.24). The method has come to be known as the electrostatic (Langmuir) probe, first used by Langmuir and Mott-Smith [112], and is based on the polarization of a plasma and the fact that plasmas do not obey Ohm's law. The current to the probe is measured as a function of the probe potential to obtain the volt–ampere characteristics, commonly known as the probe characteristic (Fig. 3.24).

Whereas the experimental procedure is simple, the theory has a tendency to become extremely complicated. Articles by Chen [8], Schott [113], and Cherrington [114] and monographs by Swift and Schwar [115] and Chung *et al.* [116] are excellent sources of detailed information for the interested reader. The only probe regime that can be reasonably understood is the region ($p \leqslant 0.1$ torr) in which the mean free path λ is very much greater than the probe size, which in turn is very much greater than the Debye length λ_D, the distance in a plasma over which significant deviations from charge neutrality (and, therefore, significant electrostatic fields) can exist.

It is easy to recognize three regions in Fig. 3.24. Let us first consider region I, the ion-saturation region. For the condition $T_e \gg T_i$ obtained in

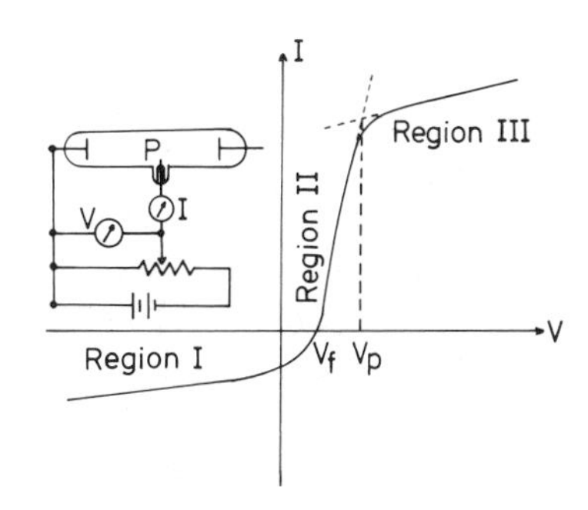

Fig. 3.24. A single probe and its typical voltage–current (V–I) characteristic.

glow discharges, the ions must enter the sheath with an energy kT_e. Since the ions are accelerated through a presheath region to this energy value, the collected ion-saturation current I_i becomes, as given by Chen [8], approximately

$$I_i = 0.61 A_p ne(kT_e/m_i)^{1/2}, \tag{37}$$

where A_p is the collection area of the probe. Thus, if T_e is known we can evaluate the plasma density n. Since ions are less affected than electrons by magnetic fields plasmas in such fields can be analyzed in this way. Only when the sheath is thin compared to the probe radius ($r_p \gg \lambda_D$) is the current determined by the space charge–diode relationship, as discussed by Chen [8], and the electron saturation current (region III) can be used to give a value for $n(kT_e)^{1/2}$.

The shape of the retarding field region (II) obviously is related to the distribution of electron energies $f(E)$ and, hence, gives kT_e when the distribution is Maxwellian, i.e.,

$$I_e = -\tfrac{1}{4} A_p ne(8kT_e/\pi m_e)^{1/2} \exp(-eV_\phi/kT_e), \tag{38}$$

where $V_\phi = V_p - V$ is the retarding potential. For $V_p > V_f$ (Fig. 3.24), the ion current is small and a plot of $\ln I_e$ as a function of V_ϕ will give T_e. For the non-Maxwellian condition of glow discharges the distribution is still isotropic; the electron current is

$$I_e(V_\phi) = A_p ne(2\pi e/m_e^2) \int_{eV_\phi}^{\infty} Ef(E)(1 - eV_\phi/E)\, dE, \tag{39}$$

where $E = \frac{1}{2}mv_e^2$ is the electron energy. Here $f(E)$ is normalized so that a Maxwellian distribution would be $f(E) = (m_e/2\pi kT_e)^{3/2} \exp(-E/kT_e)$. In this case, $f(E)$ can be obtained directly from the second derivative of the probe characteristic, i.e.,

$$f(E) = [m_e^2/A_p ne 2\pi e^3][d^2 I_e(V_\phi)/dV_\phi^2]\Big|_{E = eV_\phi} \tag{40}$$

This is the widely used probe technique for determining the electron energy distribution function.

The plasma potential V_p can be determined by locating the junction between regions II and III (Fig. 3.24) or by measuring the floating potential V_f at which the current nulls out and calculating V_p. However, as pointed out by Koenig and Maissel [10] and Coburn and Kay [117], V_f cannot be used to calculate V_p in rf discharges used for film deposition. Vossen [11] has shown that in these systems the measured potentials depend upon the relative area of the excitation electrode and of other grounded surfaces in

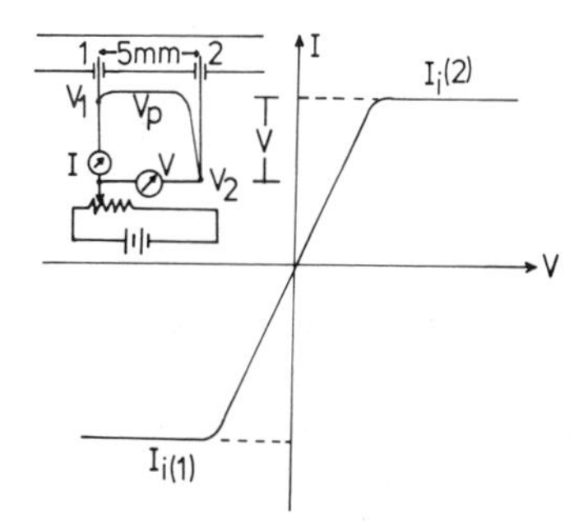

Fig. 3.25. A symmetrical double probe and its typical V–I characteristic.

contact with the discharge. Kay *et al.* [30] have discussed these problems in some detail.

Double probes (Fig. 3.25) are generally used for rf and microwave discharges in which a reference electrode is usually not available (see Inspektor *et al.* [48]. Since both probes are insulated from the ground the probes float with the plasma and are unaffected by changes in V_p. As shown in Fig. 3.25 both probes will be at a negative potential with respect to the plasma since no net current can flow to the system. The circuit is balanced when the electron current flowing to one probe is balanced by the ion current flowing to the other. The currents are, therefore, limited by the ion-saturation currents for each probe. For this reason double probes are generally preferred when magnetic fields are present. As shown in Fig. 3.25, the V–I characteristic is found by overlapping the ion-saturation characteristic of each probe. For the ideal situation in which the electron saturation regions are flat, the current is given by

$$I = I_i \tanh(eV/2kT_e). \tag{41}$$

The electron temperature is determined by taking the slope of the curve at the origin (Fig. 3.25), i.e.,

$$dI/dV\bigg|_{I=0} = (e/kT_e)I_{i(1)}I_{i(2)}/[I_{i(1)} + I_{i(2)}]. \tag{42}$$

Once kT_e is known, the plasma density can be computed from the ion saturation current. From Fig. 3.25 it is obvious that only a small fraction of the energy distribution of the electrons in the plasma, the high-energy tail, is sampled by the symmetrical double-probe arrangement. However, this difficulty has been overcome by the use of asymmetrical systems. Swift [118] has shown that the complete distribution is sampled if the area ratio is sufficiently large. Groh and Reuschling [119] have described an electronic analyzer for computing T_e (10^4–10^6 K) and n_e (10^{11}–10^{13} cm^{-3}) directly from double-probe characteristics of electrodeless glow discharges.

The foregoing discussion has neglected negative ions by assuming that

$$n_-/n_e \ll (m_- T_e/m_e T_-)^{1/2}. \quad (43)$$

As Lergon *et al.* [120] have shown, glow discharges of halogen-bearing gases or oxygen have large concentrations of negative ions. In such cases the contribution due to negative ions must be included.

Probes have been used for diagnostics of sputter coating discharges and plasma deposition systems. Such applications are reviewed by Eser *et al.* [121], Thornton [122], and Clements [123]. Probe surface contamination affects probe characteristics and data interpretation. Clements [123] has pointed out that the effects of contaminants on a probe surface can be minimized by biasing the probes with an rf voltage. Oliver *et al.* [124] excited a double floating probe system with low-amplitude ($V_b \ll kT_e/e$) rf whose frequency was between the plasma–ion and plasma–electron resonant frequencies and showed that surface oxide layers produced no effect on the measured values of T_e and n (even though dc measurements indicated changes of a factor of 2).

Niinomi and Yanagihara [125] used probes that were heated up to 1000 K with sheathed heaters inserted into them to prevent polymer films from depositing on the surface of the probe. Under film formation in an argon/benzene plasma the measured $f(E)$ showed considerable departure from a Maxwellian distribution. Yamaguchi *et al.* [126] have described the effects of plasma-polymerized styrene films on the V–I characteristics of single and double probes. Inspektor *et al.* [48] have used floating double probes to characterize microwave plasmas in hydrocarbons under pyrocarbon deposition conditions and observed that T_e and n were not similar in their behavior with respect to external parameters such as gas pressure and composition, applied power, and probe positions in the microwave plasmas. Mosburg *et al.* [127] have shown that, with proper care, electrostatic probes can be used for the measurement of T_e and n in rf discharges in silane, even though a film of semiconducting Si is continuously deposited on the probe surface. In fact they suggest that the measurement may be helped by a continuous renewal of the surface. Probe analysis of other silicon containing glow discharges have been reported by a number of authors, among them Grossman *et al.* [128], Gieres [129], and Avni *et al.* [18]. Grossman *et al.* [128] found that T_e was independent of the rf frequency, but n was higher at a higher frequency, thus indicating a frequency dependence of the degree of ionization. In dc discharges in which the probe was located in the negative glow Gieres [129] found that the electron energy distribution cannot be described by a uniform temperature.

As Venugopalan [130] pointed out, the use of electrostatic probes is fraught with dangers of not only contamination but also catalytic activity. Many of the probe materials such as Pt, W, and Ni are well-known catalysts for a number of reactions even under plasma conditions [29]. This is a field that needs to be investigated in some detail, particularly under deposition conditions.

C. Mass Spectrometric Sampling

Mass spectrometric sampling of a plasma is an *intrusive* technique in which the particle fluxes generally depend upon the sampling sheath thickness and voltage. Nevertheless, it has been applied to identify and measure particle concentrations and energies in glow discharges. In many instances mechanistic insight has been obtained for processes in the glow and for reactions (usually ion–molecule collisions) in a spatial or temporal afterglow.

For neutral species *molecular beam* methods developed by Foner and Hudson [131] are still in vogue. The neutral beam is modulated between a sampling orifice located in a wall of the discharge tube and the ionization chamber of the mass spectrometer and only the modulated component of the mass spectrometer output is recorded. This technique permits all neutral species, including radicals, to be detected with sufficient sensitivity. In the absence of modulation the sensitivity for detecting free radicals is lowered by several orders of magnitude. Hayashi *et al.* [132] used photoionization mass spectrometry for observing C_mF_n radicals produced in an rf plasma; the ionizer used noble gas resonance lines (Ar 11.6 and 11.8 eV, Kr 10.03 and 10.64 eV) or the Lyman α line (10.2 eV) of hydrogen.

For ionic species a somewhat more complex experimental configuration is necessary. The ion beam that is extracted from the plasma through a sampling orifice must be focused into the mass spectrometer. If conventional magnetic sector instruments are used, either the plasma itself or the drift tube of the mass spectrometer must be operated substantially away from ground potential so as to achieve the large and variable kinetic energy of ions required by these instruments. Wagner and Brandt [133] have described the use of such an instrument for studying the positive-ion chemistry in Si/SF_6 system.

In recent years most glow discharge studies have been performed using the *quadrupole*, first described by Paul *et al.* [134] and also known as a *mass filter*. During the past 10 years or so there have been numerous publications on ion sampling from glow discharge deposition and polymerization systems using quadrupole mass spectrometers. We may refer here to papers by Kay *et al.* [135], Smolinsky and Vasile [136], Turban *et al.* [137], and Carmi *et al.*

[32]. Wagner and Veprek [138] have described the application of mass filters for kinetic and chemical relaxation studies in Si/H system. Quadrupoles require only low-energy ions (10 eV) and can be operated without serious problems if the ion beam has a significant energy spread. Their main disadvantage is low mass resolution. The ability to ionize and collect neutral species is usually incorporated in these mass spectrometers so that much neutral sampling work done along with ion sampling. For example, Coburn and Kay [139] and Purdes *et al.* [140] describe the monitoring of ions and neutrals from plasmas through an aperture in the substrate table, behind which was mounted a differentially pumped quadrupole mass analyzer. Ion–molecule collisions in the sheath region and downstream from the sampling orifice affect the accuracy of ion densities measured this way. These effects have been discussed by a number of authors, among them Smith and Plumb [141], Helm [142], and Hasted [143].

In glow discharges used for deposition the main concern is plasma–surface interaction rather than ion density. Ions formed in the sheath or downstream from the sampling orifice will have a lower energy than ions coming directly from the bulk plasma with an energy characteristic of the plasma potential V_p. The collisional processes in the sheath region must be included to obtain the ion flux incident on surfaces. For determining the importance of collisional processes in the ion extraction Coburn [144] and Coburn and Kay [117] used an energy spectrometer between the sampling orifice and the mass spectrometer (see Fig. 3.26). Komiya *et al.* [145] have reported on the use of an ion energy analyzer in their mass spectrometric sampling of the incident ions on a substrate during deposition by hollow-cathode discharge. Franklin *et al.* [146], Vasile and Smolinsky [147], Rowe [148], and many others have used retarding potential techniques. Energy distributions obtained with deflection-energy spectrometers are much more precise than those measured with retarding potentials. The latter, however, is compatible with line-of-sight sampling of the neutrals.

In their studies of ion–molecule reactions Boehme and Goodings [149], Shahin [150], and many others combined mass spectrometric sampling of positive ions with Langmuir-type probe diagnostics. Drevillon *et al.* [151] and Gieres [129] applied such a combined diagnostics to characterize a dc glow discharge deposition system for silicon. For some years Lergon and Mueller [152] have used wall probes for negative ion extraction in plasmas of strongly electron-attaching gases such as SF_6. A typical wall probe consists of a piece of platinum melted into the Pyrex wall of a discharge tube and has an orifice size of 50 μm. Lergon *et al.* [120] have shown that the measurement of negative ion is a very sensitive method for the detection of deposited wall material.

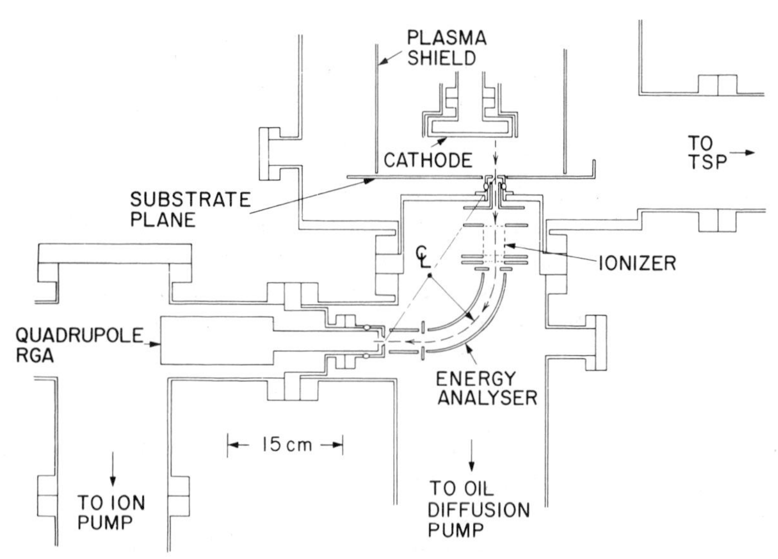

Fig. 3.26. Experimental set-up for the mass and energy analysis of ions bombarding the substrate plane. (From Coburn [144].)

D. Magnetic Resonance Techniques

Magnetic resonance methods depend on the gyromagnetic properties of the system under investigation. When these properties are associated with electrons, electron paramagnetic resonance (EPR) is observed. Free atoms and free radicals are paramagnetic and the EPR technique has, therefore, been used for studying these species. Several reviews of gas-phase EPR have appeared, among them we cite Westenberg [153], Carrington *et al.* [154], Brown [155], and Miller [86]. The last one is addressed particularly to the problems encountered in plasma diagnostics. Since the detection sensitivity for free atoms and radicals is about 10^{11} cm^{-3} the technique is suitable for glow discharges.

In general, the absorption of microwave energy by free atoms and radicals placed in strong magnetic fields is studied. A free atom or free radical in a magnetic field has its energy levels split into their (otherwise degenerate) components, which correspond to different orientations of the radical's magnetic moment along the magnetic field. This is, of course, well known as Zeeman splitting. In an EPR experiment the free atoms and free radicals must be contained in a microwave cavity (Fig. 3.8b) and the applied magnetic field adjusted such that the energy spacing between the Zeeman levels is equivalent to the resonant microwave frequency ($\simeq$9 GHz) of the cavity.

When this resonance occurs, photons are absorbed by the sample and the loss of power from the microwave field is detected. The energy $h\nu$ of one of the otherwise degenerate Zeeman components M_J is given by the equation

$$h\nu = g_J \mu_B H M_J, \tag{44}$$

where H is the magnetic field, ν is the level's frequency relative to its zero field value, g_J is the g factor, which is determined by the species' magnetic moment, μ_B the Bohr magneton, and M_J is the projection of the angular momentum J along the field direction. Term symbols give all the necessary information for calculating g_J (see Bethe and Salpeter [156]); spectroscopic selection rule gives $M_J = \pm 1$ so that the resonance field H_r is given by

$$H_r = h\nu_s / g_J \mu_B, \tag{45}$$

where ν_s is the microwave frequency of the spectrometer. This simple theory allows the prediction of absolute positions of atomic resonances usually in the absence of hyperfine structure.

A number of atomic species such as $H(^2S_{1/2})$, $D(^2S_{1/2})$, $N(^4S_{3/2}, {}^2D_{5/2})$, $O(^3P_2, {}^3P_1)$, $F(^2P_{3/2}, {}^2P_{1/2})$, etc., have been detected in glow discharges in the gas phase. The fluorine atom which is believed to play an important role in the plasma etching process, however, has a nuclear spin and can exhibit hyperfine structure. In such a case previous studies of the species usually will suggest the line positions and shifts due to changes in ν_s can be estimated readily from Eq. (45). Among the free radicals detected we mention OH, SH, CF, SF, NO, and ClO, all of which are in $^2\Pi$ states, and O_2, SO, and SeO in the $^1\Delta$ states. Since their g_J values can be calculated [86, 154], the resonance fields can be predicted.

Besides identification EPR spectra have also been used to obtain absolute concentrations of free radicals. The reader is referred to papers by Westenberg [157] and Breckenridge and Miller [158] for details and precautions. To cite a case in support, consider O_2, which has transitions that are magnetic dipole in nature and exhibits a five-line spectrum, four of which correspond to $^1\Delta_g$ state and the fifth (strongest) to the ground state $X^3\Sigma_g^-$. The integrated EPR signals of a given species are linearly proportional to the concentration of that species. Since ground-state O_2 is a stable gas its concentration can be measured, say, by a pressure measurement. Thus, by measuring the intensity of $O_2(X^3\Sigma_g^-)$ in a given spectrometer, the strength of the lines of any other atomic or molecular species will give directly its absolute concentration. For heteronuclear species that have electric dipole transitions NO is used in place of O_2 as a calibrant gas.

Whereas the possibility exists for measurements of the absolute concentrations of free radicals in gas phase, there have been no widespread application of the EPR technique for *in situ* experiments in glow discharge

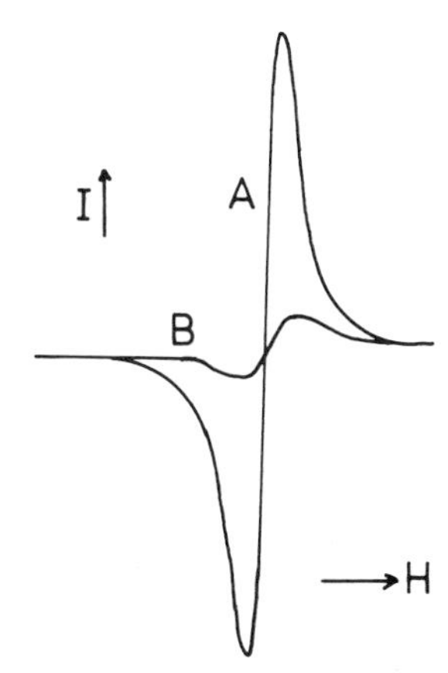

Fig. 3.27. ESR spectra of plasma-free radicals adsorbed on alumina. (A) 3.5 v/o $SiCl_4$ in Ar and (B) 3.5 v/o $SiCl_4$ + 20 v/o H_2 in Ar. (After Avni *et al.* [18].)

deposition systems. In an investigation Avni *et al.* [18] used an indirect method in which the radicals in different locations of pyrocarbon and silicon plasmas were adsorbed and stabilized on preheated alumina and then transferred to an EPR tube for measurement. Since the radicals are adsorbed on solid, the resolution of the spectrum is poor and radicals cannot be identified (Fig. 3.27). In the condensed phase, the magnetic moment of the species is determined usually by an unpaired spin (S) so that species with $S = 0$ are not detected by EPR.* During the transfer, the material invariably comes in contact with air and moisture. Morita *et al.* [159], Tkatschuk *et al.* [160], and Scott *et al.* [161] have shown that exposure of films to air causes a rapid decrease in the intensity of the original signal and even the appearance of new signals. This has been reconfirmed by Grenda and Venugopalan [162] in an *in situ* experiment that also demonstrated the dependence of the signal on the film structure.

Nevertheless, the signal intensity serves as an indicator for the radical concentration in the discharge. As Scott *et al.* [163] pointed out, the line shape is usually neither Lorentzian nor Gaussian and the relative area A under the absorption curve must be approximated by the product of the peak-to-peak amplitude and the square of the peak-to-peak linewidth. The number N_p of spins in the film is calculated from the N_s value of spin in a standard (such as DPPH) by using the equation

$$N_p = N_s(A_p G_s M_s / A_s G_p M_p), \tag{46}$$

* In the gas phase, however, there is the possibility of unquenched electronic angular momentum L, which can give rise to a magnetic moment independent of the spin. The combination of orbital magnetism and electron spin can yield better Zeeman effects than are typically encountered in condensed phases.

where G is the amplifier gain and M the modulation amplitude (in gauss or tesla). It must be pointed out that the presence of unpaired spins in films is due not only to the incorporation of gaseous free radicals formed in the discharge but also to the formation of radicals from the polymer material through the process of bond rupture caused by UV radiation in the plasma (Section II.D). This was demonstrated by Yasuda [164] and Morosoff *et al.* [165] by using ESR probes made of glass rods and polyethylene-coated tubes.

The application of the EPR method to the study of trapped radicals in frozen matrices is complicated because of magnetic anisotropies, matrix shifts, power saturation, etc., which result in severe line broadening, line distortion, and unusual intensity distributions. Despite these complexities the published literature for the past three decades is rich in EPR work on condensates from glow discharges.

IV. REVIEW OF RESULTS

In theory it is conceivable that almost any substance can be decomposed to a certain degree in a suitable glow discharge. In fact, several hundreds of substances have been subjected to glow discharges. While a review of the results from the standpoint of starting materials is desirable for mechanistic purposes, interest in glow discharge deposition has evolved because of the mechanical and electrical properties of the films obtained. For this reason in this review we consider the deposition of three categories of films: metallic, semiconducting, and insulating.

A. Metallic Films

The deposition of metallic films in or near the plasma region was reported by McTaggart [166], who decomposed the vapors of the alkali and alkaline earth metal halides in the presence of a noble gas or H_2 in a microwave discharge operated by a waveguide resonator. Although the observed dissociation rates could not be correlated with bond energies in the starting materials, the dissociation of alkali metal halides (MX) appeared to proceed as

$$MX + e^- = M + X^-.$$

The alkaline earth metal halides (MX_2), on the other hand, yielded deposits that consisted of equimolar mixtures of the metals and dihalides, i.e.,

$$MX_2 + e^- = MX + X^-,$$

$$2MX = M + MX_2.$$

Glow discharge sputter deposition is preferable for producing relatively pure metallic films. This widely employed technique has been reviewed by a number of authors, i.e., Vossen and Cuomo [44], from a process viewpoint. As early as 1852, Grove used this method of metallic film deposition, whereby a dc glow discharge is established between a plate of source material (cathode) and the substrate (anode) in a chemically inert gas. The early literature on the subject is well documented by Glockler and Lind [167]. Basically, removal of the cathode material results from the bombardment of the cathode by energetic positive ions of the discharge. The process is referred to as cathodic sputtering and is the result of energy and momentum transfer from the bombarding ions.

A binary collision is characterized by the transfer function $4m_i m_t/(m_i + m_t)^2$, where m_i and m_t are the masses of the incident ion and target species, respectively. Sputtered atoms come from the surface layers of the target so that we would expect the sputtering yield S to be proportional to the energy deposited in a thin layer near the surface, NL in Fig. 3.13. Sigmund [168] has given the following expression for the sputtering yield

$$S = (3\alpha/4\pi^2)[4m_i m_t/(m_i + m_t)^2]E/U_0, \quad (47)$$

where E is the energy of the incident ion; U_0 the surface binding energy of the material being sputtered; and α a monotonic increasing function of m_t/m_i, which has values of 0.17 for $m_t/m_i = 0.1$ increasing up to 1.4 for $m_t/m_i = 10$.

In practice, the sputtering yield rises rapidly from an apparent threshold that is approximately equal to the heat of sublimation and is usually below 100 eV. Above 100 eV it becomes a slowly varying function of E until it reaches a broad maximum at energies that are higher than would normally be encountered in a glow discharge. Selected data are given in Table 3.1 for Ar^+ ions at various energies. Vossen and Cuomo [44] have collated data for a number of ions. The interested reader is referred to articles by Winters [3, 169] for further details.

So far we have considered cathodic sputtering by positive ions. In glow discharge systems, bombarding ions are by no means monoenergetic. Sputtering yields given in Table 3.1 are, therefore, useful only to give a rough indication of the deposition rate of sputtered films. Furthermore, the flux of neutrals is about 10^4 times greater than the arrival rate of sputtered material. Besides ground-state neutrals a further source of bombardment is due to excited neutrals, of which metastables of the sputtering gas would be most abundant (Fig. 3.13). A major source of charged-particle bombardment at the anode is due to electrons the energy spectrum of which was measured by Ball [170] and Chapman *et al.* [171]. They found that the fast electrons emitted from the target by ion and other impact can have a major influence on the structure and properties of the growing film on the substrate (anode).

Negative ions sputtered from the target are accelerated across the discharge and collide with the anode. As Hanak and Pellicane [172] and Cuomo *et al.* [173] showed the sputtering produced by these negative ions not only reduced the amount of target material deposited on the anode, but, in some instances, produced steady-state sputter removal of the anode material itself. Finally, photons produced during ion or electron bombardment on any surface and from relaxation of excited atoms in the glow can affect the growth of a film.

If the substrate is different from the anode, a bias potential (usually negative) is applied to the substrate holder, so that the growing film is subject only to positive ion bombardment. This is known variously as bias sputtering or ion plating.* Coburn [144] identified the mass and energy of ions bombarding the substrate, with and without bias, in a dc discharge in argon with a Cu cathode. Although the mass spectra from rf discharges were very similar, there is a considerable difference in the ion energy distribution from the two types of discharge. Data such as these must be carefully interpreted, as the ion current magnitude cannot simply be equated with ion abundance; the quadrupole spectrum tends to underestimate higher mass ions, and its sensitivity also depends on ion energy. Sputtered particles that were ejected from the cathode as positive ions could not reach the sampling aperture because of the large retarding field for positive ions in the cathode fall region of the discharge. Therefore, the observed sputtered material was ejected from the cathode as neutral atoms and subsequently ionized in the discharge. A similar observation was made in the 1920s by von Hippel [176] and Baum [177] from their spectroscopic work. Coburn [144] measured the $^{63}Cu^+$ ion current as a function of the cathode position with and without bias voltage (Fig. 3.28), the former representing the deposition profile on the anode plane.

In an rf diode sputtering system Coburn and Kay [117] found that the energy distribution of low mass ions were broadened significantly by rf modulation during their passage through the plasma–substrate sheath. Collisional processes in the sheath region were found to influence the energy distribution of singly charged ions in the gas phase, but to a much smaller extent than in the target sheath. By using essentially the same mass/energy analysis technique, Komiya *et al.* [145] found the relative intensity ratio of $Cr^+ : Ar^+ : Cr^{++}$ as 80 : 16 : 4 in a hollow cathode discharge used for Cr deposition. The energy distribution of Cr^+, which they measured (Fig. 3.29),

* Initially, the term "ion plating" was used by Mattox [174] in reference to a process in which the deposition source was a thermal evaporation filament, instead of a sputtering target, and the substrates were connected to a dc sputtering target. Later, it was applied by Mattox [175] to any process in which the substrate is subjected to purposeful ion bombardment during film growth in a glow discharge environment.

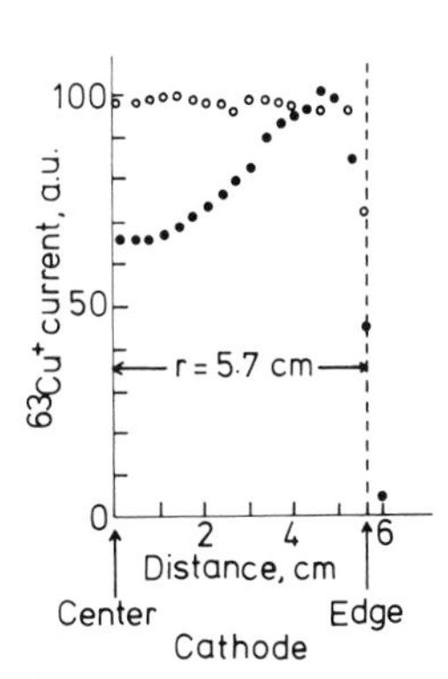

Fig. 3.28. Copper ion current versus cathode position with no bias voltage (●) and $V_b = -50$ V (○) on the sampling aperture. Sputtering voltage is 1000 V, anode–cathode spacing is 5 cm, and $p_{Ar} = 81$ μ. (From Coburn [144].)

indicated that Cr^+ is produced in a region that is at a potential of several to ten volts higher than that of the vacuum wall. More recently, Ziemann and Kay [178] and Ziemann *et al.* [179] have shown that the process of film deposition by bias sputtering in a low-pressure (1-mtorr) supported dc discharge can also be analyzed in terms of the energy delivered to the growing film by the bombarding gas ions per arriving film atom.

Greene and Sequeda-Osorio [63] have correlated sputtering parameters and results of optical emission intensity measurements for Cu (324.75 nm) in an Ar discharge. They also give a brief review of earlier work on spectroscopic diagnostics of glow discharge sputtering. As shown in Fig. 3.30 the total intensity, which is the area under the I versus t curve, is proportional to both the total change in target mass and the total thickness of deposited film. Since $I \times t$ is proportional to the film thickness, the sticking coefficient must have remained constant within the sputtering parameters studied ($p = 90$ mtorr, $V = 2$–3.5 kV). However, for very high sputtering rates or for heated substrates, we would expect a deviation from linearity as the sticking coefficient decreases. The fact that the emission intensity is proportional to

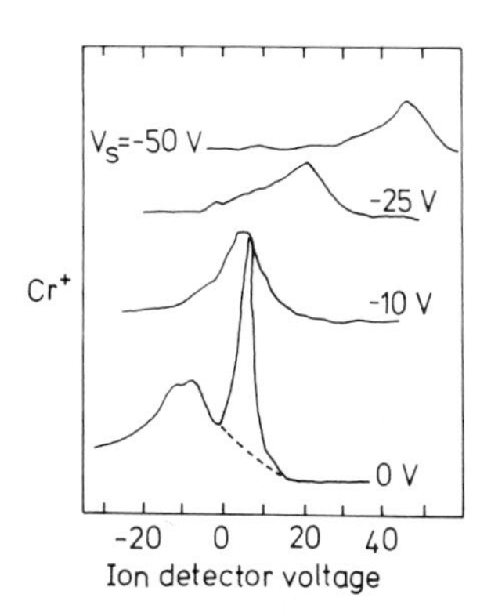

Fig. 3.29. Energy distribution of Cr^+ for various values of substrate voltage V_s. (From Komiya *et al.* [145].)

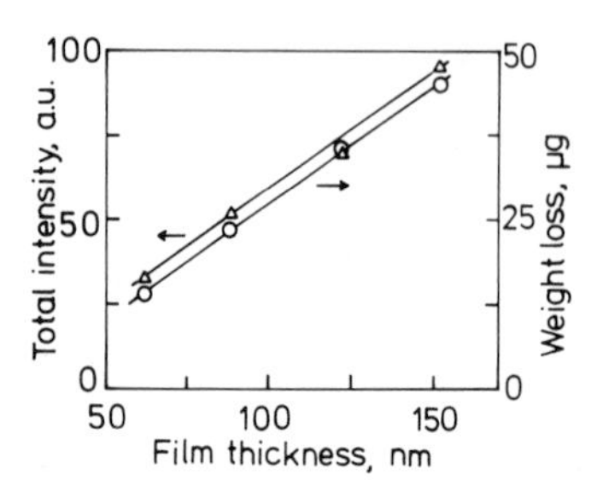

Fig. 3.30. Total emission intensity ($I \times t$) versus deposited film thickness and target weight loss. (After data of Greene and Sequeda-Osorio [63].)

the sputtering rate implies that the probability of target atom excitation in the discharge was constant under their experimental conditions. A possible explanation is that the maximum concentration of sputtered atoms occurs in the cathode fall region where the electron energy is high and the relative probability of electron collision excitation is small, that is, the high energy tail of the cross section for electron impact excitation of neutral atoms. The results for In shown in Fig. 3.31 are, however, typical of optical emission data obtained by varying the Ar pressure at constant target voltage. The divergence between the cathode fall peak and the negative glow peak appears to indicate an increasing loss of sputtered material due to gas-phase scattering of sputtered atoms in transit from the target to the substrate.

Greene [64] has determined the spatial distribution of sputtered Cu atoms by absorption spectroscopy and showed that the deposition rate at the substrate for a parallel-plate reactor is proportional to $1/p(dN/dx)$, where dN/dx is the anode-side slope of the sputtered atom distribution curve.

Greene [64] has also investigated the effect of negative substrate bias V_b on the sputtering rate of Cu as a function of its number density in the discharge. Figure 3.32 shows that N_{max} varied linearly with the target sputtering rate for $V_b < 100$ V; for $V_b > 100$ V, N_{max} increased at a faster rate due to resputtering from the film contributing a measurable fraction of the

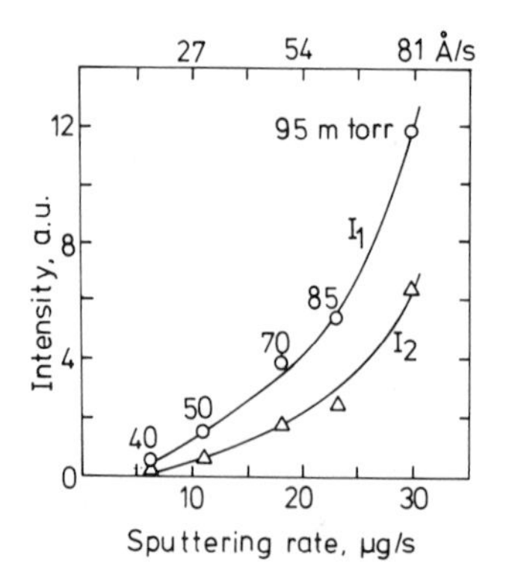

Fig. 3.31. The sputtering rate of an In target in a dc diode glow discharge as a function of the maximum In emission intensity in the cathode fall I_1 and the negative glow I_2 at 410.4 nm. The argon pressure was varied at a constant target voltage of -2.5 kV. (From Greene [64].)

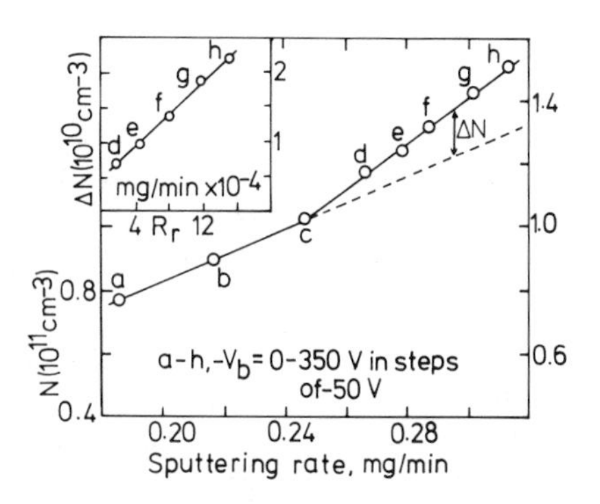

Fig. 3.32. The sputtering of a Cu target plotted as a function of the maximum number density N of Cu atoms in the discharge. The substrate bias voltage was varied at a constant target voltage of -2.5 kV and Ar sputtering pressure of 50 mtorr. Inset shows the resputtering rate R_r during deposition of a Cu film plotted against ΔN for various values of substrate bias. (From Greene [64].)

Cu atoms to the discharge, as well as to a slight increase in the residence time of sputtered atoms in the discharge. The resputtering rate R_r was calculated by using the equation

$$R_r = R_s(1 - r)[1 - \xi] = R_s(1 - r) - R_a, \qquad (48)$$

where the accumulation rate $R_a = R_s(1 - r)\xi$, r_s is the sputtering rate, r the fraction of material that is lost from the discharge by scattering, and ξ the probability that the impinging atom will be incorporated into the film. The last equality is obtained for the case in which the thermal sticking probability is unity. The inset in Fig. 3.32 shows that the resputtering rate during deposition is linearly proportional to the excess signal ΔN for various values of V_b.

Most of our discussion until now has been concerned with the sputtering of a single metal. The sputter deposition of alloys, and multicomponent films in general (Section IV.B), is complicated because the components may have different sputtering yields, condensation coefficients, and transport properties. An example of this is found in Coburn's [144] work using an aluminum–bronze (Fe, 3a/o; Al, 15a/o; Cu, 82a/o) cathode in a dc discharge in argon. The positive-ion mass spectrum showed a relatively large abundance of Fe^+, which could not be explained. Greene *et al.* [180] used emission spectroscopy to detect not only the discharge gas but also the elements from alloy targets such as Monel K-500 (Ni, 65.33w/o; Cu, 29.31w/o; Fe, 1.01w/o) and Inconel 718 (Ni, 53.25w/o; Fe, 18.14w/o; Cr, 18.01w/o). Upon initial bombardment of the surface the constituents with the highest sputter yield (Table 3.1) are preferentially removed from the surface, enriching the surface layer in the lower sputter yield material (segregation) until a steady state is reached. The mean free paths of the sputtered material should not be very different, but the condensation coefficient of a multicomponent target will be different, causing an effective change in the composition of the depositing alloy. Resputtering will, of course, deplete the film of the component with the higher sputtering yield and will be particularly effective under bias conditions.

Chemical sputtering may occur if reactive gases are present in the deposition system. It involves the reaction of an excited neutral or ionized gas with a surface to form volatile compounds. The technique is mainly used for plasma treatment of organic surfaces and for etching in plasmas. As early as 1926, Guentherschulze [181] observed that C, Se, As, Sb, and Bi formed volatile hydrides in a cold-cathode discharge in hydrogen and that these compounds dissociated outside of the cathode dark space to deposit metallic films. However, a relatively recent study of the chemical sputtering of graphite in an oxygen plasma by Holland and Ojha [9] indicated that impacting ions sputtered C–O compounds without their dissociating in the plasma. Also, Hanak and Pellicane [172] found that etching of glass substrates rather than film deposition occurred in systems with reactive anion targets of TbF_3 and $TbCl_3$. Such glow discharge systems have not been analyzed in any detail.

Neutral metal dimers have been observed during the sputtering of metal targets. For early work in this field, the reader is referred to papers by Honig [182], Woodyard and Cooper [183], and Oechsner and Gerhard [184]. Coburn *et al.* [185] made a mass spectrometric study of the sputtered species arriving at the substrate in an rf diode sputtering system when various metal oxide (M_xO_y) targets were used. Figure 3.33 shows the extent to which species MO^+ are observed relative to the species M^+ as a function of M–O bond energy. If the species MO were ionized in the discharge with an efficiency comparable to that for the ionization of the species M, then

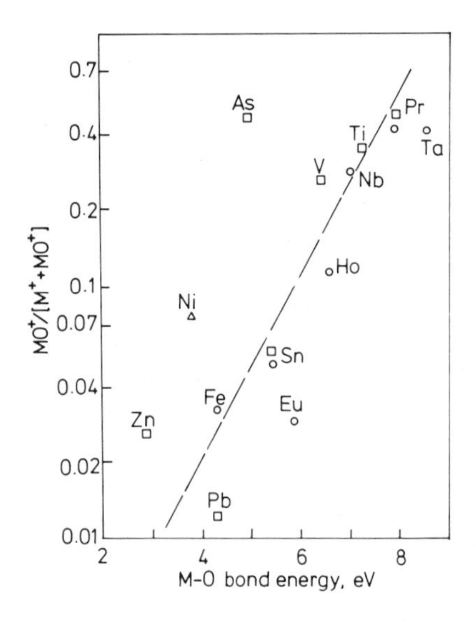

Fig. 3.33. The dependence of $MO^+/(M^+ + MO^+)$ on the M–O bond energy. Here p_{Ar} = 60 mtorr, 13.56 MHz rf 100 W, target area = 20 cm^2. (From Coburn *et al.* [185].)

molecular sputtered species MO would constitute a sizeable fraction of the total sputtered material particularly if the M–O bond energy is large.

B. Semiconducting Films

The glow discharge deposition of semiconductor films, particularly silicon, has been extensively studied. Since the early experiments of Schwartz and Heinrich [186] numerous investigators have decomposed silane (SiH_4) under a variety of conditions in an effort to prepare pure Si. However, the deposited film is largely hydrogenated amorphous silicon, sometimes also called polysilane, a-Si : H, or simply a-Si. The preparation of polycrystalline Si (pc-Si) and Ge films was accomplished by Veprek and Marecek [187], who used the *chemical transport* of Si and Ge in a glow discharge in H_2. More recently, microcrystalline silicon (μc-Si) has been deposited by the decomposition of $SiCl_4$ and $SiHCl_3$ and their mixtures with H_2 in glow discharges. During the past decade there have been numerous publications on the analysis for characterization of these films. There have also been several experiments to deplete their H or Cl content. In view of their commercial importance the preparation of doped or ion-implanted silicon films has also been a significant field of activity. LeComber and Spear [188] and Bauer and Bilger [189] have discussed some of the discharge conditions that influence the properties of the films. It is only during the last five years or so that serious efforts have been made to study the basic processes that result in different types of silicon.

1. Glow Discharges in Silicon Hydrides

Mosburg *et al.* [127] have reported probe measurements of electron temperature and density in rf discharges in silane (Fig. 3.34) and B-doped silane, even though a film of Si was continuously deposited on the probe surface. Their work suggested that the measurement may be helped by a continuous renewal of the surface. During typical depositions of thin films of a-Si, values of $kT_e = 2$–2.5 eV and $n_e = (1$–$1.5) \times 10^9$ cm^{-3} were obtained in silane plasmas at 72 mtorr and 1.2-W rf power. By using microwave interferometry* Turban *et al.* [137] measured the mean electron density $\langle n_e \rangle$ of an rf discharge in 5% SiH_4–He as a function of the rf power (Fig. 3.35). de Rosny *et al.* [190] deduced electron temperatures ranging from 1.6 to 2.5 eV from the ratio of the Si (288.2 nm) to H (656.3 nm) emission intensities and the cross sections listed by Perrin and Schmitt [191]. All these values appear

* The phase shift introduced by a plasma column of radius R gives the mean electron density $\langle n_e \rangle$ according to $\Delta\phi = 4\pi(R/\lambda)[(1 - \langle n_e \rangle/n_c)^{1/2} - 1]$, where λ is the mean-free-space wavelength and n_c the cutoff electron density.

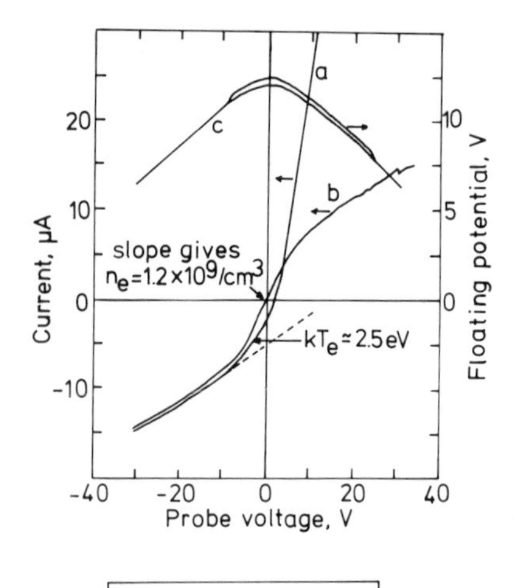

Fig. 3.34. (a) Langmuir probe and (b) double probe currents measured in pure silane at a pressure of 72 mtorr with a nominal input power of 1.2 W. (c) Mean floating potential of the double probe measured relative to the ground. (From Mosburg, Jr. *et al.* [127].)

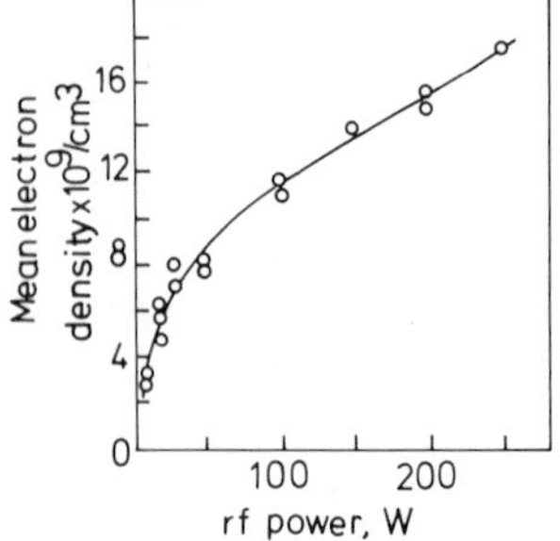

Fig. 3.35. Mean electron density as a function of the rf power determined by microwave interferometry. (From Turban *et al.* [137(a)].)

to be reasonable for the given set of experimental parameters such as gas pressure and discharge power/frequency.

Hata *et al.* [108] have used the CARS technique (Section III.A.5) for spatially resolved density measurements of SiH_4 molecules and their rotational temperatures. In addition, they have detected for the first time the elusive SiH_2 in a hydrogen-diluted silane ($SiH_4/H_2 = 1/9$) plasma. Interestingly enough, O'Keefe and Lampe [192] have also detected SiH_2*, but by the optical emission spectroscopy (OES) technique in SiH_4 and SiH_4–SiF_4 mixtures irradiated by a pulsed CO_2 laser. Since SiH_2 is considered to be the most probable species dominating a-Si deposition [67], further application of the CARS technique is expected to elucidate the role of SiH_2 in the surface reaction. For the present, OES and mass spectrometry (MS) are the two techniques widely used for species identification and concentration measurement in these systems. In the following sections we shall briefly survey recent papers on these two topics.

a. Optical Emission Spectroscopy. Griffith *et al.* [193] and Kampas and Griffith [194] identified optical emission from the species H, H_2, Si, SiH, and several impurity species in silane discharges. Perrin and Delafosse [69] found that the rotational levels of the $A^2\Delta$ state of SiH (4127 Å) are abnormally excited. Further work by Kampas and Griffith [67] and Kampas [195] using collision partners such as N_2 and Ar have revealed that H and the emitting excited states of Si and SiH are primary products of the silane decomposition. They also found that the electron concentration decreased and the average electron energy increased as [SiH_4] was increased in SiH_4–Ar mixtures. This dependence explains the work of Knights *et al.* [196], which showed that the deposition rate of a-Si from silane–noble gas mixtures is not proportional to [SiH_4].

In their later work Knights *et al.* [70] measured the vibration–rotation bands of the electronic ground state of SiH and deduced vibrational and rotational temperatures of this species to be 2000 and 484 K, respectively. This suggested that for all dissociative states of SiH_4 produced by electron impact leading to SiH formation there is a large excess of energy released in excitation of the fragments. Knights *et al.* [70] also measured the infrared absorption spectrum of SiH_4 (ν_3 band at 5μ) and used it to identify the SiH_4 absorption from the plasma. The rotational and vibrational temperature of 300 and 850 K thus deduced for SiH_4 is consistent with a nonequilibrium population driven by direct electron impact excitation of rovibration and subsequent collisional quenching.

A typical OES spectrum obtained by Matsuda *et al.* [66] and Matsuda and Tanaka [197] from a pure SiH_4 glow discharge plasma is shown in Fig. 3.36. They found that the OES intensity ratio of H* to SiH* is strongly correlated with the median wave number $\tilde{\nu}_m$ of the Si–H stretching absorption band of the deposited film over a wide parameter space of the plasma conditions including inert gas (He, Ne, and Ar) dilution of SiH_4. The paper of Matsuda and Tanaka [197] also describes the phenomenological relationship between the OES data and the properties of the deposited film.

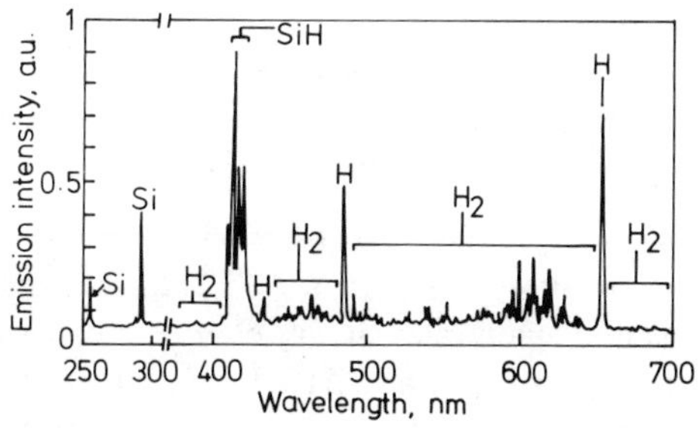

Fig. 3.36. Typical OES spectrum from a pure SiH_4 glow discharge; flow rate = 5 sccm/min; pressure = 80 mtorr; rf power = 20 W. (From Matsuda and Tanaka [197].)

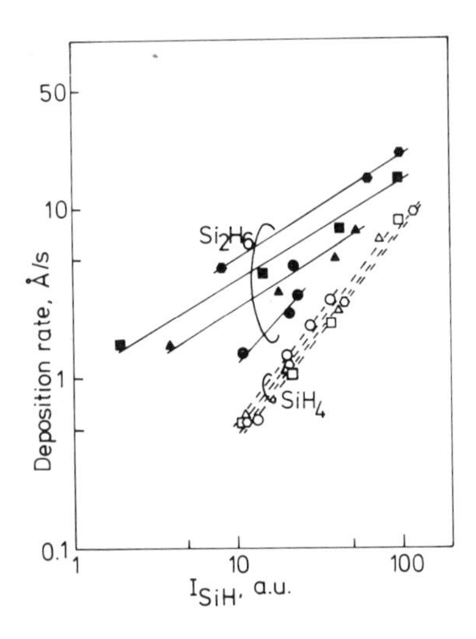

Fig. 3.37. Deposition rate of a-Si : H films plotted against the line intensity of [SiH] emissive radicals in glow discharges of SiH_4 (symbols: ○, 1 sccm; △, 5 sccm; □, 15 sccm; ○, 30 sccm) and Si_2H_6 (closed symbols: ●, 1 sccm; ▲, 5 sccm, ■, 15 sccm; •, 30 sccm). rf power and gas flow rate were varied from 1 to 80 W and 1 to 30 sccm, respectively, keeping gas pressure constant at 50 mtorr. (From Matsuda *et al.* [198(b)].)

In their later work Matsuda *et al.* [198] studied Si_2H_6 discharges as well. Figure 3.37 shows the deposition rate of a-Si films from SiH_4 and Si_2H_6 glow discharges in diode and triode reactors plotted against the line intensity of SiH observed by OES during deposition. As is evident, the deposition rate of a-Si from the SiH_4 glow discharge is proportional to the line intensity of SiH emission (4127 Å). No such simple behavior was found for Si_2H_6 discharges. From the quantitative measurement of the luminosity of the SiH emission, Matsuda *et al.* [198] estimated the density of SiH* to be 3×10^6 cm^{-3}. Since this value is 3–4 orders of magnitude lower than the density species required for explaining the observed deposition rate, it was concluded that SiH* is not a direct precursor for a-Si deposition but other species correlated with SiH* (possibly SiH neutral radicals) are responsible for the deposition from SiH_4 plasma. By measuring the deposition rate as a function of the separation between the mesh and the substrate electrode in a triode reactor, Matsuda *et al.* [198] determined that the SiH radical in Si_2H_6 plasma has a much shorter lifetime than that in SiH_4 plasma.

Optical emission spectroscopy has also been applied by Marcyk and Streetman [199] for measuring As-implanted Si, and by Zesch *et al.* [200], among others, for measuring B-dopant concentrations in a-Si.

b. Mass Spectrometry. Haller [201] observed high relative abundances of ions containing more than one Si atom ($Si_jH_k^+$) and suggested that their most rapid reactions in a silane plasma are charge transfer and formation of new Si–Si bonds upon collision with SiH_4. Drevillon *et al.* [151] found that SiH_3^+ was the dominant ion by mass spectrometric sampling of a multipole discharge in 80% SiH_4–20% H_2.

Based on a mass spectrometric analysis of the positive ions and neutrals in SiH_4–He, SiH_4–D_2, and SiD_4–He plasmas, Turban *et al.* [137] proposed the scheme shown in Fig. 3.38 for the deposition of a-Si films. They observed ion–molecule, disproportionation, insertion, and abstraction reactions involving the following active species: SiH_2, SiH_3, H, SiH_2^+, and SiH_4. Their study of the species flux to the walls of the discharge tube showed that free radicals SiH_n ($n = 0$–3) are the gaseous precursors of the film and that heterogeneous reactions of these radicals, H atoms, and ions at the walls control the composition and structure of the films. The possibility that SiH_2 and/or SiH_3 radicals play an important role in the film formation was also suggested by Knights [202] and Scott *et al.* [203]. Apparently the Si–H bonds in the film are produced partly from SiH_n radicals and partly from the direct incorporation of H atoms into the growing film. The H-atom etching of a-Si films, which perhaps initiates the Si chemical transport observed by Veprek and Marecek [187], indicates some reversibility in the deposition process (Fig. 3.38).

Matsuda and Tanaka [197] combined OES with MS for characterizing the glow discharges of SiH_4, Si_2H_6, and SiH_4–inert gas mixtures under conditions of a-Si deposition. They found that the density of ionic species, dominantly SiH_3^+, is lower than that of neutral radicals by 4–5 orders of magnitude. Their biased substrate experiments showed that ionic species gave no discernible change in the deposition rate but strongly affected the quantity of hydrogen introduced into the film in the form of H^+. The deposition rate could be well correlated with the MS data and analyzed

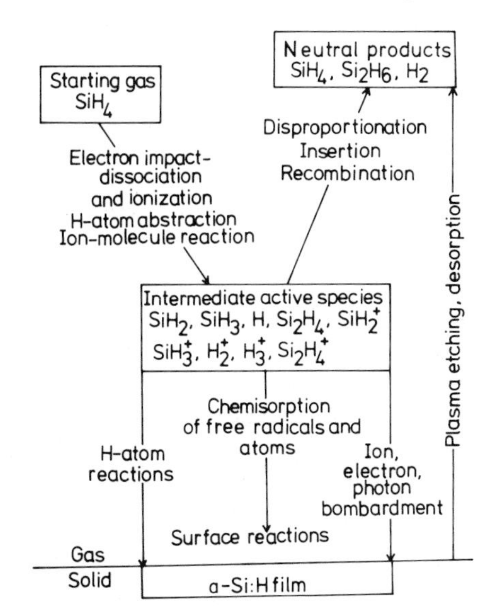

Fig. 3.38. Schematic diagram for the silane glow discharge deposition. (From Turban *et al.* [137(b)].)

quantitatively in terms of the H-to-Si ratio in the plasma. On the basis of mass spectrometric measurement, Matsuda *et al.* [198] have also demonstrated that the crystallite size is strongly affected by the amount of ionic species impinging on the growing surface of μc-Si.

2. Chemical Transport in Hydrogen Plasma

Chemical transport in a low-pressure plasma of hydrogen has been used for preparing films of several materials since the formation of pc-Si and Ge films was reported by Veprek and Marecek [187]. The transport takes place only in the direction of increasing gas temperature, but the discharge current density is nearly constant between the charge and the deposition zone. Through a series of refinements in technique Veprek *et al.* [204] have found it possible to deposit polycrystalline films of silicon at temperatures as low as 80°C. For a long time the emphasis in these investigations have been on the properties of the films and their control rather than understanding their formation via analysis of the glow discharges [17].

Wagner and Veprek [138] made a mass spectrometric study of a deuterium discharge over c-Si powder, a silane discharge, and a deuterium discharge over a-Si deposited by a silane discharge. They found that at long residence times ($>$5s), the steady-state signal heights of SiH_2^+ (silane discharge) and SiD_2^+ (Si/D_2 discharge) are the same even though the two systems approach steady state from opposite sides. The equivalence of the signal heights at long residence times means that the rates of production of silane (forward reaction),

$$Si(s) + (x/m)H_m(g) \rightleftharpoons SiH_x(g), \tag{49}$$

and the removal of silane (reverse reaction) are equal. For this to be the case the authors proposed that a "partial chemical equilibrium" (nonisothermal conditions!) must have been reached in the Si/H_2 discharge. The kinetic parameter controlling the extent of departure of the system from such a state is the ratio τ/t_{res}, where τ is the characteristic time of Reaction (49), either from the educt or product side, and t_{res} is the mean residence time of the species in the reaction zone. This work is significant in that the authors were able to delineate the conditions under which a-Si ($t_{res}/\tau < 1$) and μc-Si ($t_{res}/\tau \geqslant 10$) films are deposited.

In their later work, Wagner and Veprek [138] applied a chemical relaxation technique to show that both SiH_4 and $H_2/Si(s)$ discharges operating under the same conditions of pressure, discharge current density, temperature, and residence time display the same kinetic response as well as silane concentration when they are at "partial chemical equilibrium," i.e., $t_{res} \gg \tau$. Since the formation of H_2 and the decomposition of silane occurred on the same time scale, a single-step mechanism of silane decomposition was proposed as

$$SiH_4 = SiH_2 + H_2. \quad (50)$$

3. Glow Discharges in Chlorosilanes

The deposition of μc-Si from glow discharges in $SiCl_4$ and $SiHCl_3$ and their mixtures with H_2 and/or noble gases has been studied. Bruno *et al.* [205], Katz *et al.* [206], Gafri *et al.* [207], Grossman *et al.* [128, 208], and Grimberg *et al.* [209] have described the dependence of the deposition rate on parameters such as position of the substrate in the reactor, gas pressure and composition and discharge power and frequency. Grossman *et al.* [128] also characterized an rf plasma in Ar + 10v/o H_2 using a floating double probe; based on this characterization they interpreted the deposition results from an rf plasma in a gas mixture containing 2.5v/o $SiCl_4$ and 7.5v/o H_2 in Ar. Table 3.4 shows the values of T_e, n_i, and degree of ionization at two radio frequencies as a function of pressure. The electron temperature was found to be almost independent of the frequency, while the positive ion concentration was higher at a higher frequency. As a consequence, there is an increase in the degree of ionization at the higher frequency. Since a higher degree of ionization induces a higher ion–molecule interaction, an increase in the decomposition of $SiCl_4$ and deposition of Si is to be expected and was actually observed.

Avni *et al.* [18], Grill *et al.* [210], and Manory *et al.* [211] have made mass spectrometric sampling of microwave plasmas of $SiCl_4 + H_2 + Ar$ mixtures, which were characterized using the double floating probe system

TABLE 3.4

Electron Temperature T_e, Positive Ion Density n_i, and Degree of Ionization at Two Radio Frequencies[a,b]

	T_e (eV)		$n_i \times 10^{-11}/cm^{-3}$		Degree of ionization $\times 10^4$	
p (mbar)	0.4 MHz	27.12 MHz	0.4 MHz	27.12 MHz	0.4 MHz	27.12 MHz
0.5	4.48	4.74	0.5	3.7	0.10	0.80
1.0	5.17	5.43	0.8	7.0	0.09	0.73
3.0	5.00	4.48	1.0	9.8	0.03	0.34
5.0	3.10	2.58	1.5	10.7	0.03	0.22

[a] From Grossman *et al.* [128].

[b] Gas mixture Ar + 10 v/o H_2; substrate in the center of the glow; current in the rf coil 12 A (0.4 MHz) and 4 A (27.12 MHz).

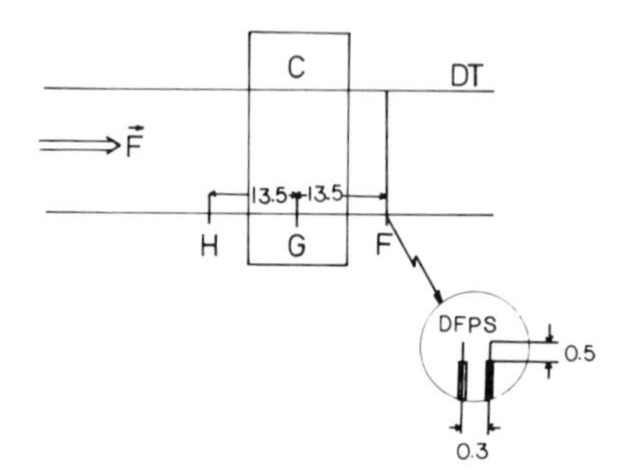

Fig. 3.39. Probe layout in the gas stream showing the sampling positions before the microwave cavity (H), in the center of the cavity (G), and beyond the cavity (F); DT, discharge tube; F, direction of gas flow; C, microwave cavity. Inset shows the double floating probe system (DFPS). (From Inspektor *et al.* [48].)

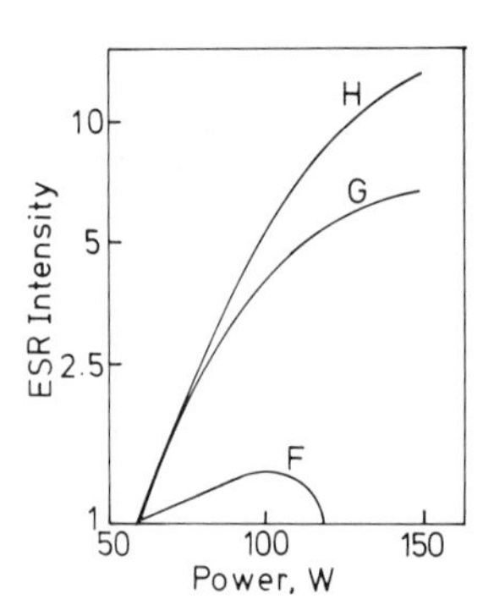

Fig. 3.40. ESR intensity of free radicals (normalized to 60 W) versus microwave power input for 3.5 v/o $SiCl_4$ + 20 v/o H_2 in Ar at 4-torr pressure. (From Avni *et al.* [18(b)].)

shown in Fig. 3.39. They recognized that the deposition rate of Si is highest and the Cl contamination in the Si deposit the lowest where the $SiCl_4$ enters the plasma. The free radicals from the plasma were adsorbed and stabilized on alumina and their concentrations were determined to be highest upstream of the plasma rather than downstream (Fig. 3.40). Based on these observations the formation of polymerized silicon species by radicals and ion–molecule interactions was proposed:

$$SiCl_2^{\cdot} \xrightarrow{SiCl_4} Si_2Cl_{2-4}^{\cdot} \xrightarrow{SiCl_4} \cdots Si_xCl_y + Cl_2, \tag{51}$$

$$SiCl_3^+ \xrightarrow{SiCl_4} Si_2Cl_5^+ \xrightarrow{SiCl_4} \cdots Si_xCl_y^+ + Cl_2. \tag{52}$$

However, they noticed a nonproportionality between the deposition process of Si and its concentration in the plasma at higher pressures and input powers. This, of course, could be attributed to the etchant properties of chlorine atoms and ions and sputtering by Ar^+ ions.

C. Insulating Films

Insulating films are invariably compounds, usually oxides grown by plasma oxygenation of metals and carbonaceous materials, containing H or

F produced by glow discharge polymerization of hydrocarbons or fluorocarbons.* They have also been produced by glow discharge sputtering.

Glow discharge sputtering of compound targets has been studied by a number of authors. By using the ion sampling and energy analysis system that was described in Section III.C, Coburn *et al.* [185] found that the intensity of sputtered neutral molecules exceeded that of sputtered neutral atoms for several oxide targets in an rf diode glow discharge. Given that a metal oxide target, and by implication any other compound target, is unlikely to be sputtered in a molecular form, it is not surprising that the stoichiometry of the resulting film will be different from that of the target, usually being deficient in the gaseous or other volatile species. However, Erskine and Cserhati [214] have shown that the stoichiometry of a quartz (SiO_2) film is the same as the target if the sputtering is done in a mixture of 95%Ar : 5%O_2. Usually this involves some sort of *reactive sputtering.* This is the type of process in which a compound is synthesized by sputtering a target (e.g., Ti) in a reactive gas (e.g., O_2 or Ar–O_2 mixtures) to form a compound (TiO_2). It is impossible for us to review the numerous reactive sputtering processes that have been described in the literature. We refer the reader to an extensive bibliography prepared by Vossen and Cuomo [44]. At very low reactive gas partial pressures and high target sputtering rates, it is well established that virtually all of the compound synthesis occurs at the substrate and that the stoichiometry of the film depends on the relative rates of arrival at the substrate of metal vapor and reactive gas.

1. Plasma Oxidation/Anodization

The formation of an oxide film on a metal or semiconductor surface immersed and floating in an oxygen plasma is called plasma oxidation. When the growth is stimulated by applying a bias to the substrate so that the film surface potential is above the floating potential the process is called plasma anodization. The topic has been reviewed by Dell'Oca *et al.* [215], O'Hanlon [216], Gourrier and Bacal [217], Ojha [218], and many others.

Earlier work on plasma oxidation by Ligenza [219], Weinreich [220], and others were done in microwave or dc discharges which were inadequate in producing pinhole-free thin oxide films at controlled rates. Furthermore, Leslie *et al.* [221] have reported that the films produced in dc discharges were contaminated by the cathode material. Greiner [222] was able to overcome these problems by placing the substrate to be oxidized on the cathode of an rf system and adjusting the cathode potential to a value that resulted in simultaneous etching and oxidation by oxygen ions during the

* According to Hauser [212] and Hosokawa *et al.* [213] carbon coatings deposited by sputtering are conducting rather than insulating.

oxide growth. Ray and Reisman [223] recorded the growth of SiO_2 only at pressures greater than 10 mtorr; the growth rate on the Si surface facing away from the discharge was significantly higher. Harper *et al.* [224], however, reported that the rf oxidation technique suffers from such drawbacks as contamination of the oxide film due to sputtering of the walls, deposition of back-sputtered material, and excessive heating of the sample during oxidation. Since particle energies may reach as high as 10^3 eV, some ion implantation is to be expected. Despite these problems, the rf process has become dominant for the fabrication of tunnel barriers for Josephson technology, for both lead alloy and niobium films.

A schematic representation of the physical processes occurring during plasma anodization is shown in Fig. 3.41. The fact that the oxide growth rate and the oxide film thickness increase on application of a positive potential led O'Hanlon and Pennebaker [225] to suggest that negative oxygen ions in the discharge were essential for the oxide growth. However, Olive *et al.* [226] found only a small decrease in the anodization rate when the barrier for the negative ion flux was increased by decreasing the dc surface potential of a tantalum substrate through the application of a combination of dc and rf potentials at constant anode current. Gourrier *et al.* [227] found from photodetachment and probe measurements that the supply of negative ions from the plasma cannot account for the observed oxidation rates of GaAs. Although O'Hanlon and Sampogna [228] were able to increase the anodization rate by biasing positively an additional grid, their probe measurements indicated essentially that the electron temperature was enhanced. By using a magnetic field Chang [229] prevented the electrons from reaching the plasma surface and found a strong reduction of the growth rate. Therefore, electron-assisted surface processes have been invoked. According to Ando and Matsumura [230] electron capture by adsorbed oxygen atoms is the most likely process, the oxygen atoms originating from the plasma or on the surface by electron-stimulated dissociation. The possibility of dissociative attachment to adsorbed oxygen molecules cannot be ruled out.

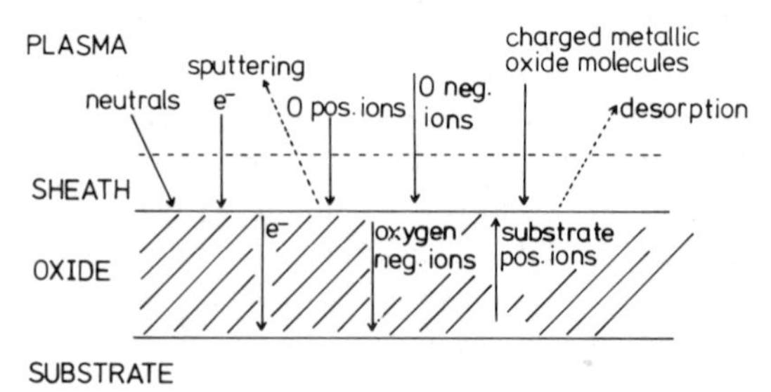

Fig. 3.41. Schematic respresentation of the physical processes occurring during plasma anodization. (From Gourrier and Bacal [217].)

During plasma anodization charge transport occurs in the oxide: electrons and oxygen anions move inward, and sample cations move outward. Only a few results for GaAs and Si have been reported on the ratio of anionic to cationic movements. By using $^{18}O_2$ as a marker, Yamasaki and Sugano [231] showed that the oxygen order is largely conserved during GaAs plasma anodization in an rf discharge. This suggested that long-range migration of oxygen anions by direct interstitial transport is not dominant. Perriere *et al.* [232] arrived at a similar conclusion from their ^{16}O plasma anodization of Si through thin ^{18}O-enriched ZrO_2 layers. By using double-layered structures of Al/GaAs, Chang [229] and Gourrier *et al.* [227] found that GaAs oxide can be formed on both sides of the Al layer, which is itself oxidized, and concluded that both oxygen ions and substrate ions move during the anodization. However, Ho and Sugano [233] and Chang *et al.* [234] found that only oxygen migration occurs during Si oxidation. Thus, the results appear to depend on the oxidized or anodized material. Fromhold [235] and Fromhold and Baker [236] have carried out calculations using a very simplified model which assumes that the rate of oxidation is limited by the transport of ionic species through the already formed layer, with the transport mechanism being the thermally activated hopping of ionic defects in the presence of electric fields due to the surface potential established by the discharge and modified by the space charge of the mobile ionic defects. The results on the oxidation of InP that Meiners [237] and Wager *et al.* [238] have obtained by using heated afterglows rather than plasmas seem to suggest a significant contribution from plasma excited neutrals as well.

2. Glow Discharge Nitriding

Nearly half a century ago, the use of a low-pressure glow discharge for nitriding was introduced by Berghouse [239] and von Bosse *et al.* [240]. The material to be nitrided is used as the cathode in a glow discharge produced in a gas mixture which contains nitrogen. The gas mixtures used are $N_2 + Ar$, $N_2 + H_2$, $NH_3 + Ar$, etc. For rapid nitride growth, high current and power densities (0.1–10 mA/cm^2 and 0.5–1 kV) are desirable so that the abnormal region of the glow discharge is preferred. Besides steel, refractory metals such as Zr, Ti, Mo, Nb, and carbides such as TiC have been nitrided in a glow discharge. Gruen *et al.* [241], Veprek [17], and Ojha [218] have reviewed plasma nitriding in some detail.

Emission spectroscopy of glow discharges in N_2 has shown the presence of N^+ and N_2^+ so that the cathode undergoes energetic ion bombardment besides adsorbing neutral gas species such as atomic nitrogen. Because a steel specimen could not be nitrided in a floating position, Hudis [242] concluded that adsorbed gas species are insufficient for nitriding. Edenhofer [243] and

Lakhtin *et al.* [244] observed that reactive sputtering of the cathode and back-diffusion of the sputtered species are responsible for the nitriding of steel. Sputtered Fe atoms react with nitrogen to form unstable FeN, which is back-scattered onto the cathode and decomposes to lower nitrides, such as Fe_2N, Fe_3N, and Fe_4N. A fraction of the nitrogen atoms released due to such a decomposition are believed to diffuse into the steel.

Hudis [242] found by mass spectrometric sampling of ions through a hole in the center of the cathode that in a nitrogen–hydrogen plasma nitrogen ions (N^+, N_2^+, etc.) comprise less than 0.1% of the cathode current while nitrogen–hydrogen molecular ions comprise 10–20%. In a nitrogen–argon plasma, however, N^+ and N_2^+ comprise 80% of the cathode current but produce very little nitriding. Nevertheless, a number of authors, among them Hudis [242], Lebrun *et al.* [245], and Content *et al.* [246] suggested that fast ions and neutrals from the glow penetrate into the cathode to a depth of several atomic lengths and that the nitrogen absorbed in this way diffuses into the metal and forms a layer with a nitride structure. A nitriding mechanism based on the formation of vacancy and nitrogen ion pairs and their subsequent diffusion inside the metal was proposed by Brokman and Tuler [247]. They used crossed electric and magnetic fields to enhance the ion current density and found that the diffusion coefficient was proportional to the current density. However, Tibbetts [248] found nitriding even when positive ions were repelled from a steel surface by a biased grid.

Szabo and Wilhelmi [249] have made a mass spectrometric study of continuous and pulsed dc discharges used for steel nitriding. They observed NH_x ($x = 0$–4) and a strong cataphoretic enrichment of hydrogen. When a freshly nitrided steel surface was sputtered in Ar, ions of Fe, $FeNH_2$, and $FeNH_3$ were detected. Therefore, they proposed the formation of a $FeNH_{2-3}$ boundary layer. This is consistent with the work of Braganza *et al.* [250], which showed that in a hydrogen glow discharge nitrided stainless steel is considerably depleted in nitrogen.

Liu *et al.* [251] found that the kinetics of Ti and Zr nitride formation are enhanced by about a factor three in the dc discharge technique compared with thermal nitriding. Konuma and Matsumoto [252], Matsumoto and Kanzaki [253], and Matsumoto *et al.* [254] have studied the nitriding of Ti and Zr in a nitrogen and 10% H_2–90% N_2 rf discharge at 5–20-torr pressure. Probe measurements revealed that the N_2 and $N_2 + H_2$ plasmas had electron energies of 8–10 and 9–11 eV, respectively, and ion densities of 10^9–10^{10} and $\simeq 10^{10}$ cm^{-3}, respectively. From these values, V_f was estimated as -80 to -100 V. Emission spectroscopy showed a number of electronic transitions, among them the $C^3\Pi_u - B^3\Pi_g$ transition of N_2, the $B^2\Sigma_u^+ - X^2\Sigma_g^+$ transition of N_2^+, and the $A^3\Pi - X^3\Sigma^-$ transition of NH, the latter in the plasma that contained 10% H_2 (Table 3.5). The N_2^+ ion was estimated to

TABLE 3.5

Electronic Transitions Observed in the Emission Spectra of Nitriding (rf) Plasmas[a]

Plasma	Species/Transitions		T_v/K
N_2	N_2, $B^3\Pi_g - A^3\Sigma_g^+$	$\Delta v = 3, 4\ (v' = 13)$	
	N_2, $C^3\Pi_u - B^3\Pi_g$	$\Delta v = -7$ *to* $+3\ (v' = 11)$	5400 ($\Delta v = -2$)
			4400 ($\Delta v = +1$)
	N_2, $D^3\Sigma_u^+ - B^3\Pi_g$	$\Delta v = -4$ to $-1\ (v' = 4)$	
	N_2^+, $B^2\Sigma_u^+ - X^2\Sigma_g^+$	$\Delta v = -2$ to $+2\ (v' = 4)$	5700 ($\Delta v = -1$)
	N_2^+, $D^2\Pi_g - A^2\Pi_u$	$\Delta v = -2$ to $+2\ (v' = 7)$	
	NO, $A^2\Sigma^+ - X^2\Pi$	$\Delta v = -5$ to $+2\ (v' = 6)$	
N_2/H_2	N_2, $C^3\Pi_u - B^3\Pi_g$	$\Delta v = -7$ to $+3\ (v' = 11)$	4000 ($\Delta v = -2$)
(9/1)			3800 ($\Delta v = +1$)
	N_2, $D^3\Sigma_u^+ - B^3\Pi_g$	$\Delta v = -4$ to $-1\ (v' = 4)$	
	N_2^+, $B^2\Sigma_u^+ - X^2\Sigma_g^+$	$\Delta v = 0\ (0 - 0, 1 - 1)$	
	NH, $A^3\Pi - X^3\Sigma^-$	$\Delta v = 0\ (0 - 0, 1 - 1)$	
	H, H_α, H_β, H_γ		

[a] Based on Matsumoto and Kanazaki [253].

have a vibrational temperature of 5700 K for $\Delta v = 1$. Besides N^+ and N_2^+, mass spectrometric sampling of the $N_2 + H_2$ plasma while nitriding showed NH_x^+ ($x = 1$–4) species. Matsumoto and co-workers found that the addition of H_2 decreased the concentration of N_2^+ and consequently the sputtering of Ti by N_2^+. Based on an ESCA analysis that revealed NH radicals on titanium surface, Matsumoto proposed mechanisms consistent with an initial nitride forming reaction involving NH with the substrate followed by the diffusion of nitrogen from nitride to metal.

The nitriding of silicon in rf plasmas has been reported. By using a double probe, Akashi *et al.* [255] found that the plasma (electron) density in N_2–He discharge was much higher than those in N_2 discharges and attributed the effective nitriding of Si in N_2–He discharges to increased activation and ionization of N_2 by the Penning effect of excited He. The SiN films were, however, uneven due perhaps to a radial distribution of N_e and T_e in the reactor tube. Matsumoto and Yatsuda [256] used N_2 and N_2–H_2 plasmas with electron densities of 10^8–10^{10} cm^{-3} and electron energies of 4–9 and 4–14 eV, respectively. The mass spectrometric samplings of their N_2 plasma showed N^+, N_2^+, NO, and Si; the addition of H_2 suppressed Si and NO. The NO formation was explained by the sputtering of SiO_2 from the discharge walls by N_2^+ and the reaction of SiO_2 with N_2 in the plasma. The hydrogen apparently suppressed NO formation and produced NH_x ($x = 1$–3), which aided in the nitriding process. Evidence from ion-beam studies of Taylor *et al.* [257] suggests that N_2^+ ions undergo charge exchange and dissociate at the surface of silicon and its oxides (SiO, SiO_2) to form hot N atoms, which react with Si producing nitrides that are similar to those of the type Si_3N_4.

3. Plasma Carburizing

In principle, the method is similar to glow discharge nitriding. Grube and Gay [258] studied the carburizing of a steel cathode in a dc cold cathode glow discharge of a hydrocarbon gas (1–20 torr) heated by an external furnace to a temperature of 1050°C. Under the experimental conditions, the hydrocarbon dissociates not only pyrolytically but also by electron bombardment in the discharge region, the latter enhancing the deposition rate of carbon on the metal surface. It has been suggested that high infusion rates of carbon into the surface, rather than high diffusion rates, are responsible for the high carburizing rates achieved in plasma carburizing.

Yoshihara *et al.* [259] have reported the preparation of SiC films by carburization in an rf discharge of a 2 : 3 mixture of SiH_4 and CH_4 by volume. Konuma *et al.* [260] used an rf plasma in methane (10–30 torr) for carburizing of metals such as Ti, Zr, Nb, and Ta. Although the carbides were identified on the metal surface by ESCA analysis, the discharge produced oily products that were identified by mass spectroscopy to be polyethylene and a mixture of higher-chain hydrocarbons with mass numbers as high as 450 amu. Although these glow discharges have not been analyzed under carburizing conditions, there is a considerable body of evidence from glow discharge polymerization work (Section IV.C.4) that CH_x radicals may be responsible for carbide formation.

4. Glow Discharge Polymerization*

When a glow discharge is produced in the vapor of an organic compound or its mixture with a noble gas, such as Ar, films of a polymeric nature are deposited on the surfaces exposed to the glow/afterglow. The observation of this phenomenon dates back to de Wilde [262] and Thenard and Thenard [263]. For nearly a century, however, such films were considered undesirable by-products and efforts were made to prevent film formation. For example, Suhr [264] designed reactors with externally heated walls to prevent film formation. About the 1950s the advantageous features of these organic films (e.g., flawless thin coatings, good adhesion to the substrate, chemical inertness, and low dielectric constant) were recognized. Since then much

* In contrast to conventional polymerization, i.e., molecular polymerization, polymer formation in glow discharge may be characterized as elemental or atomic polymerization. As Yasuda [261] pointed out, the term "plasma polymerization" should strictly be applied to polymerization that occurs in a plasma state. The term "glow discharge polymerization" has a wider meaning in the sense that includes plasma-induced polymerization.

applied research on the use of the process has been done and several hundred papers published. It is impossible for us to review all this work here. The interested reader is referred to review articles by Kolotyrkin *et al.* [265], Mearns [266], Millard [267], Havens *et al.* [268], Bell [31, 269], Yasuda [261], Kay *et al.* [30], Shen and Bell [270], and Boenig [271]. Gazicki and Yasuda [272] have summarized the electrical properties of these films and their control by regulating the glow discharge parameters.

Besides hydrocarbons, organic compounds containing nitrogen, fluorine, oxygen, and silicon have been used as starting materials for glow discharge polymerization. The overall mechanism has been represented schematically, as in Fig. 3.42, by Poll *et al.* [273]. It suggests that monomers can be converted into reactive and nonreactive products through processes occurring in the plasma (II and IV) as well as entering into polymer formation (I). The reactive products may also contribute to polymer deposition (III) or be converted to nonreactive products (V). The degradation of the polymer to form nonreactive products (VI) is included.

It should be noted that the predominance of a mechanism will depend not only on the chemical structure of the starting materials but also on the conditions of the discharge. The nature of the gaseous product(s) plays a significant role in determining the extent of the ablation process. For example, Kay [274] has reported that ablation of the polymer deposit occurs by shutting off the hydrogen, which causes the deposition of polymer in an otherwise non-polymerizing CF_4 discharge. Evidently H_2 removes H as HF from the discharge system, thus reducing the etching effect by the fluorine plasma. The importance of the ablation process is also demonstrated by the poor polymer formation reported by Yasuda [275] in glow discharges of oxygen-containing compounds. In short, the deposition observed in glow discharges is the resultant of the competitive ablation and polymerization processes. Further, it is necessary to distinguish between rate of polymerization and rate of deposition. In the literature these two terms are used interchangeably; unfortunately, they do not represent the same thing.

Most of the published work is concerned with the deposition rate on processing factors such as the reactor geometry, discharge mode, frequency and power, gas pressure, and flow rate of the organic compound and carrier

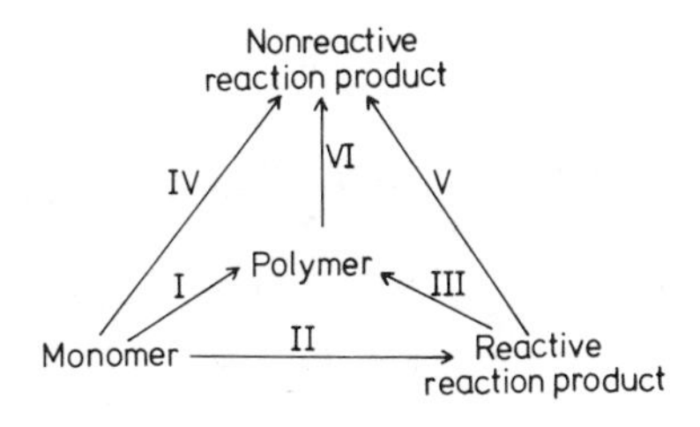

Fig. 3.42. Overall mechanism of glow discharge polymerization. (After Poll *et al.* [273].)

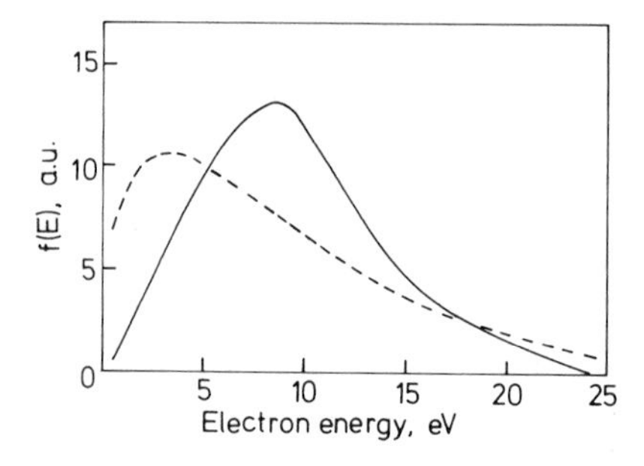

Fig. 3.43. Electron energy distribution function of a polymerizing benzene plasma under conditions of good film deposition: C_6H_6/Ar (——), Maxwell (- - -). (From Niinomi and Yanagihara [125]. Reprinted with permission from *ACS Symp. Ser.* **108**, 87. Copyright 1979 American Chemical Society.)

gas, if any. While high deposition rates result from the imposition of a magnetic field [25], pulsed discharges reduce the rate of polymer formation considerably [276]. Both Yasuda [261] and Boenig [271] have summarized the effect of these external parameters on film deposition.

Diagnostics of a polymerizing plasma were reported by Niinomi and Yanagihara [125], Yanagihara *et al.* [277], and Yanagihara and Niinomi [278], who measured T_e, n_e, and $f(E)$ for an rf plasma sustained in an Ar/C_6H_6 mixture using heated probes. Under conditions of good-quality film deposition, they found that $f(E)$ is non-Maxwellian (Fig. 3.43). Unlike in pure Ar or C_6H_6 vapor the value of n_e decreased by four decades when a small amount of C_6H_6 (100–200 SCCM) was added to the Ar plasma; however, n_e increased by three decades with increasing benzene flow rate (Fig. 3.44). The electron temperature (Fig. 3.45) showed a similar flow rate dependency to that of n_e, although not on a logarithmic scale. The reaction conditions that deposited good-quality polymer film usually corresponded to a plateau region of low pressure (0.5 torr), high benzene flow rate (500 SCCM), and n_e and T_e values of $\simeq 2.25 \times 10^9$ cm^{-3} and 4.25×10^4 K,

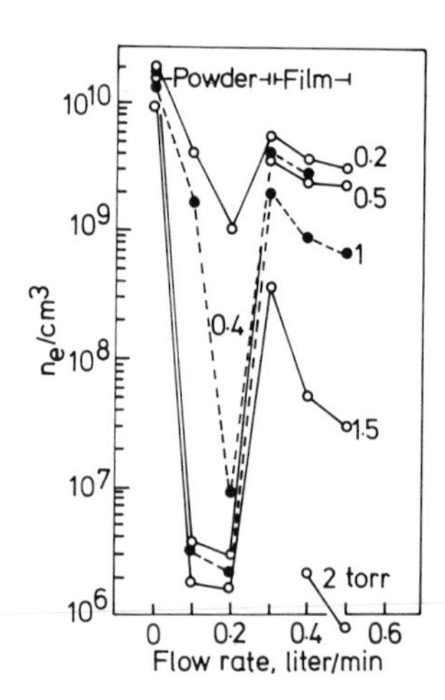

Fig. 3.44. Relationship between n_e and benzene flow rate at a fixed Ar flow rate of 150 sccm/min. (From Niinomi and Yanagihara [125]. Reprinted with permission from *ACS Symp. Ser.* **108**, 87. Copyright 1979 American Chemical Society.)

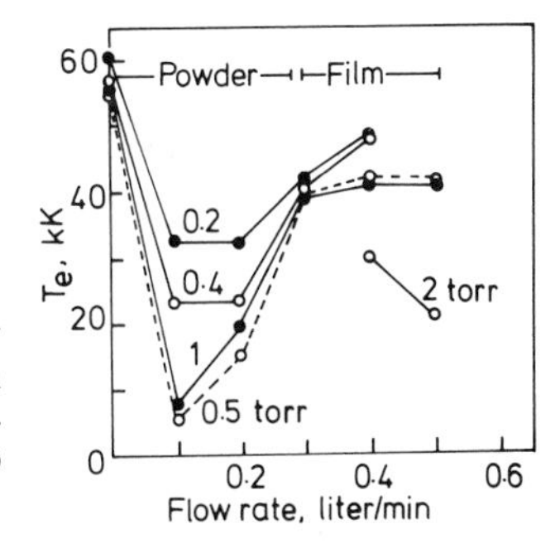

Fig. 3.45. Relationship between T_e and benzene flow rate at a fixed Ar flow rate of 150 sccm/min. (From Niinomi and Yanagihara [125]. Reprinted with permission from *ACS Symp. Ser.* **108**, 87. Copyright 1979 American Chemical Society.)

respectively. The reaction conditions for the lowest value of n_e and T_e corresponded to those producing a lot of powder.

Niinomi and Yanagihara [125] monitored the concentration of positive ions and neutrals under the preceding conditions. The mass spectrum of the positive plasma ions exhibited almost the same trends as the mass spectrum of neutral species present in the plasma (Fig. 3.46) and the pyrolysis mass spectrum of plasma-polymerized benzene reported by Venugopalan *et al.* [279]. Principal plasma ions assigned from the observed m/z values are listed in the Table 3.6. Although the mechanism of the formation of the observed ionic species is uncertain, the formula of C_7 to C_{12} oligomers (the second column, coded as 1st additives) can be made up by simply adding C_6H_6 to a corresponding fragment formula, which was found in abundance in the plasma, e.g.,

$$C_6H_5^+ + C_6H_6 \longrightarrow C_{12}H_{11}^+. \tag{53}$$

For oligomers of C_{13} to C_{18} as seen in the third column, a similar ion–molecule reaction mechanism appears to be valid. For the higher homologs, however, hydrogen loss seems to occur for structural stabilization. Although ion–molecule reactions play a role in plasma polymerization (in the gas

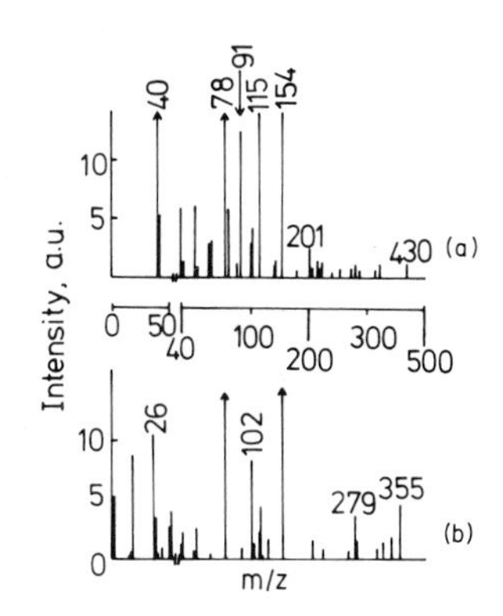

Fig. 3.46. Mass spectrum of (a) the positive plasma ions and (b) neutral species in a plasma under conditions of good film deposition. Benzene and Ar flow rates were 400 and 300 sccm/min; rf power = 40 W; pressure = 0.5 torr. (From Niinomi and Yanagihara [125]. Reprinted with permission from *ACS Symp. Ser.* **108**, 87. Copyright 1979 American Chemical Society.)

TABLE 3.6

Positive Ions from an RF Plasma in Ar/C_6H_6 under Conditions of Good Film Deposition[a]

	Oligomer ions					
Fragment	1st	2nd	3rd	4th	5th	6th
		$C_{13}H_{12}^+$			$C_{31}H_{22}^+$	$C_{37}H_{34}^+$
CH_2^+	$C_7H_8^+$	$C_{13}H_{14}^+$		$C_{25}H_{19}^+$		
	$C_7H_{10}^+$					
	$C_8H_7^+$					$C_{38}H_{36}^+$
$C_2H_2^+$	$C_8H_8^+$	$C_{14}H_{14}^+$				
$C_2H_3^+$	$C_8H_9^+$	$C_{14}H_{15}^+$		$C_{26}H_{25}^+$		
		$C_{14}H_{16}^+$				
					$C_{33}H_{23}^+$	$C_{39}H_{31}^+$
$C_3H_2^+$	$C_9H_8^+$	$C_{15}H_{14}^+$	$C_{21}H_{18}^+$			
$C_3H_3^+$	$C_9H_9^+$		$C_{21}H_{21}^+$			
		$C_{16}H_{10}^+$				
$C_4H_2^+$		$C_{16}H_{12}^+$				
$C_4H_3^+$	$C_{10}H_9^+$	$C_{16}H_{15}^+$	$C_{22}H_{20}^+$	$C_{28}H_{20}^+$		
$C_4H_4^+$						
					$C_{35}H_{23}^+$	
		$C_{17}H_8^+$	$C_{23}H_{16}^+$			
$C_5H_3^+$			$C_{23}H_{18}^+$			
$C_5H_4^+$	$C_{11}H_{10}^+$	$C_{17}H_{16}^+$	$C_{23}H_{21}^+$	$C_{29}H_{29}^+$		
					$C_{36}H_{24}^+$	
$C_6H_3^+$		$C_{18}H_{15}^+$	$C_{24}H_{15}^+$			
$C_6H_4^+$		$C_{18}H_{16}^+$	$C_{24}H_{24}^+$			
$C_6H_5^+$	$C_{12}H_{11}^+$	$C_{18}H_{17}^+$		$C_{30}H_{29}^+$		
$C_6H_6^+$						
$C_6H_7^+$						

[a] From Niinomi and Yanagihara [125]. Reprinted with permission from *ACS Symp. Ser.* **108**, 87. Copyright 1979 American Chemical Society.

phase), there is no evidence that the mechanism of the formation of a thin polymer film of a substrate is the same as that occurring in the gas phase.

Yanagihara *et al*. [280] and Yanagihara and Yasuda [281] have applied similar diagnostic techniques to rf glow discharges in CH_4, C_2H_4, CF_4, and C_2F_4.

In an rf discharge in methane, Smolinsky and Vasile [282] and Vasile and Smolinsky [147, 283] found that polymer deposits were formed more rapidly on the electrodes than on the walls and they detected a greater number of one and two carbon ions, especially $C_2H_3^+$, $C_2H_2^+$, CH_3^+, CH_2^+, and CH^+, in the dark space adjacent to the electrodes than in the space adjacent to the walls. Since the neutral C_2H_4 and C_2H_2 did not influence the rate they concluded that ions arriving on the surface are more important than neutrals in terms of the rate of polymerization. Following later works on extraction of positive ions from rf plasmas in C_2H_4 [284] and in C_2H_2 [285], they concluded that the polymerization in hydrocarbon plasmas is propagated by positive ions. This is in agreement with the earlier work of Westwood [286] and Thompson and Mayhan [287], and the work of Khait *et al.* [38], which showed that the deposition rate on a biased electrode in the plasma is nearly linearly proportional to the biasing voltage, indicating that the nature of the deposition by polymerization is dependent upon charge carriers.

Inspektor *et al.* [48, 288] and Carmi *et al.* [32] have made a detailed analysis of microwave cavity plasmas of methane, propylene, and their mixtures with argon in which pyrocarbon is deposited. They used three sampling and probe positions (Fig. 3.39) that were located immediately before the cavity (H), in the center of the cavity (G), and immediately beyond the cavity (F). By using a double floating probe system (DFPS) they determined T_e, n_e, and n_i at these locations. With respect to position F, the addition of Ar to CH_4 increased T_e (Fig. 3.47) suggesting that a heating effect was caused by Ar, which is a better heat conductor than CH_4. Also at this position the values of n_i and n_e in the plasma of Ar + CH_4 mixture were higher than the values for the CH_4 plasma alone. The saturation currents of the positive ions on both probes and the electron current density increased with increasing Ar/CH_4 ratio and was attributed to the charge transfer reaction from Ar to CH_4:

$$Ar^+ + CH_4 \rightarrow Ar + CH_4^+; \qquad \Delta H_f = -2.78 \text{ eV}; \qquad \sigma = 2 \times 10^{-15} \text{ cm}^2 \tag{54}$$

$$Ar^+ + CH_4 \rightarrow Ar + CH_3^+ + H; \qquad \Delta H_f = -1.36 \text{ eV}; \qquad \sigma = 1.04 \times 10^{-14} \text{ cm}^2 \tag{55}$$

According to Inspektor *et al.* [48] the values of n_i and n_e in a pure methane (no Ar added) increased in going from position H to G, but decreased at position F as pressure or power was increased (Figs. 3.48 and 3.49). This is to be compared with the values of n_i and n_e with added Ar

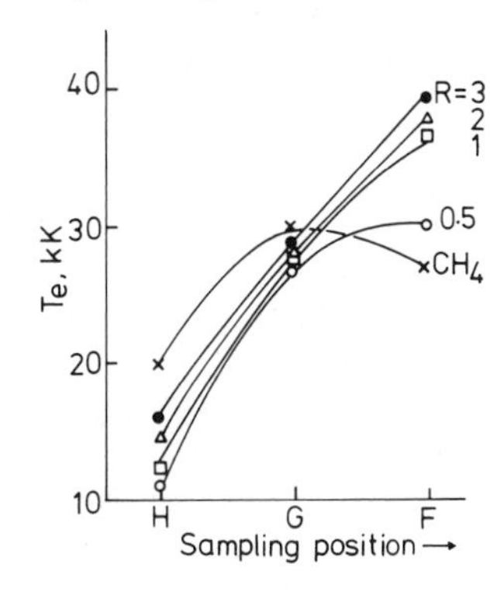

Fig. 3.47. Electron temperature in positions, H, G, and F at a total gas pressure of 3.0 torr. Methane flow rate is 47 ml/min; R = Ar/CH_4 ratio; microwave power = 100 W. (From Inspektor *et al.* [48].)

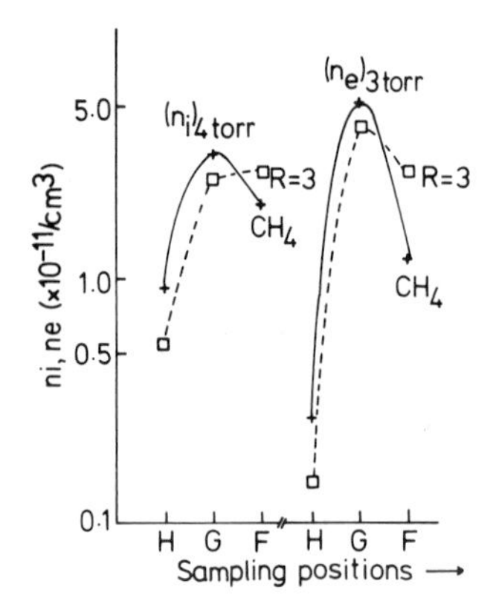

Fig. 3.48. Ion (n_i) and electron (n_e) densities in positions H, G, and F. Experimental parameters as in Fig. 3.47. (From Inspektor *et al.* [48].)

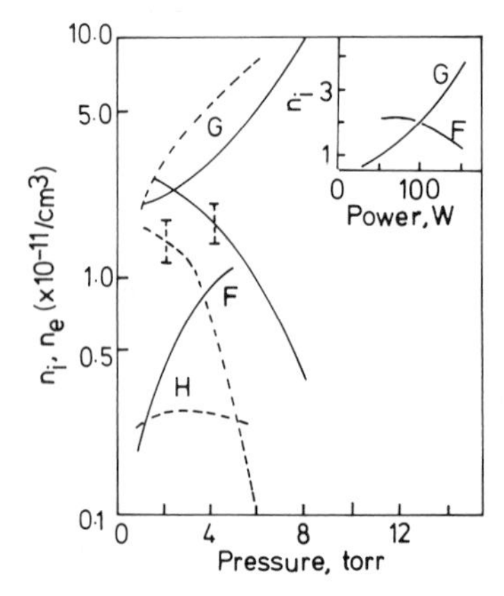

Fig. 3.49. Ion density (——) and electron density (- - -) (measured by electric conductivity method) versus methane pressure in position H, G, and F. Inset shows ion density versus microwave power at positions G and F. Experimental parameters as in Fig. 3.47. (From Inspektor *et al.* [48].)

shown in Fig. 3.50. By using three different techniques, viz., probe (DFPS), Stark broadening, and electric conductivity,* at position G in the Ar/C_3H_6 mixture, Inspektor *et al.* [48] found that the value of n_e increased with increasing pressure. They also noted that for a given pressure, the value of n_e depended on the diagnostic technique and could be different by as much as an order of magnitude (Table 3.7). In view of this the discrepancies between n_i and n_e values in Figs. 3.48–3.50 were considered to originate from different diagnostic techniques and ignored. It was concluded that the decrease of n_i and n_e in going from position G to F was due to the formation of highly

* If we assume that the current in the microwave plasma is carried mainly by electrons and consider only the dc component of the electric field strength E, the electron conductivity $\sigma_e = e\mu_e n_e$, where σ_e is in ohm^{-1} per centimeter. The electron density is directly evaluated from the measured electron current density j_e by $n_e = (j_e/E)e\mu_e$, where E is in volts per centimeter and j_e is given in amperes per square centimeter.

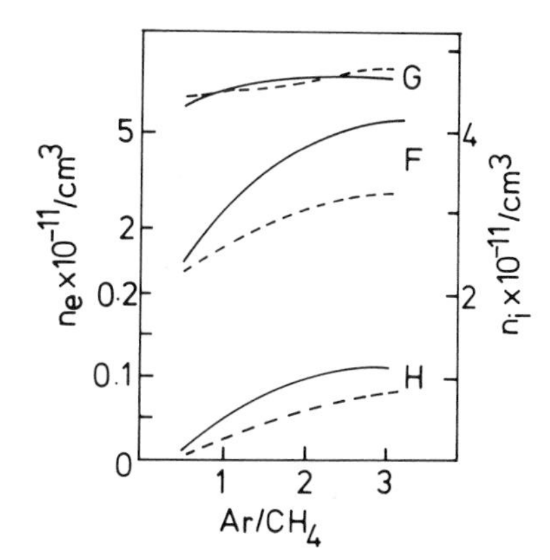

Fig. 3.50. Ion (——) and electron (- - -) density versus Ar/CH_4 ratio at 3 torr in positions H, G, and F. Experimental parameters as in Fig. 3.47. (From Inspektor *et al.* [48].)

unsaturated hydrocarbon polymer ions that are deposited on the walls and lose their charge.

The results of mass spectrometric sampling of a propylene–argon plasma by Carmi *et al.* [32] are shown in Fig. 3.51. Note the nearly linear decrease of $[C_3H_6]$ along the reactor, i.e., the reaction time in the flowing plasma, which is representative of a first-order kinetics in the plasma. It was suggested that the reaction is initiated by the ionization of C_3H_6 by either electrons, Ar^+, or metastable Ar:

$$C_3H_6 \xrightarrow{e,\ Ar^+,\ Ar^*} C_3H_6^+. \tag{56}$$

TABLE 3.7

Electron Density in an $Ar + C_3H_6$ Plasma Measured by Different Methods[a,b]

	$n_e \times 10^{-11}/cm^{-3}$			
			Stark broadening	
Pressure (torr)	DFPS	σ_e	H_β	$H_{n=11}$
1	2.4	1.5	15.0	4.2
3	3.5	3.0	17.8	5.6
5	4.5	3.7	20.5	8.0
7	7.8	6.1	23.2	—
11	10.5	9.4	30.3	—

[a] From Inspektor *et al.* [48].

[b] Measurements made in position G (Fig. 3.39); power = 100 W; $Ar/C_3H_6 = 3$.

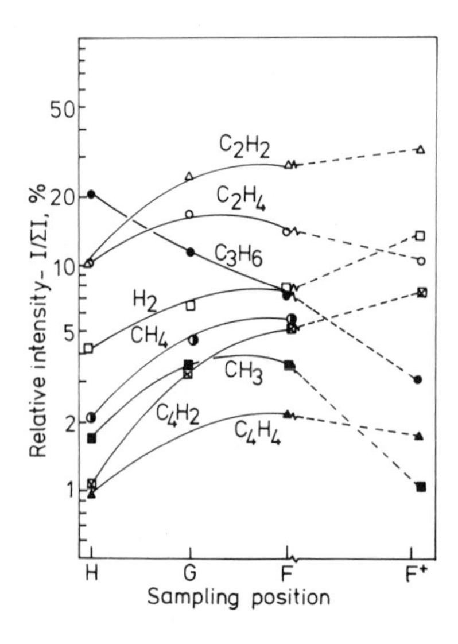

Fig. 3.51. Relative intensity of species along the plasma reactor; $R = 5$, $p = 3$ torr, $P = 150$ W, $Q(C_3H_6) = 7.5$ sccm. (From Carmi *et al.* [32].)

Since C_3H_6 is always the species in great excess, the following propylene chain was proposed:

$$C_3H_6^+ \xrightarrow{C_3H_6} C_6H_{10-6}^+ \xrightarrow{C_3H_6} C_9H_{14-9} \xrightarrow{C_3H_6} \cdots C_pH_q^+, \qquad p > q. \quad (57)$$

Fragmentation of the $C_3H_6^+$ ion to $C_2H_2^+$ rather than CH_4^+ is thermodynamically favored and supported by mass spectrometric data. Therefore, an acetylene chain was also proposed; i.e.,

$$C_2H_2^+ \xrightarrow{C_3H_6} C_5H_{4-6}^+ \xrightarrow{C_3H_6} \cdots C_rH_s^+, \qquad r > s. \quad (58)$$

Since ethylene is produced in high concentration the following ethylenic chain was also included:

$$C_2H_4^+ \xrightarrow{C_3H_6} C_5H_{5-8}^+ \xrightarrow{C_3H_6} \cdots C_uH_r, \qquad u > r. \quad (59)$$

All three chain reactions were proposed to proceed in parallel and result in pyrocarbon deposits.

Inspektor *et al.* [48] defined a degree of polymerization *Pm* as the normalized sum of peak intensities I_x of the high pyrocarbons, each multiplied by its carbon content x, i.e.,

$$Pm = \frac{1}{3} \sum x[(\sum Ix)/I_{C_3H_6}], \qquad x > 3. \quad (60)$$

The variation of *Pm* with and without argon is shown in Fig. 3.52 as a function of pressure. The polymerization is higher in the afterglow (F) rather

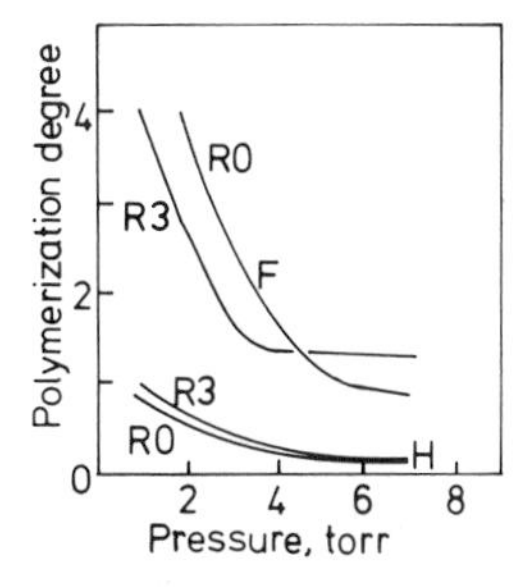

Fig. 3.52. Degree of polymerization at various Ar/C_3H_6 ratios RX versus pressure at positions H and F. (From Inspektor *et al.* [48].)

than in the foreglow (H) and is decreased in the afterglow by an increase in pressure and/or the addition of argon. These results are consistent with the measured decreases in n_e and n_i in the afterglow (Fig. 3.50).

Besides chain propagation by cationic species free-radical polymerization mechanisms have been proposed. The basis for this is the presence of unpaired spins, which has been reported by a number of authors who studied ESR spectroscopy of the plasma-polymerized films. If we assume that the trapped radicals in the films are related to the free radicals produced in the plasma it is easily recognized that the free-radical concentrations in the plasma are 10^3–10^5 higher than the ion concentrations. Further, Morosoff *et al.* [165] have observed high radical concentrations at the surfaces of polymers exposed to a plasma. Surface free radicals are known to react with gas-phase free radicals and unsaturated molecules such as ethylene and propylene. Since these evidences are not derived directly from the analysis of glow discharge we simply refer the reader to articles in which Bell [31, 269] relates the rate at which a polymer is formed with the experimentally controlled variables in glow discharge polymerization using free radicals as the primary species propagating chain growth, both in the gas phase and on the surface of the deposited polymer. Vinogradov *et al.* [289] have discussed an activation growth model for the formation of $(CH_x)_n$ polymer in CH_4 discharges taking the radicals as building blocks while the charged particles promote the activation of the surface sites for the growing polymer film. The exact polymer deposition mechanism can only be delineated by experiments yet to be performed, in which either the ions or the radicals formed can be isolated from the substrate surface where deposition takes place.

Both saturated and unsaturated fluorocarbons deposit fluorine-containing films in glow discharges. The general conclusion from ESCA analysis of numerous deposits is that plasma species from saturated fluorine compounds react mostly with the surface creating a somewhat permanently attached fluorinated layer. Unsaturated fluorocarbons, on the other hand, readily polymerize in a glow discharge, the main result being deposition of polymeric materials and not their grafting. Both neutral and ionic species

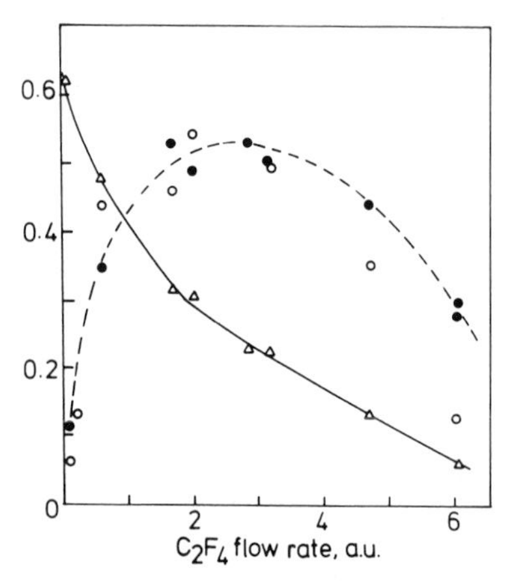

Fig. 3.53. Correlation between polymer deposition rate and unsaturated fluorocarbon species in plasma determined by mass spectrometry; ●, summation of unsaturated species as function of C_2F_4 flow rate at constant pressure; ○, polymer deposition rate determined by oscillating frequency quartz microbalance; △, summation of saturated species, mainly CF_3^+. (From Kay and Dilks [290].

have been monitored by Kay *et al.* [135] and by Vasile and Smolinsky [285] using *in situ* mass spectrometric analysis of C_2F_4 glow discharges. In contrast to hydrocarbon systems, particularly C_2H_4, C_2F_4 discharges produced a high yield of gaseous products as well as polymer deposits. Furthermore, the intensities of the signals relating to unsaturated species are directly related to the polymer deposition rate (Fig. 3.53). Millard and Kay [291] have determined that the polymerization rate in a C_3F_8 plasma is related to the intensity of the CF_2 band (265 nm) (Fig. 3.54). Since the presence of $(CF_2)_n$ species was observed in the effluents of plasmas excited in several different fluorocarbons and since their gas-phase concentration was directly related to the deposition rate, Kay *et al.* [30, 292] and Kay and Dilks [290] proposed that oligomerization of CF_2 may occur to form C_2F_4, C_3F_6, C_4F_8, etc. Evidently the primary precursors to polymerization have the general formula $(CF_2)_n$, which may include both cyclic alkanes and mono-olefins. Since the deposited film has $F/C < 2$, the plasma species, such as photons, ions, and metastables, must interact with the polymer to eliminate F and even small fragments by ablation.

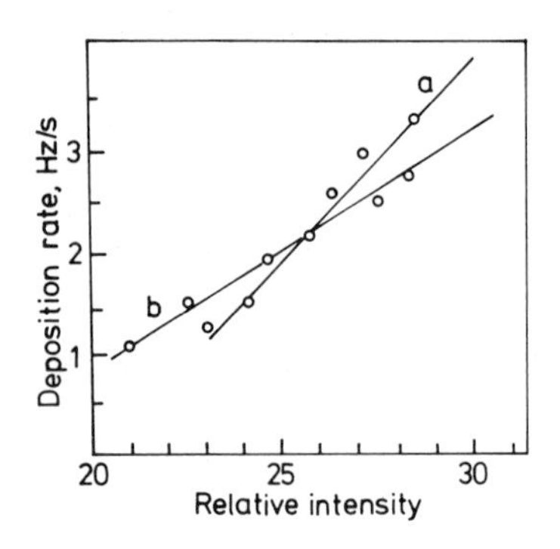

Fig. 3.54. Rate of polymerization of C_3F_8 in a parallel plate reactor versus (a) CF_2 band emission at 265 nm and (b) F atom emission at 686 nm. (From Millard and Kay [291]. Reprinted by permission of the publisher, The Electrochemical Society, Inc.)

Cramarossa *et al.* [293], d'Agostino [77], and d'Agostino *et al.* [294, 295] have used actinometric emission spectroscopy and mass spectrometry to analyze rf discharges in several fluorocarbons and their mixtures with H_2 or O_2, the latter because of their usefulness in plasma etching. They related emission intensities of both Ar and N_2 actinometers (see Section III.A.2) to the electron excitation functions $f(n_e)$, a function of the electron densities n_e at energies $E > E_{th}$ (where E_{th} is 11.5 and 13.7 eV for N_2 and Ar, respectively). Table 3.8 gives the normalized value of $f(n_e)$, the measured radical densities and polymer deposition rates. They explained the polymerization process on the basis of a mechanism that involves CF and CF_2 radicals as building blocks as well as an activation of the polymer surface by means of charged particle bombardment.

The deposition processes occurring in a fluorocarbon plasma are critically dependent upon the effective F/C ratio, low ratios ($<$2–3 depending on the bias) favoring polymerization and high ratios ($>$2–3 depending on the bias) favoring etching. Kay *et al.* [292] and Kay and Dilks [290] have combined the processes of etching and polymerization for producing metal-containing fluorocarbon polymer deposits. The films were prepared by using a mixture of argon and perfluoropropane (C_3F_8) in a capacitively coupled diode reactor in which the substrate was mounted on the grounded electrode. Germanium, molybdenum, and copper were used for the rf powered electrode, which at 13.56 MHz attained an overall negative potential due to the greater mobility of the electrons than the ions in the plasma and functioned as the cathode. Therefore, the positive ions arrive at the cathode with increased kinetic energy, and cathode material is removed from its surface by competitive physical sputtering in which momentum transfer to the surface is involved as well as chemical plasma etching through the formation of volatile species that subsequently desorb and enter the gas

TABLE 3.8

Normalized Values of Electron Excitation Functions $f(n_e)$, Radical Densities, and Polymer Deposition Rates r_p in Pure C_2F_4 and in Mixtures of C_nF_{2n+2} + 20% of Additives[a]

Feed	$f(n_e)$	CF	CF_2	r_p[b]
CF_4–H_2	1.0	0.13	0.004	1.3
CF_4–C_2F_4	0.41	0.20	0.20	1.2
C_2F_6–H_2	0.45	0.10	0.20	2.9
C_3F_8–H_2	0.20	0.08	0.05	2.1
C_2F_4	0.06	1.0	1.0	0.9

[a] From d'Agostino *et al.* [295].
[b] In units of g/cm^2 min.

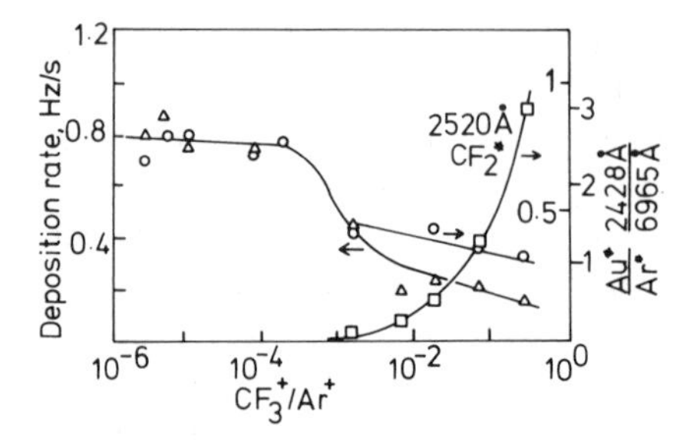

Fig. 3.55. Deposition rate, CF_2 band emission, and relative Au atomic emission (Au/Ar) as a function of the gas composition measured by mass spectrometry (CF_3^+/Ar^+). (After Kay and Hecq [296].)

phase. However, copper is a typical example of a material that forms nonvolatile fluorides; therefore, in this case material can be removed only by physical sputtering. Indeed, mass spectra of the plasma effluents showed that the relative concentration of $(CF_2)_n$ in the gas phase is directly proportional to the polymer deposition rate that is in the decreasing order Ge $>$ Mo $\gg$ Cu. This order also reflects the efficiencies of the cathode etching process to lower the F/C ratio in the plasma and, thus, increase the concentration of unsaturated species and, therefore, the polymerization rate.

Kay and Hecq [296] have used both mass spectrometry and optical emission for the diagnostics of a C_3F_8/Au system in which the deposition rate was simultaneously measured by a quartz crystal microbalance as a function of different Ar/C_3F_8 mixtures. Their data, shown in Fig. 3.55, indicates that the deposition rate is nearly constant up to $CF_3^+/Ar^+ \leqslant 2 \times 10^{-4}$. Also, the normalized Au optical emission intensity paralleled the deposition rate curve. However, the CF_2 emission starts when the deposition rate drops, suggesting that the decrease in gold deposition is due to the onset of polymer formation on the Au cathode.

5. Plasma-Activated Chemical Deposition

Films have been deposited by initiating chemical reactions in a gas with an electrical discharge. The technique is often referred to as plasma-activated (or plasma-assisted) chemical vapor deposition (PCVD). Such films are typically carbides, nitrides, oxides, and oxynitrides of elements such as Al, B, Ge, Si, and Ti. Empirical descriptions of processes and properties of films are to be found in articles by Kirk [297], Hollahan and Rosler [298], Rand [299], Reinberg [300], Veprek [17], Ojha [218], and Bonifield [301]. Table 3.9 is a representative, but not all-inclusive, list of the films produced and the starting materials used in rf/microwave discharges. The cited references are also good sources of information on the correlation between deposition and properties of the films and the macrovariables of the glow discharges. Practically none of these glow discharges have been analyzed for understanding the mechanism of film deposition.

TABLE 3.9

Inorganic Films Deposited by PCVD Technique

Film	Reactants	Reference
	Carbides	
GeC	GeH_4/CH_4	[302]
SiC	SiH_4/CH_4	[303]
		[304]
		[305]
	SiH_4/C_2H_4	[302]
		[306]
	SiH_4/C_2H_2	[307]
	SiH_4/CF_4	[298]
	$Si(CH_3)_4/H_2$	[308]
	$Si(CH_3)_4/H_2/Ar$	[206, 309]
	$SiCl_2(CH_3)_2/H_2/Ar$	[206, 309]
TiC	$TiCl_4/C_2H_2/H_2/Ar$	[310]
	Nitrides	
AlN	Al/N_2	[311]
		[312]
	$Al/N_2/Cl_2$	[313]
	$AlCl_3/N_2$	[314]
		[315]
BN	$B/N_2/H_2$	[16]
	$B_2H_6/H_2/NH_3$	[316]
	BBr_3/N_2	[317]
	B_2H_6/NH_3	[215]
P_3N_5	P/N_2	[318]
	PH_3/N_2	[204]
SiN	SiH_4/N_2	[319]
		[320]
		[321]
		[322]
	SiH_4/NH_3	[303]
		[323]
		[324]
	Nitrides (*continued*)	
		[325]
		[326]
		[327]
		[328]
	$SiH_4/NH_3/N_2$	[329]
	$SiCl_4/NH_3$	[330]
	SiI_4/N_2	[19]
	Oxides	
Al_2O_3	$AlCl_3/O_2$	[331]
		[312]
B_2O_3	$B(OC_2H_5)_3/O_2$	[332, 333]
GeO_2	$Ge(OC_2H_5)_4/O_2$	[332, 333]
	$GeCl_4/O_2$	[334]
		[335]
SiO_2	$Si(OC_2H_5)_4$	[336]
	$Si(OC_2H_5)_4/O_2$	[337]
		[332, 333]
		[338]
	$SiCl_4/O_2$	[334]
		[335]
	SiH_4/O_2	[303]
	SiH_4/N_2O	[324]
		[297]
		[339]
T_xO_y	$Ti(OC_2H_5)_4/O_2$	[332, 333]
	Oxynitrides	
$Si_xO_yN_z$	SiO/N_2	[340]
	$SiH_4/NO/NH_3$	[340]

There have been few diagnostic investigations of glow discharges used in PCVD work. For example, Haque *et al.* [341] used *in situ* mass spectrometry to study chemical transport in a carbon deposition system involving CO and CO_2; Veprek *et al.* [85] used matrix-isolation spectroscopy to study a carbon deposition system of hydrocarbons; Yoshihara *et al.* [308] made a mass spectral analysis of a SiH_4/CH_4 system used for SiC deposition; and

Ron *et al.* [330] applied ESR spectroscopy to conclude that a radical mechanism governs the dissociation of $SiCl_4$ and NH_3 in the formation of Si_3N_4. Based on such limited studies, PCVD processes have been described thus far as chemical vapor deposition by free radicals produced in the glow discharge. Clearly, for each of the systems listed in Table 3.9 investigations are badly needed to elucidate the role of the glow discharge electrons, ions, photons, and other excited species in the deposition process.

D. Proposed Models

The heterogeneous reactions such as boriding, carburizing, and nitriding of the respective atoms, molecules, or ions with metal surfaces are mainly exothermic. This is substantiated by Samsonov and Vinitskii's [342] data for the Gibbs energy change (ΔG), which are collated in Table 3.10. The computations, however, assume that the boron, nitrogen, and carbon were supplied as atoms to the metal surface. In a glow discharge, the dissociation of B_2H_6 or BCl_3, N_2 or NH_3, or hydrocarbon gases to B, N, or C atoms, respectively, takes place with a certain degree of atomization or ionization, both being functions of the macro and micro variables of the plasma. From the thermodynamic point of view, the interaction between atoms and/or ions with the metal surface results in chemical compounds, the extent of their deposition, i.e., thickness beneath the surface, being diffusion controlled. Figure 3.56 shows the various chemical compounds, such as oxides, borides, carbides, and silicides, which were detected on 440-C steels by Brainard and Wheeler [343] by using the XPS technique. The 440-C steels were pre-oxidized to form a Spinel interface ($FeO-Fe_2O_3$) on which Mo_2B_5, Mo_2C, and $MoSi_2$ were deposited by rf sputtering. A review of the literature, for example, the papers by Avni [344], Mattox [345], and Thornton [46], however, indicates that not enough data is available to conclude that the

TABLE 3.10

Gibbs Energy of Formation at 773 K for Refractory Compounds[a]

Boriding	ΔG (kcal/mole)	Nitriding	ΔG (kcal/mole)	Carburizing	ΔG (kcal/mole)
$Ti + 2B = TiB_2$	−68.0	$Ti + \frac{1}{2}N_2 = TiN$	−63.0	$Ti + C = TiC$	−41.9
$2Ti + 5B = Ti_2B_5$	−99.5	$V + \frac{1}{2}N_2 = VN$	−43.5	$V + C = VC$	−11.3
$V + 2B = VB_2$	−60.0	$4Cr + \frac{1}{2}N_2 = 2Cr_2N$	−26.5	$7Cr + 3C = Cr_7C_3$	−46.4
$Cr + 2B = CrB_2$	−30.2	$B + \frac{1}{2}N_2 = BN$	−18.5	$2Fe + C = Fe_2C$	+2.0
		$Al + \frac{1}{2}N_2 = AlN$	−59.8	$4B + C = B_4C$	−10.9
		$3Si + 2N_2 = Si_3N_4$	−117.8	$Si + C = SiC$	−11.5

[a] Data collected from Samsonov and Vinitskii [342].

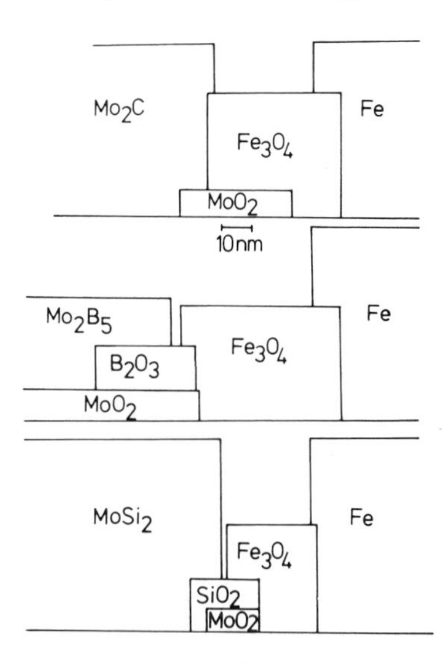

Fig. 3.56. Representation of interfacial regions of rf sputtered coatings on oxidized 440-C substrates; bias = −300 V. (From Brainard and Wheeler [343].)

adherence of a deposited film on a surface is controlled by a chemical bond rather than a van der Waals bond.

Film deposition in a glow discharge system is a dynamic, irreversible, kinetic process that begins with homogeneous reactions in both plasma bulk (PB) and layer (PL) and terminates through heterogeneous reactions in the neutralization layer (NL). Table 3.11 lists the various kinetic processes leading to the formation of the coating films. The rate-determining step in this dynamic deposition process is probably controlled by the heterogeneous plasma surface interactions (PSI). Before describing the various kinetic models it is worthwhile to emphasize the irreversibility of the process, that is to say, a solid product is obtained from a gas monomer supplied to the plasma.

Consider the plasma process in which solid μc-Si is deposited on a substrate from the monomer of tetrachlorosilane vapor:

$$SiCl_4(g) + Ar + H_2 \xrightarrow{\text{plasma}} \mu\text{c-Si} : H : Cl(s) + HCl(g) + Ar. \tag{61}$$

Two main processes may contribute to the reversibility of the reaction, viz., the formation of tetrachlorosilane gas from solid microcrystalline silicon:

(1) sputtering of Si atoms or clusters by Ar^+ bombardment, and
(2) plasma etching by hydrogen and chlorine, releasing chlorosilane back to the PL.

The calculated and measured values of fluxes and rates of sputtering, etching, and deposition are given in Table 3.12 for Ar^+, Cl_2, and F_2. The rate of μc-Si

TABLE 3.11

Kinetics of Reactions in Plasma Surface Interactions (PSI)

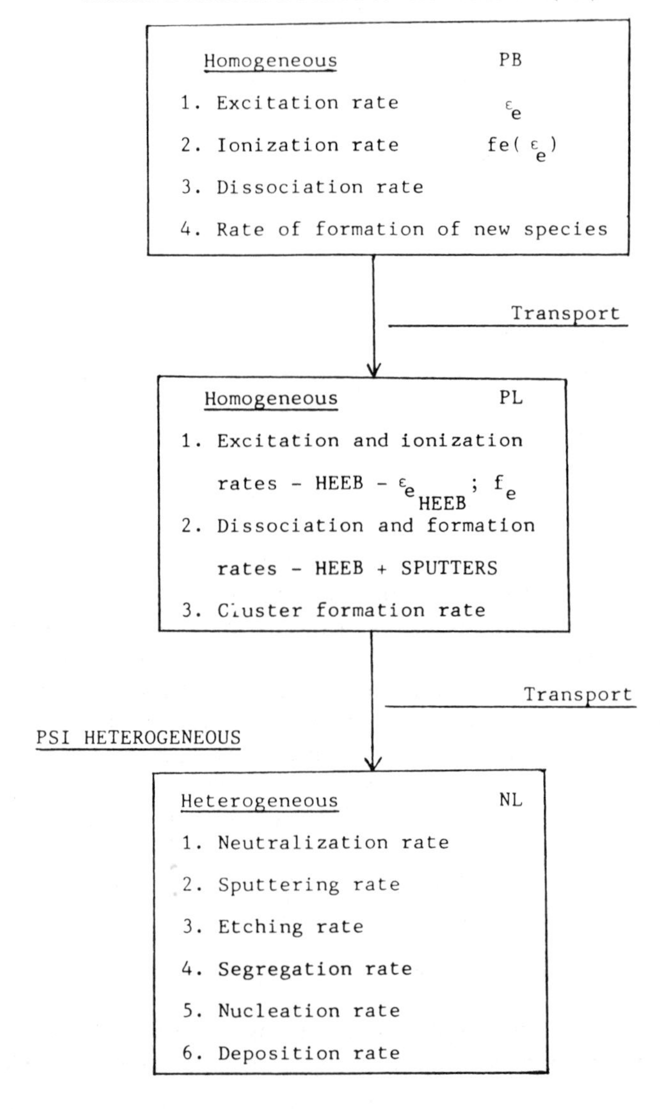

TABLE 3.12

Fluxes and Rates of Sputtering, Etching, and Deposition of Silicon in a Microwave (2.45-GHz) Plasma[a]

	Sputter	Etch[b]		Deposition[c]		
Plasma	Ar	$Ar + F_2$	$Ar + Cl_2$	$Ar + SiF_4$	$Ar + SiCl_4$	$Ar + SiCl_4 + H_2$
Flux = $10^{16}/cm^2$ s	1.6	9.5	1.9	1.0[d]	1.2[d]	2.8[d]
Rate (nm/s)	—	0.08	0.03	1.2	1.22	1.55

[a] Kinetic energy of $Ar^+ = 600$ eV; $n_i(Ar^+) = 10^{11}/cm^3$; $J_{Ar^+} = 3.4 \times 10^{16}/cm^2$ s; $J_{F_2} = J_{Cl_2} = 10^{16}/cm^2$ s; $5 \times 10^{-2} \leqslant P \leqslant 2.0$ torr; Si substrate grounded.

[b] Flux data from Winters [3] and Vinogradov *et al.* [35]; rate data on single crystal from Chapman [7] and Vinogradov *et al.* [35].

[c] Microcrystalline Si film deposited at 0.5 and 2.0 torr from 5 v/o SiF_4 in Ar (Avni, unpublished), 5 v/o SiF_4 in Ar (data from Avni [344] and Avni *et al.* [18]), and 5 v/o $SiCl_4$ + 15 v/o H_2 in Ar (data from Avni [344] and Avni *et al.* [18]).

[d] Calculated from the $I_{Si}/\Sigma I$ normalized concentration as measured by mass spectrometry (data from Avni [344] and Avni *et al.* [18]).

deposition is higher by more than one order of magnitude of the combined rates of sputtering and chemical etching. This has been taken as evidence of a low probability for reversible reactions at pressures above 0.5 torr in an rf discharge.

Tibbitt *et al.* [346] have proposed reinitiation processes based on earlier models by Denaro *et al.* [347] and Poll *et al.* [273] in which the electrons, radiation, and ion fluxes of the plasma reactivate the deposited polymer to form active radicals on the surface that, in turn, may react with the incoming gas particles, such as fresh monomer, its fragments, or polymerized molecules. These reactions release what Poll *et al.* [273] termed nonreactive reaction products, such as H_2 and Cl_2. In these reinitiation processes the irreversibility described earlier is still maintained.

Denaro *et al.* [347, 348], Poll *et al.* [273], Lam *et al.* [349], and Tibbitt *et al.* [346] have proposed theoretical models for evaluating the deposition rate in plasma polymerization processes. Bell [31] has reviewed these models, which are based on the following observations and assumptions:

(1) The propagation of homogeneous and heterogeneous polymerization is promoted mainly by radical reactions that are listed in Table 3.13.

(2) The rate of polymer deposition is expressed in terms of the rate of monomer consumed.

(3) The extent of gas-phase termination in the plasma system is very small and essentially all of the radicals formed in the plasma state are adsorbed on the deposited polymer surface.

TABLE 3.13

Reaction Mechanism for Plasma Polymerization of Unsaturated Hydrocarbons[a]

	Initiation
1.	$e + M_g \rightarrow M'_g + H_2 + e$
2.	$e + M_g \rightarrow M'_g + 2H + e$
3.	$e + M_g \rightarrow 2R_g + e$
4.	$e + H_2 \rightarrow 2H + e$
	Propagation (Homogeneous)
5.	$H + \left\{\begin{matrix} M_g \\ M'_g \end{matrix}\right\} \rightarrow R_{g_1}$
6.	$R_{g_n} + \left\{\begin{matrix} M_g \\ M'_g \end{matrix}\right\} \rightarrow R_{g_{n+1}}$
	Adsorption
7.	$S + \left\{\begin{matrix} M_g \\ M'_g \end{matrix}\right\} \rightarrow \left\{\begin{matrix} M_s \\ M'_s \end{matrix}\right\}$
8.	$S + H \rightarrow H_s$
9.	$S + R_{g_n} \rightarrow R_{s_n}$
	Propagation (Heterogeneous)
10.	$R_{s_n} + \left\{\begin{matrix} M_g \\ M'_g \end{matrix}\right\} \rightarrow R_{s_{n+1}}$
11.	$R_{s_n} + \left\{\begin{matrix} M_s \\ M'_s \end{matrix}\right\} \rightarrow R_{s_{n+1}}$
	Termination
12.	$H_s + H \rightarrow H_2$
13.	$R_{g_m} + H \rightarrow P_{g_m}$
14.	$R_{g_m} + R_{g_n} \rightarrow P_{g_{m+n}}$
15.	$R_{g_m} + R_{s_n} \rightarrow P_{s_{m+n}}$
16.	$R_{s_m} + R_n \rightarrow P_{s_{m+n}}$
	Reinitiation
17.	$e + P_{g_{m+n}} \rightarrow R_{g_m} + R_{g_n}$
18.	$P_s \xrightarrow{e, h\nu, I^+} R_{s_m} + R_{s_n}$
19.	$H + P_{g_n} \rightarrow R_{g_n} + H_2$
20.	$H + P_{s_n} \rightarrow R_{s_n} + H_2$

[a] From Tibbitt *et al.* [346]. Reprinted with permission from *Macromol.* **10**, 647. Copyright 1977 American Chemical Society.

(4) The concentrations of adsorbed monomer and free radicals are proportional to the gas phase concentrations of these species.

(5) The rate of hydrogen atoms formed (hydrocarbon systems) is equivalent to the rate of free radicals produced.

Tibbitt *et al.* [346] gave the following analytical expression for the polymer deposition rate dG/dt:

$$(dG/dt) = [2ca/(a - b)][M_g]_0^2\{(1 - \exp[-(a + b)\tau])/\exp[(a + b)\tau]\}. \quad (62)$$

With reference to Table 3.13, we write

$$a = k_i[e] = k_1[e] = K_2[e] = k_4[e] = 2.1 \times 10^{-2} \quad s^{-1};$$

$$b = (2/L)k_a; \qquad k_a = k_8[S] = 7.5 \times 10^{-3} \quad cm\ s^{-1};$$

$$c = (L/2)(k_{pg} + k_{ps}); k_{ps} = k_{10}K_R = 5.5 \times 10^{-18} \quad cm^4\ s^{-1};$$

$$k_{pg} = k_5 = k_6 = 3.5 \times 10^{-18} \quad cm^3\ s^{-1};$$

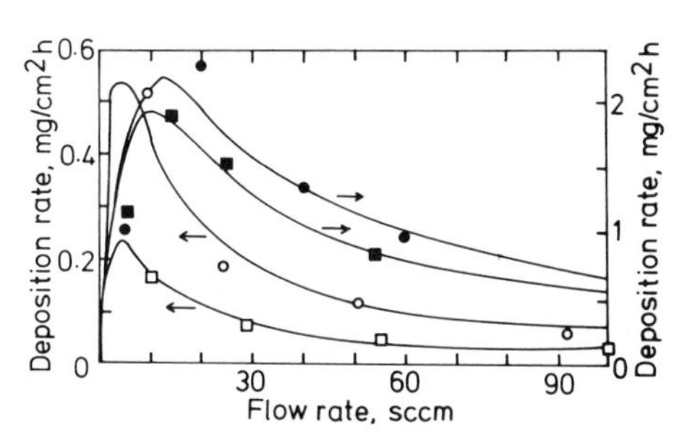

Fig. 3.57. Calculated and experimental polymer deposition rates for olefins as a function of monomer flow rate; ●, ethylene; ■, butadiene; ○, propylene; and □, isobutylene. (From Tibbitt *et al.* [346]. Reprinted with permission from *Macromol.* **10**, 647. Copyright 1977 American Chemical Society.)

and $\tau = V/Q$, where V is the plasma volume, Q the flow rate, and L the interelectrode separation. The quality of the fit of Eq. (62) to experimental data is shown in Fig. 3.57 for a number of hydrocarbons. Since the calculated values agree very well with the experimental results, a discussion of the assumptions in arriving at Eq. (62) is given below.

1. Homogeneous and Heterogeneous Propagation

Two types of reactions, viz., positive ion–molecule reactions and radical–molecule reactions have been discussed in connection with mechanisms in plasma systems of flowing gases. Which of these is rate determining in homogeneous and heterogeneous processes in a given plasma system has not yet been delineated.

Avni [344] and Avni *et al.* [18] have introduced the concept according to which free radicals and positive and negative ions are formed continuously along Z in the plasma as a result of the interactions of the monomer gas with energetic electrons of the plasma. Therefore, the concentration per unit volume (cm^{-3}) of free radicals n_R positive ions n_i and negative ions n_N and their spatial gradients are variables that must be considered in the kinetics of the reactions with monomer molecules. Other variables controlling the kinetics are the reactivities of the radicals and ions, i.e., the cross sections σ_R, σ_i, and σ_N of these reactions. The probability P of a reaction is the product of n and σ in a given location Z along the plasma. It follows that if

$$n_R \times \sigma_R > n_i \times \sigma_i, \qquad \text{then} \quad P_R > P_i; \tag{63}$$

i.e., the probability of radical–molecule interactions is higher than the probability of ion–molecule reactions. Conversely, if

$$n_i \times \sigma_i > n_R \times \sigma_R, \qquad \text{then} \quad P_i > P_R. \tag{64}$$

Bell [31] has suggested a value of 10^{-18} cm^3/molecule s for the rate constant for radical–monomer interactions, such as Reactions (5) and (6) in Table 3.13. According to Chatham *et al.* [350] ion–molecule charge transfer reaction rates are typically of the order of 10^{-10}–10^{-12} cm^3/molecule s.

Thus, for concentrations $n_R = 10^3 n_i$, $P_i > P_R$. Avni *et al.* [18, 344] have shown that for the polymerization of hydrocarbons $P_i > P_R$ in rf plasmas with Ar, while $P_R \geqslant P_i$ in rf plasmas without Ar. The same type of behavior was reported by Avni *et al.* [18, 344] for $SiCl_4$ + Ar ($P_i > P_R$) and for $SiCl_4 + H_2$ + Ar ($P_R \geqslant P_i$) rf plasma systems. Thus, depending on the plasma system either $P_R \geqslant P_i$ or $P_i \geqslant P_R$ can be selected and controlled. Such a selection, however, is not feasible for the theoretical condition where $P_R = P_i$.

2. Deposition Rate Expressed as the Rate of Monomer Consumed

This assumption has been investigated by Avni *et al.* [18, 344] by using mass spectrometric sampling at different locations along the plasma. An overall reaction rate k_o for the dissociation of the monomer M $(I_M/\sum I)$ and the formation of products was evaluated from data such as those shown in Fig. 3.51 for a C_3H_6 + Ar plasma and in Fig. 3.58 for a $SiCl_4$ + Ar plasma with and without H_2. Because of the large excess of monomer in plasma chemical reactions of the type

$$M + S^{(+)(\cdot)} \rightarrow \text{products},$$

where $S^{(+)(\cdot)}$ is some species in the plasma (ion or radical), were considered to be pseudo-first-order in M and S. Since at all positions (H, G, and F in Fig. 3.39) in the plasma steady-state conditions exist and the flow velocities between H and G and G and F are known, the conventional kinetic equation applies:

$$\int_{\tau_o \text{ at H}}^{\tau_o \text{ at G}} d\ln(I_M/\textstyle\sum I) = k_o \tau_o. \tag{65}$$

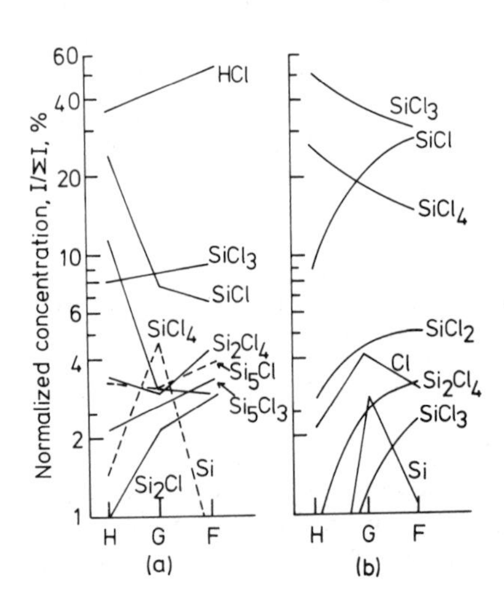

Fig. 3.58. Normalized concentration of chlorosilane species for plasma sampled in three positions at 100 W and 4.0 torr; (a) 3.5 v/o $SiCl_4$ + 20 v/o H_2 + Ar and (b) 3.5 v/o $SiCl_4$ + Ar. (From Avni *et al.* [344(b)].)

TABLE 3.14

Overall Reaction Rate Constants k_0 for the Dissociation of C_3H_6, SiH_4, and $SiCl_4$ [a,b]

	Plasma			
Gas mixture (vol %)	C_3H_6	SiH_4	$SiCl_4$	$k_0 = k_0(Ar + H_2)(NH_3)/k_0(Ar)$
16 C_3H_6 + Ar	228	—	—	0.26
66 C_3H_6 + Ar	60	—	—	0.26
5.0 SiH_4 + Ar	—	330	—	0.29
5.0 SiH_4 + 15 H_2 + Ar	—	95	—	0.29
3.5 $SiCl_4$ + Ar	—	—	15	40.0
3.5 $SiCl_4$ + 20 H_2 + Ar	—	—	600	40.0
3.5 $SiCl_4$ + 15 NH_3 + Ar	—	—	400	27.0

[a] From Avni *et al.* [344].
[b] Input power = 100 W; positions H–G; pressure = 1.0 torr; overall reaction time (t_0) = 2.2 ms.

Therefore, the value of k_o could be evaluated from the slope of the plot of $I_M/\sum I$ versus sampling position shown in Figs. 3.51 and 3.58, and the values of τ_o for the distances H–G and G–F at a given flow rate and pressure. Table 3.14 lists the calculated values of k_o for the dissociation of C_3H_6, SiH_4, and $SiCl_4$.

Avni *et al.* [18] used a different model for species S, which were short-lived in the plasma. This allowed the calculation of a local rate constant k_L using the equation

$$k_L \tau_L P = \ln(I/\sum I), \tag{66}$$

where τ_L is the ion or radical residence time and P the reactant partial pressure. The ionic residence time was evaluated by using its mobility and the Langevin relationship, while the radical residence time was found from its velocity and collision frequency. By making use of Eqs. (65) and (66) the monomer consumption in the plasma dM/dt could be calculated and related to the polymer deposition rate dG/dt. Representative data are given in Table 3.15 for μc-Si, Si_3N_4, PyC, FeB, and SiC, as the ratio $\bar{v}_{eff}/v$ in different rf plasmas. The constancy of this ratio for different plasmas at different operating conditions verifies the assumption and allows dM/dt to be correlated with dG/dt.

Equation (17) (Section II.D) gives the rate of ionization and excitation in the PL. When compared with phenomena in the PB there are strongly enhanced rates of excitation and ionization (Fig. 3.15) in the PL due to the secondary electrons (HEEB). The flux of HEEB j_{es} is given by Khait *et al.* [38] as

$$j_{es} = \gamma(\varepsilon_{is}) j_{is} = \tfrac{1}{4}\gamma a_i n u_i, \tag{67}$$

TABLE 3.15

Monomer Consumption dM/dt and Deposition Rate dG/dt in RF Plasmas with Reactor Cross Section of 80 cm^2

Monomer in gas mixtures (vol %)	Monomer feed (mg/cm^2 s)	Monomer dissociation[a] $I/\Sigma I$ (%)	dM/dt (mg/cm^2 s)	dG/dt (mg/cm^2 s)	$\bar{v}_{eff}/v^g$ dG/dM
$5SiCl_4 + 20H_2/Ar$[b]	0.30	85	0.25	0.10	0.40
$3.5SiCl_4 + 15NH_3/Ar$[c]	0.19	80	0.15	0.07	0.46
$16C_3H_6/Ar$[d]	0.35	82	0.29	0.14	0.48
$5BCl_3 + 15H_2/Ar$[e]	0.40	75	0.30	0.10	0.34
$1(CH_3)_4Si + 5H_2/Ar$[f]	0.28	78	0.22	0.08	0.36

[a] Measured by a quadrupole mass spectrometer sampling the plasma at different locations [18, 344].

[b] Microcrystalline Si deposited on ATJ graphite (grounded); $P = 2$ torr (data from Havens *et al.* [268]).

[c] Si_3N_4 deposited on AISI400 steel (floated); $P = 4$ torr (data from Ron *et al.* [329]).

[d] Pyrocarbon deposited on ATJ graphite, biased at -100 V; $P = 7$ torr (data from Inspektor *et al.* [48]).

[e] Boridation of AISI400 steel (grounded FeB); $P = 3$ torr (data from Raveh *et al.* [351]).

[f] SiC deposited on Ti alloys (grounded); $P = 4$ torr (data from Katz *et al.* [38]).

[g] See text and Khait *et al.* [38].

where γ is the coefficient of emission of electrons from the target surface, j_{is} and ε_{is} the ion flux and energy bombarding a substrate surface, u_i the mean velocity of the ions, n the monomer particle concentration in the plasma (cm^{-3}), and a_i ($=n_i/n$) the degree of ionization in the plasma. By introducing Eq. (67) into Eq. (17), Khait *et al.* [38] showed that

$$dn_i^*/dt = \bar{v}ns(1 + [ba_i u_i \sigma_i \sqrt{v\sigma_{es}}]). \tag{68}$$

Here b is the electron trajectory due to elastic and inelastic collisions, given by Eq. (22), and $\bar{v}$ is a characteristic velocity of the resulting ionization or excitation rate dn_i^* in the PL, given by

$$\bar{v} = \tfrac{1}{4}q_{es}(\varepsilon_{es})a_i(P, E)\gamma(\varepsilon_{is}, T_s)u_i(T_g, P, E) \tag{69}$$

with σ_i and σ_{es} the cross sections for ionization and electron–neutral collisions, respectively, T_s and T_g, the substrate and plasma temperatures, respectively, P the total gas pressure, and E the electric field strength in the sheath.

In the following discussion of closed (nonflowing gases) and open (flowing gases) systems the PL is considered to govern both the polymerization (hydrocarbons, chlorosilanes, etc.) and deposition rates.

a. Deposition Rate in Closed Plasma Systems. These systems have been discussed by Poll *et al.* [273]. Nonflowing gases are characterized by $v_f = 0$ or $Q = 0$. Following Khait *et al.* [38], we consider ΔL, the length of PL (Fig. 3.13). Let Δt_R be the time interval in which η_0 (the quantity of monomer material in the gas mixture to be polymerized, i.e., $\eta_0 = n_M/n_{mix}$) at $t = 0$ is transformed after polymerization to a final η_f, and S_0, the PL cross section, is transformed into S'_0, the cross section of the substrate where deposition occurs (after the time interval, Δt_R). At any time t, the fraction $\delta n_m(t)$ of material particles deposited can be expressed as:

$$\delta n_m(t) = (1/m_0) \int_0^t \frac{dG}{dt}\, dt, \tag{70}$$

where m_0 is the mass of material deposited on the substrate. If we assume that only a fraction $\alpha < 1$ of monomer is excited, ionized, and polymerized by $\bar{\varepsilon}_{es}$, then the rate of deposition dG/dt is

$$(dG/dt)_{vf=0} = m_0 \alpha (dn_m^{*(+)}/dt). \tag{71}$$

By applying Eqs. (68) and (69) for PL, we get

$$(dG/dt)_{v_f=0} = \tfrac{1}{4} m_0 \alpha q_{es} \gamma a_i n u_i S'_0. \tag{72}$$

According to Poll *et al.* [273] at $v_f = 0$ or $Q = 0$, the relative concentration of monomer decreases with time according to

$$\eta_f(t) = \eta_0 \exp(-t/\Delta t_R). \tag{73}$$

The time interval for deposition Δt_R is given by

$$\Delta t_R = \Delta L/\bar{v}\alpha, \tag{74}$$

where $\bar{v}$ is the characteristic velocity of the resulting ionization, excitation, and/or polymerization. Thus,

$$(dG/dt)_{vf=0} = [n_m(0)/\Delta t_R] m_0 \exp(-t/\Delta t_R). \tag{75}$$

It follows that the rate of deposition in a plasma system with nonflowing gases decreases exponentially with the duration of operation. As the data in Table 3.15 indicate this result can be applied to the consumption rate of the monomer $(dM/dt)_{v=0}$. Koizlik [352], Poll *et al.* [273], and Luhleich *et al.* [353] have confirmed experimentally the exponential decay of dG/dt in a plasma with nonflowing gases (closed system).

b. Deposition Rate in Open Plasma Systems. With $v_f > 0$, Δt_R is the time interval in which $\eta_0 \to \eta_f$, $S_0 \to S'_0$ and the flow velocity $v_f \to v'_f$. At steady state

$$(1/m_0)\, dG/dt + [n_m(0) - n_m]/\Delta t_R = 0, \tag{76}$$

and the rate of deposition is

$$dG/dt = [m_0 n_m/\Delta t_R]\beta\psi(v), \tag{77}$$

where $\beta = S_0/S_0'$ is a geometric factor and $\psi(v)$ an efficiency coefficient of excitation, ionization, and polymerization of particles having a flow velocity v_f'.

In the PL with HEEB interaction

$$\Delta t_R/\beta = \Delta L/\beta\alpha\bar{v}, \tag{78}$$

where $\bar{v}$ is again the characteristic velocity of the resulting dn_i^* or polymerization in the PL. The coefficient $\psi(v)$ behaves such that at velocity $v_f = 0$, $\psi(v) = 1$, representing the maximum ionization, excitation, or polymerization and deposition rate. As $v_f \to \infty$, $\psi(v) \to 0$, i.e.,

$$\psi(v/\bar{v}) = 1 - \exp(-\bar{v}\alpha\beta/v_f). \tag{79}$$

Under these conditions the rate of deposition is given by

$$dG/dt = m_0\eta_f(dn_i^*/dt)_{v_f}\,\Delta V_{PL}\bar{v}_{eff}, \tag{80}$$

where ΔV is the plasma layer volume and $\bar{v}_{eff}$ in dimensions of velocity includes the coefficients α, β, and $\psi(v/\bar{v})$, i.e.,

$$\bar{v}_{eff} = \alpha\beta\psi(v/\bar{v})v_f\bar{v}/[\alpha\bar{v}\psi(v/\bar{v})\beta] + v_f. \tag{81}$$

Equation (81) cannot be solved analytically. It can be numerically evaluated as follows:

The experimental efficiency of deposition is the ratio between the deposition rate and the rate of monomer consumption, that is,

$$[(dG/dt)/(dM/dt)]_{PL} = \bar{v}_{eff}/v_f. \tag{82}$$

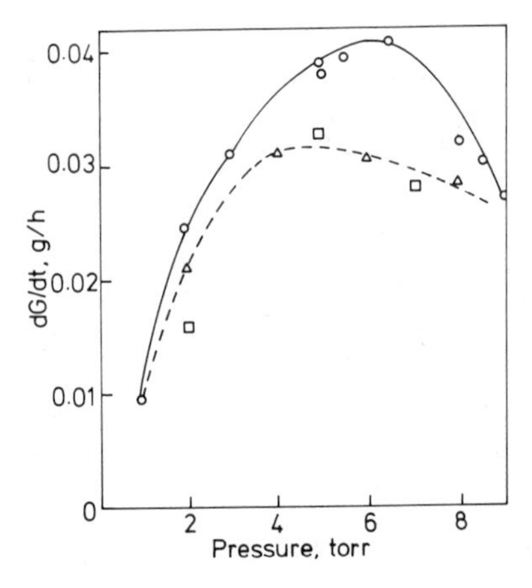

Fig. 3.59. Rate of deposition dG/dt versus pressure in position F; $R = 5$; ——, 7.6 sccm C_3H_6; – – –, 25.0 sccm C_3H_6; □, values predicted by the theoretical model; substrate grounded; rf power = 0.6 kW; time of deposition = 5 h. (From Khait *et al.* [38].)

Table 3.15 gives the ratios of $\bar{v}_{eff}/v_f$ as evaluated from experimental values of dG/dt and monomer consumption dM/dt. For the various rf plasmas listed in Table 3.15 the ratio $\bar{v}_{eff}/v_f$ varies between 0.34 for a boridation process to 0.48 for the pyrocarbon deposition from C_3H_6 + Ar plasma. Depending on the flow velocity v_f the effective velocity $\bar{v}_{eff}$ can be selected for optimizing the rates of deposition and monomer consumption. Figures 3.59–3.61 show the experimental values of dG/dt and the calculated values for $\bar{v}_{eff}/v_f = 0.46$ in the pyrocarbon deposition from mixtures of propylene and argon in an rf plasma. The agreement between the predicted and measured values suggests the validity of the theoretical model of Khait *et al.* [38] for flow systems.

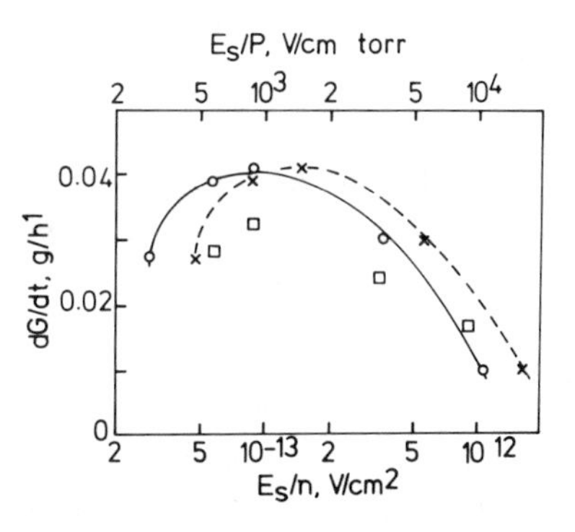

Fig. 3.60. Rate of deposition dG/dt versus calculated E_s/P (——) and E_s/n (– – –) with pressure parameter for substrate grounded and for experimental conditions as in Fig. 3.59; □, predicted values. (From Khat *et al.* [38].)

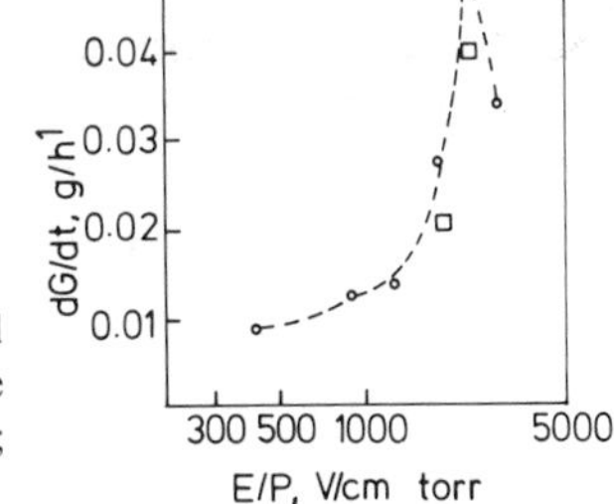

Fig. 3.61. Rate of deposition dG/dt versus calculated E_s/P with R as parameter at $P = 4.0$ torr and with the substrate grounded. Experimental conditions as in Fig. 3.59; □, predicted values. (From Khait *et al.* [38].)

V. CONCLUDING REMARKS

The understanding of film deposition processes in plasma systems is still in its infancy. The deposition models described in the preceding section have a phenomenologically qualitative character. There is insufficient information on plasma surface interactions (PSI) chiefly due to the fact that few experiments have been done with known fluxes of radicals on well-defined surfaces and little is known about synergetic effects in plasmas. To remedy this situation basic research in the following areas is needed:

(1) Recombination reactions at a surface where radicals recombine to form stable molecules.

(2) Synergetic effects where photon, electron, and ion bombardment change the reactivity of the incident radical with the surface.

(3) Adsorption characteristics of the incident species as a function of surface coverage and plasma conditions, such as electrical variables, bias voltages on the substrate, and partial pressures.

(4) Measurements of the rate of conversion of reactants to products on a surface.

(5) The growth mechanism in the case of polymer deposition.

It would be most desirable to perform well-defined surface experiments inside a plasma system. At present this seems to be very difficult if not impossible. A step in the right direction would be to apply both *in situ* and off-line techniques. For example, an insight to the PSI can be obtained by *in situ* measurements of the plasma layer (Table 3.11). Optical spectroscopy such as emission, absorption, laser-induced fluorescence, laser optogalvanic effect, and laser induced Raman effect can analyze the particles' flux gradients to the surface and monitor those sputtered away and/or etched from the surface. The information gained would be the reaction rates of excitation, ionization, dissociation, formation of polymeric species, and clusters. Off-line measurements of the deposited films by techniques such as Auger, LEED, UPS, XPS, SIMS, infrared and Raman spectroscopy, and EPR will supply data on surface coverage, sticking coefficients, nucleation, chemical bondings, and rates of diffusion and segregation. The correlation between the *in situ* with the off-line measurements could form the basis for developing appropriate theoretical models on the film deposition processes in glow discharges.

REFERENCES

1. M. Venugopalan, ed., "Reactions under Plasma Conditions," Vols. I and II, Wiley, New York, 1971.
2. E. Bauer and H. Poppa, *Thin Solid Films* **12**, 167 (1972).
3. H. F. Winters, *in* "Plasma Chemistry III" (S. Veprek and M. Venugopalan, eds.), pp. 69–125, Springer-Verlag, Berlin, 1980.
4. J. D. Cobine, "Gaseous Conductors." Dover, New York, 1958.
5. A. von Engel, "Ionized Gases." Oxford Univ. Press, London and New York, 1965.
6. E. Nasser, "Fundamentals of Gaseous Ionization and Plasma Electronics." Wiley, New York, 1971.
7. B. Chapman, "Glow Discharge Processes." Wiley, New York, 1980.

8. F. F. Chen, *in* "Plasma Diagnostic Techniques" (R. H. Huddlestone and S. L. Leonard, eds.), pp. 113–200, Academic Press, New York, 1965.
9. L. Holland and S. M. Ojha, *Thin Solid Films* **38**, L17 (1976); L. Holland and S. M. Ojha, *Vacuum* **26**, 53 (1976).
10. H. R. Koenig and L. I. Maissel, *IBM, J. Res. Dev.* **14**, 276 (1970).
11. J. L. Vossen, *J. Electrochem. Soc.* **126**, 319 (1979).
12. H. S. Butler, and G. S. Kino, *Phys. Fluids* **6**, 1346 (1963).
13. G. S. Anderson, W. N. Mayer, and G. K. Wehner, *J. Appl. Phys.* **33**, 2991 (1962).
14. R. S. Rosler and G. M. Engle, *Solid State Technol.* (December), 88 (1979).
15. M. Sokolowski, *J. Cryst. Growth* **46**, 136 (1979).
16. M. Sokolowski, A. Sokolowska, A. Michalski, Z. Romanowski, A. Rusek-Mazurek, and M. Wronikowski, *Thin Solid Films* **80**, 249 (1981); M. Sokolowski, A. Sokolowska, A. Rusek, Z. Romanowski, B. Gokieli, and M. Gazewska, *J. Cryst. Growth* **52**, 165 (1981).
17. S. Veprek, *Curr. Top. Mater. Sci.* **4**, 151 (1980); S. Veprek, *Pure Appl. Chem.* **54**, 1197 (1982).
18. (a) R. Avni, U. Carmi, R. Manory, E. Grossman, and A. Grill, *Proc. Int. Symp. Plasma Chem., 6th, Montreal, 1983*, 820 (1983); R. Avni, U. Carmi, I. Rosenthal, and A. Inspektor, *Proc. Int. Symp. Plasma Chem., 6th, Montreal, 1983*, 522 (1983); (b) R. Avni, U. Carmi, I. Rosenthal, R. Manory, and A. Grill, *Thin Solid Films* **107**, 235 (1983).
19. M. Shiloh, B. Gayer, and F. E. Brinckman, *J. Electrochem. Soc.* **124**, 295 (1977).
20. L. Holland, *in* "Science and Technology of Surface Coating" (B. N. Chapman and J. C. Anderson, eds.), pp. 369–385, Academic Press, New York, 1974.
21. F. M. Penning, *Physica* **3**, 873 (1936).
22. J. A. Thornton and A. S. Penfold, *in* "Thin Film Processes" (J. L. Vossen and W. Kern, eds.), pp. 75–113, Academic Press, New York, 1978.
23. D. B. Fraser, *in* "Thin Film Processes" (J. L. Vossen and W. Kern, eds.), pp. 115–129, Academic Press, New York, 1978.
24. R. K. Waits, *in* "Thin Film Processes" (J. L. Vossens and W. Kern, eds.), pp. 131–174, Academic Press, New York, 1978.
25. N. Morosoff, W. Newton, and H. Yasuda, *J. Vac. Sci. Technol.* **15**, 1815 (1978).
26. F. Kaufman, *Adv. Chem. Ser.* **80**, 29 (1969).
27. M. Venugopalan and R. A. Jones, "Chemistry of Dissociated Water Vapor and Related Systems," Wiley, New York, 1968.
28. M. Capitelli and E. Molinari, *in* "Plasma Chemistry II" (S. Veprek and M. Venugopalan, eds.), p. 59, Springer-Verlag, Berlin, 1980.
29. M. Venugopalan and S. Veprek, *in* "Plasma Chemistry IV" (S. Veprek and M. Venugopalan, eds.), pp. 1–58, Springer-Verlag, Berlin, 1983.
30. E. Kay, J. Coburn, and A. Dilks, *in* "Plasma Chemistry III" (S. Veprek and M. Venugopalan, eds.), pp. 1–42, Springer-Verlag, Berlin, 1980.
31. A. T. Bell, *in* "Plasma Chemistry III" (S. Veprek and M. Venugopalan, eds.), pp. 43–68, Springer-Verlag, Berlin, 1980.
32. U. Carmi, A. Inspektor, and R. Avni, *Plasma Chem. Plasma Process.* **1**, 233 (1981).
33. J. W. Coburn, *Plasma Chem. Plasma Process.* **2**, 1 (1982).
34. D. L. Flamm and V. M. Donnelly, *Plasma Chem. Plasma Process.* **1**, 317 (1981).
35. G. K. Vinogradov, P. I. Nevzorov, L. S. Polak, and D. I. Slovetsky, *Vacuum* **32**, 529 (1982).
36. D. Bohm, *in* "The Characteristics of Electrical Discharges in Magnetic Fields" (A. Guthrie and R. K. Wakerling, eds.), pp. 77–86, McGraw-Hill, New York, 1949.
37. H. D. Hagstrum, *Phys. Rev.* **123**, 758 (1961).
38. Y. L. Khait, A. Inspektor, and R. Avni, *Thin Solid Films* **72**, 249 (1980).
39. H. D. Hagstrum, *in* "Inelastic Ion-Surface Collisions" (N. H. Tolk, J. C. Tully, W. Heiland, and C. W. White, eds.), Academic Press, New York, 1977.

40. D. M. Gruen, A. R. Krauss, S. Susman, M. Venugopalan, and M. Ron, *J. Vac. Sci. Technol.* **A1**, 924 (1983).
41. M. Kaminski, "Atomic and Ionic Impact Phenomena on Metal Surfaces," Springer-Verlag, Berlin, 1965.
42. Y. L. Khait, U. Carmi, and R. Avni, *Proc. Int. Symp. Plasma Chem., 6th, Montreal, 1983*, 729 (1983).
43. N. Laegried and G. K. Wehner, *J. Appl. Phys.* **32**, 365 (1961).
44. J. L. Vossen and J. J. Cuomo, *in* "Thin Film Processes" (J. L. Vossen and W. Kern, eds.), pp. 11–73, Academic Press, New York, 1978.
45. J. Roth, *in* "Sputtering by Particle Bombardment" (R. Behrisch, ed.), Vol. II, Chapter 3, Springer-Verlag, Berlin, 1982.
46. J. A. Thornton, *in* "Deposition Technologies for Films and Coatings" (R. F. Bunshah, ed.), Chapters 2 and 5, Noyes Publ., Park Ridge, New Jersey, 1982.
47. R. V. Stuart and G. K. Wehner, *J. Appl. Phys.* **33**, 2345 (1962).
48. A. Inspektor, U. Carmi, R. Avni, and H. Nickel, *Plasma Chem. Plasma Process.* **1**, 377 (1981).
49. J. Lothe and G. M. Pound, *in* "Molecular Processes on Solid Surfaces" (E. Dragulis, R. D. Gretz, and R. I. Jaffe, eds.), McGraw-Hill, New York, 1969.
50. D. J. Ehrlich, R. M. Osgood, and T. F. Deutsch, *Appl. Phys. Lett.* **38**, 946 (1981).
51. S. R. J. Brueck and D. J. Ehrlich, *Phys. Rev. Lett.* **48**, 1678 (1982).
52. D. J. Ehrlich and J. Y. Tsao, *J. Vac. Sci. Technol.* **1B**, 969 (1983).
53. J. H. de Boer, *in* "Molecular Processes on Solid Surfaces" (E. Dragulis, R. D. Gretz, and R. I. Jaffe, eds.), McGraw-Hill, New York, 1969.
54. R. H. Huddlestone and S. L. Leonard, eds., "Plasma Diagnostic Techniques." Academic Press, New York, 1965.
55. W. Loechte-Holtgreven, ed., "Plasma Diagnostics," North-Holland Publ., Amsterdam, 1968; W. Loechte-Holtgreven and J. Richter, *in* "Plasma Diagnostics" (W. Lochte-Holtgreven, ed.), Chapter 5, pp. 250–346, North-Holland Publ., Amsterdam, 1968.
56. I. M. Podgornyi, "Topics in Plasma Diagnostics," Plenum, New York, 1971.
57. A. Eubank and E. Sindoni, eds., "Course on Plasma Diagnostics and Data Acquisition Systems," Editrice Compositori, Bologna, 1975.
58. F. Cabannes and J. Chapelle, *in* "Reactions under Plasma Conditions" (M. Venugopalan, ed.), Vol. I., pp. 367–469, Wiley, New York, 1971.
59. H. R. Griem, "Plasma Spectroscopy," McGraw-Hill, New York, 1964.
60. K. Bergsted, E. Ferguson, H. Schluter, and H. Wulff, *Proc. Int. Conf. Ion. Phenom. Gases, 5th, Munich, 1962*, **1**, 437 (1962).
61. E. B. Turner, *in* "Plasma Diagnostic Techniques" (R. H. Huddlestone and S. L. Leonard, eds.), Chapter 7, pp. 319–358, Academic Press, New York, 1965.
62. M. F. Kimmitt, A. C. Prior, and V. Roberts, *in* "Plasma Diagnostic Techniques" (R. H. Huddlestone and S. L. Leonard, eds.), Chapter 9, pp. 399–430, Academic Press, New York, 1965.
63. J. E. Greene and F. Sequeda-Osorio, *J. Vac. Sci. Technol.* **10**, 1144 (1973).
64. J. E. Greene, *J. Vac. Sci. Technol.* **15**, 1718 (1978).
65. W. R. Harshbarger, T. A. Miller, P. Norton, and R. A. Porter, *Appl. Spectrosc.* **31**, 201 (1977); W. R. Harshbarger, R. A. Porter, and P. Norton, *J. Electron. Mater.* **7**, 429 (1978).
66. A. Matsuda, K. Nakagawa, K. Tanaka, M. Matsumura, S. Yamasaki, H. Okushi, and S. Iizima, *J. Non-Cryst. Solids* **35/36**, 183 (1980).
67. F. J. Kampas and R. W. Griffith, *Appl. Phys. Lett.* **39**, 407 (1981); F. J. Kampas and R. W. Griffith, *J. Appl. Phys.* **52**, 1285 (1981).
68. M. Taniguchi, M. Hirose, T. Hamasaki, and Y. Osaka, *Appl. Phys. Lett.* **37**, 787 (1980).

69. J. Perrin and E. Delafosse, *J. Phys. D.* **13**, 759 (1980).
70. J. C. Knights, J. P. M. Schmitt, J. Perrin, and G. Guelachvili, *J. Chem. Phys.* **76**, 3414 (1982).
71. A. Bradley and J. P. Hammes, *J. Electrochem. Soc.* **110**, 15 (1963).
72. T. Williams and M. W. Hayes, *Nature (London)* **216**, 614 (1967).
73. R. d'Agostino, F. Cramarossa, S. de Benedictis, and G. Ferraro, *J. Appl. Phys.* **52**, 1259 (1981).
74. R. d'Agostino, V. Colaprico, and F. Cramarossa, *Plasma Chem. Plasma Process.* **1**, 365 (1981).
75. M. Kaufman, *Pure Appl. Chem.* **48**, 155 (1976).
76. J. W. Coburn and M. Chen, *J. Appl. Phys.* **51**(6), 3134 (1980).
77. R. d'Agostino, *Proc. Int. Conf. Phenom. Ionized Gases, 16th, Duesseldorf, 1983*, 251 (1983).
78. H. J. Tiller, D. Berg, and R. Mohr, *Plasma Chem. Plasma Process.* **1**, 247 (1981).
79. D. E. Ibbotson, D. L. Flamm, and V. M. Donnelly, *J. Appl. Phys.* **54**, 5974 (1983).
80. G. S. Selwyn and E. Kay, *Proc. Int. Symp. Plasma Chem., 6th, Montreal, 1983*, 716 (1983).
81. R. E. Klinger and J. E. Greene, *Appl. Phys. Lett.* **38**, 620 (1981).
82. A. M. Bass and H. P. Broida, eds., "Formation and Trapping of Free Radicals," Academic Press, New York, 1960.
83. G. J. Minkoff, "Frozen Free Radicals," Wiley (Interscience), New York, 1960.
84. D. M. Gruen, S. L. Gaudioso, R. L. McBeth, and J. L. Lerner, *J. Chem. Phys.* **60**, 89 (1974).
85. S. Veprek, D. L. Cocke, and K. A. Gingerich, *Chem. Phys.* **7**, 294 (1975).
86. T. A. Miller, *Plasma Chem. Plasma Process.* **1**, 3 (1981).
87. V. E. Bondybey and T. A. Miller, *J. Chem. Phys.* **66**, 3337 (1977).
88. T. A. Miller and V. E. Bondybey, *J. Chem. Phys.* **77**, 695 (1980).
89. R. A. Stern, *Phys. Fluids* **21**, 1287 (1977).
90. V. S. Burakov, P. Y. Misyakov, P. A. Naumenko, S. V. Nechaev, G. T. Razdobarin, V. V. Semenov, L. V. Sokolova, and I. P. Folomkin, *JETP Lett.* **26**, 403 (1977).
91. V. S. Burakov, *J. Appl. Spectrosc. (USSR)* **29**, 1504 (1978).
92. P. Bogen and E. Hintz, *Comments Plasma Phys. Controlled Fusion* **4**, 115 (1978).
93. A. Elbern, D. Rusbueldt, and E. Hintz, *Proc. Int. Symp. Plasma-Wall Interaction*, 475 (1977); A. Elbern, E. Hintz, and B. Schweer, *J. Nucl. Mat.* **76–77**, 143 (1978).
94. A. Elbern, *Appl. Phys.* **15**, 111 (1978).
95. M. J. Pellin, R. B. Wright, and D. M. Gruen, *J. Chem. Phys.* **74**, 6448 (1981).
96. R. B. Wright, M. J. Pellin, and D. M. Gruen, *Surf. Sci.* **110**, 151 (1981); R. B. Wright, C. E. Young, M. J. Pellin, and D. M. Gruen, *J. Vac. Sci. Technol.* **20**, 510 (1982).
97. J. M. Cook, T. A. Miller, and V. E. Bondybey, *J. Chem. Phys.* **69**, 2562 (1978).
98. D. H. Katayama, T. A. Miller, and V. E. Bondybey, *J. Chem. Phys.* **72**, 5469 (1980).
99. T. A. Miller and V. E. Bondybey, *Chem. Phys. Lett.* **50**, 275 (1977).
100. V. E. Bondybey and T. A. Miller, *J. Chem. Phys.* **69**, 3597 (1978); V. E. Bondybey and T. A. Miller, *J. Chem. Phys.* **70**, 138 (1979).
101. T. A. Miller, V. E. Bondybey, and J. H. English, *J. Chem. Phys.* **70**, 2919 (1979); T. A. Miller, V. E. Bondybey, and B. R. Zegarski, *J. Chem. Phys.* **70**, 4982 (1979).
102. T. J. Sears, T. A. Miller, and V. E. Bondybey, *J. Am. Chem. Soc.* **102**, 4864 (1980).
103. A. C. Eckbreth, P. A. Bonczyk, and J. F. Verdieck, *Appl. Spectrosc. Rev.* **13**, 15 (1977); A. C. Eckbreth, P. A. Bonczyk, and J. F. Verdieck, *Prog. Energy Combust. Sci.* **5**, 253 (1979).
104. S. R. J. Brueck and S. Pang, *Proc. Int. Symp. Plasma Chem., 6th, Montreal, 1983*, 714 (1983).
105. R. Walkup, R. W. Dreyfus, and P. H. Avouris, *Proc. Int. Symp. Plasma Chem., 6th, Montreal, 1983*, 701 (1983).
106. W. G. Rado, *Appl. Phys. Lett.* **11**, 123 (1967).
107. J. W. Nibler, *in* "Non-Linear Raman Spectroscopy and Its Chemical Applications," NATO Advanced Study Institute Series (W. Kiefer and D. A. Long, eds.), pp. 261–280, D. Reidel Publ., Boston, 1982.

108. (a) N. Hata, A. Matsuda, K. Tanaka, K. Kajiyama, N. Moro, and K. Sajiki, *Jpn. J. Appl. Phys.* **22**, L1 (1983); (b) N. Hata, A. Matsuda, and K. Tanaka, *Proc. Int. Conf. Amorphous and Liquid Semiconductors, 10th, Tokyo, 1983*, (1983); N. Hata, A. Matsuda, and K. Tanaka, *Proc. Int. Conf. Ion and Plasma Assisted Techniques, 4th, Kyoto, 1983*, (1983).
109. J. W. Nibler, J. R. McDonald, and A. B. Harvey, *Opt. Commun.* **18**, 371 (1976).
110. W. M. Shaub, J. W. Nibler, and A. B. Harvey, *J. Chem. Phys.* **67**, 1883 (1977).
111. M. Pealat, J. P. E. Taran, J. Taillet, M. Bacal, and A. M. Bruneteau, *J. Appl. Phys.* **52**, 2687 (1981); M. Pealat, J. P. E. Taran, J. Taillet, M. Bacal, and A. M. Bruneteau, *Proc. Int. Symp. Plasma Chem., 5th, Edinburgh, 1981*, 476 (1981).
112. I. Langmuir and H. M. Mott-Smith, *Gen. Electr. Rev.* **27**, 449 (1924).
113. L. Schott, *in* "Plasma Diagnostics" (W. Lochte-Holtgreven, ed.), pp. 668–731, North-Holland Publ., Amsterdam, 1968; L. Schott, *in* "Reactions under Plasma Conditions" (M. Venugopalan, ed.), Vol. I, pp. 515–542, Wiley, New York, 1971.
114. B. E. Cherrington, *Plasma Chem. Plasma Process.* **2**, 113 (1982).
115. J. D. Swift and M. J. R. Schwar, "Electrical Probes for Plasma Diagnostics," Amer. Elsevier, New York, 1969.
116. P. M. Chung, L. Talbot, and K. J. Touryan, "Electric Probes in Stationary and Flowing Plasmas: Theory and Applications," Springer-Verlag, Berlin, 1975.
117. J. W. Coburn and E. Kay, *J. Appl. Phys.* **43**, 4965 (1972).
118. J. D. Swift, *Br. J. Appl. Phys. Ser. 2* **2**, 134 (1969).
119. K. H. Groh and R. Reuschling, *Proc. Int. Symp. Plasma Chem., 5th, Edinburgh, 1981*, 493 (1981).
120. H. G. Lergon, M. Venugopalan, and K. G. Mueller, *Plasma Chem. Plasma Process.* **4**, 107 (1984); **5**, 89 (1985).
121. E. Eser, R. E. Ogilvie, and K. A. Taylor, *J. Vac. Sci. Technol.* **15**, 199 (1978).
122. J. A. Thornton, *J. Vac. Sci. Technol.* **15**, 188 (1978).
123. R. M. Clements, *J. Vac. Sci. Technol.* **15**, 193 (1978).
124. B. M. Oliver, R. M. Clements, and P. R. Smy, *J. Appl. Phys.* **41**, 2117 (1970).
125. M. Niinomi and K. Yanagihara, *ACS Symp. Ser.* **108**, 87 (1979).
126. S. Yamaguchi, G. Sawa, and M. Ieda, *J. Appl. Phys.* **48**, 2363 (1977); S. Yamaguchi, G. Sawa, and M. Ieda, *ACS Symp. Ser.* **108**, 115 (1979).
127. E. R. Mosburg, Jr., R. C. Kerns, and J. R. Abelson, *J. Appl. Phys.* **54**, 4916 (1983).
128. E. Grossman, R. Avni, and A. Grill, *Thin Solid Films* **90**, 237 (1982); E. Grossman, A. Grill, and R. Avni, *Plasma Chem. Plasma Process.* **2**, 341 (1982).
129. G. Gieres, *Proc. Int. Symp. Plasma Chem., 6th, Montreal, 1983*, 456 (1983).
130. M. Venugopalan, *Plasma Chem. Plasma Process.* **3**, 275 (1983).
131. S. N. Foner and R. L. Hudson, *J. Chem. Phys.* **21**, 1374 (1953); S. N. Foner and R. L. Hudson, *Adv. Chem. Ser.* **36**, 34 (1962).
132. T. Hayashi, M. Miyamura, and S. Komiya, *Jpn. J. Appl. Phys.* **21**, L755 (1982); T. Hayashi, M. Kikuchi, T. Fujioka, and S. Komiya, *Proc. Int. Ion Eng. Congr.–ISIAT & IPAT, Kyoto, September 1983*, 1611 (1983).
133. J. J. Wagner and W. W. Brandt, *Plasma Chem. Plasma Process.* **1**, 201 (1981).
134. W. Paul, H. P. Reinhard, and U. von Zahn, *Z. Phys.* **152**, 143 (1958).
135. E. Kay, J. W. Coburn, and G. Kruppa, *Vide* **183**, 89 (1976).
136. G. Smolinsky and M. J. Vasile, *Int. J. Mass Spectrom. Ion Phys.* **24**, 11, 311 (1977).
137. (a) G. Turban, Y. Catherine, and B. Grolleau, *Thin Solid Films* **60**, 147 (1979); (b) G. Turban, Y. Catherine, and B. Grolleau, *Plasma Chem. Plasma Process.* **2**, 61 (1982).
138. J. J. Wagner and S. Veprek, *Plasma Chem. Plasma Process.* **2**, 95 (1982); J. J. Wagner and S. Veprek, *Plasma Chem. Plasma Process.* **3**, 219 (1983).

139. J. W. Coburn and E. Kay, *Solid State Technol.* (December), 49 (1971).
140. A. J. Purdes, B. F. T. Bolker, J. D. Bucei, and T. C. Tisone, *J. Vac. Sci. Technol.* **14**, 98 (1977).
141. D. Smith and I. C. Plumb, *J. Phys. D.* **6**, 1431 (1973).
142. H. Helm, *J. Phys.* B **7**, 170 (1974).
143. J. B. Hasted, *Int. J. Mass Spectrom. Ion Phys.* **16**, 3 (1975).
144. J. W. Coburn, *Rev. Sci. Instrum.* **41**, 1219 (1970).
145. S. Komiya, K. Yoshikawa, and S. Ono, *J. Vac. Sci. Technol.* **14**, 1161 (1977).
146. J. L. Franklin, S. A. Studniarz, and P. K. Ghosh, *J. Appl. Phys.* **39**, 2052 (1968).
147. M. J. Vasile and G. Smolinsky, *Int. J. Mass Spectrom. Ion Phys.* **12**, 133; **13**, 381 (1973).
148. B. Rowe, *Int. J. Mass Spectrom. Ion Phys.* **16**, 209 (1975).
149. D. K. Boehme and J. M. Goodings, *Rev. Sci. Instr.* **37**, 362 (1966).
150. M. M. Shahin, *Adv. Chem. Ser.* **80**, 48 (1969).
151. B. Drevillon, J. Huc, A. Lloret, J. Perrin, G. de Rosny, and J. P. M. Schmitt, *Appl. Phys. Lett.* **37**, 646 (1980).
152. H. G. Lergon and K. G. Mueller, *Z. Naturforsch. A* **32**, 1093 (1977).
153. A. A. Westenberg, *Science* **164**, 381 (1969).
154. A. Carrington, D. H. Levy, and T. A. Miller, *Adv. Chem. Phys.* **18**, 149 (1970).
155. J. Brown, *MTP Int. Rev. Sci.* **4**, 235 (1972).
156. H. A. Bethe and E. E. Salpeter, "Quantum Mechanics of One and Two Electron Atoms," Springer-Verlag, Berlin, 1957.
157. A. A. Westenberg, *J. Chem. Phys.* **43**, 1544 (1965).
158. W. H. Breckenridge and T. A. Miller, *J. Chem. Phys.* **56**, 475 (1972).
159. S. Morita, T. Mizutani, and M. Ieda, *Jpn. J. Appl. Phys.* **10**, 1275 (1971); S. Morita, G. Sawa, and M. Ieda, *J. Macromol. Sci.* **10**, 501 (1976).
160. B. W. Tkatschuk, L. J. Ganuik, and E. P. Laurs, *Khim. Vys. Energ.* **11**, 350 (1977).
161. T. W. Scott, K.-C. Chu, and M. Venugopalan, *J. Polym. Sci. Poly. Chem. Ed.* **17**, 267 (1979).
162. M. S. Grenda and M. Venugopalan, *J. Polym. Sci. Poly. Chem. Ed.* **18**, 1611 (1980).
163. T. W. Scott, K.-C. Chu, and M. Venugopalan, *J. Polym. Sci. Poly. Chem. Ed.* **16**, 3213 (1978).
164. H. Yasuda, *in* "Plasma Chemistry of Polymers" (M. Shen, ed.), pp. 21, 39, Marcel Dekker, New York, 1976.
165. N. Morosoff, B. Crist, M. Bumgarner, T. Hsu, and H. Yasuda, *J. Macromol. Sci.* **10**, 451 (1976).
166. F. K. McTaggart, *Aust. J. Chem.* **18**, 937, 949 (1965); F. K. McTaggart. "Plasma Chemistry in Electrical Discharges," pp. 219–221, Elsevier, Amsterdam, 1967; F. K. McTaggart, *Adv. Chem. Ser.* **80**, 176 (1969).
167. G. Glockler and S. C. Lind, "The Electrochemistry of Gases and Other Dielectrics," pp. 400–429, Wiley, New York, 1939.
168. P. Sigmund, *Phys. Rev.* **184**, 383 (1969).
169. H. F. Winters, *Adv. Chem. Ser.* **158**, 1 (1976).
170. D. J. Ball, *J. Appl. Phys.* **43**, 3047 (1972).
171. B. N. Chapman, D. Downer, and L. J. M. Guimaraes, *J. Appl. Phys.* **45**, 2115 (1974).
172. J. J. Hanak and J. P. Pellicane, *J. Vac. Sci. Technol.* **13**, 406 (1976).
173. J. J. Cuomo, R. J. Gambino, J. M. E. Harper, J. D. Kutsis, and J. C. Webbe, *J. Vac. Sci. Technol.* **15**, 281 (1978).
174. D. M. Mattox, *Electrochem. Technol.* **2**, 295 (1964).
175. D. M. Mattox, *J. Vac. Sci. Technol.* **10**, 47 (1973).
176. A. von Hippel, *Ann Phys.* **80**, 672 (1926).
177. T. Baum, *Z. Phys.* **40**, 686 (1927).

178. P. Ziemann and E. Kay, *J. Vac. Sci. Technol.* **21**, 828 (1982).
179. P. Ziemann, K. Koehler, J. W. Coburn, and E. Kay, *J. Vac. Sci. Technol.* **B1(1)**, 31 (1983).
180. J. E. Greene, F. Sequeda-Osorio, and B. R. Natarajan, *J. Appl. Phys.* **46**, 2701 (1975); J. E. Greene, B. R. Natarajan, and F. Sequeda-Osorio, *J. Appl. Phys.* **49**, 417 (1978).
181. A. Guentherschulze, (1926), *Z. Phys.* **36**, 563 (1926).
182. R. E. Honig, *J. Appl. Phys.* **29**, 549 (1958).
183. J. R. Woodyard and C. B. Cooper, *J. Appl. Phys.* **35**, 1107 (1964).
184. H. Oechsner and W. Gerhard, *Phys. Lett.* **40a**, 211 (1972).
185. J. W. Coburn, E. Taglauer, and E. Kay, *Jpn. J. Appl. Phys. Suppl. 2, pt. 1*, 501 (1974).
186. R. Schwartz and F. Heinrich, *Z. Anorg. Allg. Chem.* **221**, 277 (1935).
187. S. Veprek and V. Marecek, *Solid State Electron.* **11**, 683 (1968).
188. P. G. LeComber and W. E. Spear, *in* "Amorphous Semiconductors" (M. H. Brodsky, ed.), pp. 251–284, Springer-Verlag, Berlin, 1979.
189. G. H. Bauer and G. Bilger, *Thin Solid Films* **83**, 223 (1981).
190. G. de Rosny, E. R. Mosburg, Jr., J. R. Abelson, G. Devand, and R. C. Kerns, *J. Appl. Phys.* **54**, 2272 (1983).
191. J. Perrin and J. P. M. Schmitt, *Chem. Phys.* **67**, 167 (1982).
192. J. F. O'Keefe and F. W. Lampe, *Appl. Phys. Lett.* **42**, 217 (1983).
193. R. W. Griffith, F. J. Kampas, P. E. Vanier, and M. D. Hirsch, *J. Non-Cryst. Solids* **35/36**, 391 (1980).
194. F. J. Kampas and R. W. Griffith, *Solar Cells* **2**, 385 (1980).
195. F. J. Kampas, *J. Appl. Phys.* **54**, 2276 (1983).
196. J. C. Knights, R. A. Lujan, M. P. Rosenblum, R. A. Street, D. K. Biegelsen, and J. A. Reimer, *Appl. Phys. Lett.* **38**, 331 (1981).
197. A. Matsuda and K. Tanaka, *Thin Solid Films* **92**, 171 (1982).
198. (a) A. Matsuda, T. Kaga, H. Tanaka, and K. Tanaka, *Proc. Int. Conf. Amorphous and Liquid Semiconductors, 10th, Tokyo, 1983*, (1983); (b) A. Matsuda, T. Kaga, H. Tanaka, L. Malhotra, and K. Tanaka, *Jpn. J. Appl. Phys.* **22**, L115 (1983); (c) A. Matsuda, K. Kumagai, and K. Tanaka, *Jpn. J. Appl. Phys.* **22**, L34 (1983).
199. G. T. Marcyk and B. G. Streetman, *J. Vac. Sci. Technol.* **14**, 1165 (1977).
200. J. G. Zesch, R. A. Lujan, and V. R. Deline, *J. Non-Cryst. Solids*, **35/36**, 273 (1980).
201. I. Haller, *Appl. Phys. Lett.* **37**, 282 (1980).
202. J. C. Knights, *J. Non-Cryst. Solids* **35/36**, 159 (1980).
203. B. A. Scott, M. H. Brodsky, B. C. Green, P. B. Kirby, R. M. Placenik, and E. E. Simonyi, *Appl. Phys. Lett.* **37**, 725 (1980).
204. S. Veprek, Z. Iqbal, H. R. Oswald, and A. P. Webb, *J. Phys. C.* **14**, 295 (1981); S. Veprek, Z. Iqbal, J. Bunner, and M. Scharli, *Philos. Mag. B* **43**, 527 (1981).
205. G. Bruno, P. Capezzuto, V. Colaprico, R. d'Agostino, and F. Cramarossa, *Proc. Int. Symp. Plasma Chem., 4th, Zurich, 1979*, 433 (1979); G. Bruno, P. Capezzuto, F. Cramarossa, and R. d'Agostino, *Thin Solid Films* **67**, 103 (1980); G. Bruno, P. Capezzuto, F. Cramarossa, V. Angelli, R. Murri, L. Schiavulli, G. Fortunato, and F. Evangelisti, *Proc. Int. Symp. Plasma Chem., 5th, Edinburgh, 1981*, 652 (1981).
206. M. Katz, D. Itzhak, A. Grill, and R. Avni, *Thin Solid Films* **72**, 497 (1980).
207. O. Gafri, A. Grill, D. Itzhak, A. Inspektor, and R. Avni, *Thin Solid Films* **72**, 523 (1980).
208. E. Grossman, R. Avni, and A. Grill, *Proc. Int. Symp. Plasma Chem., 5th, Edinburgh, 1981*, 658 (1981).
209. A. Grimberg, R. Avni, and A. Grill, *Thin Solid Films* **96**, 163 (1982).
210. A. Grill, E. Grossman, R. Manory, U. Carmi, and R. Avni, *Proc. Int. Symp. Plasma Chem., 6th, Montreal, 1983*, 843 (1983).
211. R. Manory, A. Grill, U. Carmi, and R. Avni. *Plasma Chem. Plasma Process.* **3**, 235 (1983).

212. J. J. Hauser, *J. Non-Cryst. Solids* **23**, 21 (1977).
213. N. Hosokawa, A. Konishi, H. Hiratsuka, and K. Annoh, *Thin Solid Films* **73**, 115 (1981).
214. J. C. Erskine and A. Cserhati, *J. Vac. Sci. Technol.* **15**, 1823 (1978).
215. C. J. Dell'Oca, D. L. Pulfrey, and L. Young, *in* "Physics of Thin Films" (M. H. Francombe and R. W. Hoffmann, eds.), Vol. 6, Academic Press, New York, 1971.
216. J. F. O'Hanlon, *in* "Oxides and Oxide Films" (A. K. Vijh, ed.), Vol. 5, p. 105, Marcel Dekker, New York, 1977.
217. S. Gourrier and M. Bacal, *Plasma Chem. Plasma Process.* **1**, 217 (1981).
218. S. M. Ojha, *in* "Physics of Thin Films" (G. Hass, M. H. Francombe, and J. L. Vossen, eds.), Vol. 12, p. 237, Academic Press, New York, 1982.
219. L. R. Ligenza, *J. Appl. Phys.* **36**, 2703 (1965).
220. O. A. Weinreich, *J. Appl. Phys.* **37**, 2924 (1966).
221 J. D. Leslie, V. Keeth, and K. Knorr, *J. Electrochem. Soc.* **125**, 44 (1978).
222. J. H. Greiner, *J. Appl. Phys.* **42**, 5151 (1971); J. H. Greiner, *J. Appl. Phys.* **45**, 32 (1974).
223. A. K. Ray and A. Reisman, *J. Electrochem. Soc.* **128**, 2424, 2461, 2466 (1981).
224. J. M. E. Harper, M. Heiblum, I. L. Speidell, and J. J. Cuomo, *J. Appl. Phys.* **52**, 4118 (1981).
225. J. F. O'Hanlon and W. B. Pennebaker, *Appl. Phys. Lett.* **18**, 554 (1971).
226. G. Olive, D. L. Pulfrey, and L. Young, *Thin Solid Films* **12**, 427 (1972).
227. S. Gourrier, A. Mircea, and M. Bacal, *Thin Solid Films* **65**, 315 (1980).
228. J. F. O'Hanlon and M. Sampogna, *J. Vac. Sci. Technol.* **10**, 450 (1973).
229. R. P. H. Chang, *Thin Solid Films* **56**, 89 (1979).
230. K. Ando and K. Matsumura, *Thin Solid Films* **52**, 173 (1978).
231. K. Yamasaki and T. Sugano, *J. Vac. Sci. Technol.* **17**, 959 (1980).
232. J. Perriere, J. Siejka, and R. P. H. Chang, *Thin Solid Films* **95**, 309 (1982).
233. V. Q. Ho and T. Sugano, *Proc. Int. Conf. Solid State Devices, 3rd, Tokyo, 1979*, (1979); V. Q. Ho and T. Sugano, *Thin Solid Films* **95**, 315 (1982).
234. R. P. H. Chang, C. C. Chang, and S. Darack, *Appl. Phys. Lett.* **3**, 999 (1980).
235. A. T. Fromhold, Jr., *J. Electrochem. Soc.* **124**, 538 (1977); A. T. Fromhold, Jr., *Thin Solid Films* **95**, 297 (1982).
236. A. T. Fromhold, Jr., and J. M. Baker, *J. Appl. Phys.* **51**, 6377 (1980).
237. L. G. Meiners, *J. Vac. Sci. Technol.* **21**, 655 (1982).
238. J. F. Wager, W. H. Makky, C. W. Wilmsen, and L. G. Meiners, *Thin Solid Films* **95**, 343 (1982).
239. B. Berghouse, German patent 669, 639 (1932).
240. J. von Bosse, K. Richter, and E. F. Kruppa, Swiss patent 172, 436 (1932).
241. D. M. Gruen, S. Veprek, and R. B. Wright, *in* "Plasma Chemistry I" (S. Veprek and M. Venugopalan, eds.), pp. 96–99, Springer-Verlag, Berlin, 1980.
242. M. Hudis, *J. Appl. Phys.* **44**, 1489 (1973).
243. B. Edenhofer, *Heat Treat. Met.* **1**, 23 (1974).
244. Yu. M. Lakhtin, Ya. D. Kogan and V. N. Saposhnikov, *Metalloved. Term. Obrab. Met.* **6**, 2 (1976).
245. J. P. Lebrum, H. Michel, and M. Janpoism, *Mem. Sci. Rev. Metall.* **69**, 727 (1972).
246. M. Content, H. Michel, and M. Gantois, *Trait. Therm.* **88**, 57 (1974).
247. A. Brokman and F. R. Tuler, *J. Appl. Phys.* **52**, 468 (1981).
248. G. G. Tibbetts, *J. Appl. Phys.* **45**, 5072 (1974).
249. A. Szabo and H. Wilhelmi, *Proc. Int. Symp. Plasma Chem., 6th, Montreal, 1983*, 346 (1983).
250. C. Braganza, S. Veprek, E. Wirz, H. Stuessi, and M. Textor, *Proc. Int. Symp. Plasma Chem., 4th, Zurich, 1979*, 100 (1979).
251. M. B. Liu, D. M. Gruen, A. R. Krauss, A. H. Reis, Jr., and S. W. Peterson, *High Temp. Sci.* **10**, 53 (1978).

252. M. Konuma and O. Matsumoto, *J. Less-Common Metals* **52**, 149; **55**, 97 (1977).
253. O. Matsumoto and Y. Kanazaki, *Proc. Int. Symp. Plasma Chem., 6th, Montreal, 1983*, 340 (1983).
254. O. Matsumoto, M. Konuma, and Y. Kanzak, *J. Less-Common Metals* **84**, 157 (1982).
255. K. Akashi, Y. Taniguchi, and R. Takada, *Proc. Int. Symp. Plasma Chem., 5th, Edinburgh, 1981*, 376 (1981).
256. O. Matsumoto and Y. Yatsuda, *Proc. Int. Symp. Plasma Chem., 5th, Edinburgh, 1981*, 382 (1981).
257. J. A. Taylor, G. M. Lancaster, A. Ignatiev, and J. W. Rabalais, *J. Chem. Phys.* **68**, 1776 (1978).
258. W. I. Grube and J. G. Gay, *Metall. Trans. A* **9A**, 1421 (1978).
259. H. Yoshihara, H. Mori, M. Kiuchi, and T. Kadota, *Jpn. J. Appl. Phys.* **17**, 1693 (1978).
260. M. Konuma, Y. Kanzaki, and O. Matsumoto, *Proc. Int. Symp. Plasma Chem., 4th, Zurich, 1979*, 174 (1979).
261. H. Yasuda, *in* "Thin Film Processes" (J. L. Vossen and W. Kern, eds.), pp. 360–398, Academic Press, New York, 1978.
262. P. de Wilde, *Ber. Dtsch. Chem. Ges.* **7**, 4658 (1874).
263. P. Thenard and A. Thenard, *C. R. Acad. Sci.* **78**, 219 (1874).
264. H. Suhr, *in* "Techniques and Applications of Plasma Chemistry" (J. R. Hollahan and A. T. Bell, eds.), pp. 63–66, Wiley, New York, 1974.
265. V. M. Kolotyrkin, A. B. Gilman, and A. K. Tsapuk, *Russ. Chem. Rev.* **36**, 579 (1967).
266. A. M. Mearns, *Thin Solid Films* **3**, 201 (1969).
267. M. Millard, *in* "Techniques and Applications of Plasma Chemistry" (J. R. Hollahan and A. T. Bell, eds.), pp. 177–213, Wiley, New York, 1974.
268. M. R. Havens, M. R. Biolsi, and K. G. Mayhan, *J. Vac. Sci. Technol.* **13**, 575 (1976).
269. A. T. Bell, *in* "Plasma Chemistry of Polymers" (M. Shen, ed.), pp. 1–13, Marcel Dekker, New York, 1976.
270. M. Shen and A. T. Bell, *in* "Plasma Polymerization" (M. Shen and A. T. Bell, eds.), *ACS Symp. Ser.* **108**, 1–33 (1979).
271. H. V. Boenig, "Plasma Science and Technology," Cornell Univ. Press, Ithaca, New York, 1982.
272. M. Gazicki and H. Yasuda, *Plasma Chem. Plasma Process.* **3**, 279 (1983).
273. H.-U. Poll, M. Arzt, and K.-H. Wickleder, *Eur. Poly. J.* **12**, 505 (1976).
274. E. Kay, *Int. Round Table Plasma Poly. Treat., IUPAC Symp. Plasma Chem., Limoges, France* (1977).
275. H. Yasuda, *Int. Round Table Plasma Poly. Treat., IUPAC Symp. Plasma Chem., Limoges, France* (1977).
276. H. Yasuda and T. Hsu, *J. Polym. Sci. Poly. Chem. Ed.* **5**, 81 (1977).
277. K. Yanagihara, T. Katuta, K. Arai, and M. Niinomi, *Kobunshi Ronbunshu* **38**, 741 (1981).
278. K. Yanagihara and M. Niinomi, *Proc. Int. Ion Eng. Congr., Kyoto, September 1983*, 1475 (1983).
279. M. Venugopalan, I.-S. Lin, and M. S. Grenda, *J. Polym. Sci., Polym. Chem. Ed.* **18**, 2731 (1980).
280. K. Yanagihara, K. Arai, and M. Niinomi, Preprint *ACS Nat. Meet. Div. Org. Coating Plastics Chem., Kansas City, September 1982*, 304 (1982); K. Yanagihara, K. Arai, M. Niinomi, and H. Yasuda, Preprint *ACS Nat. Meet. Div. Org. Coating Plastics Chem., Kansas City, September 1982*, 217 (1982).
281. K. Yanagihara and H. Yasuda, *J. Polym. Sci. Polym. Chem. Ed.* **20**, 1833 (1982).
282. G. Smolinsky and M. J. Vasile, *Int. J. Mass Spectrom. Ion Phys.* **12**, 47 (1973); G. Smolinsky and M. J. Vasile *Int. J. Mass Spectrom. Ion Phys.* **16**, 137 (1975).

283. M. J. Vasile and G. Smolinsky, *Int. J. Mass Spectrom. Ion Phys.* **18**, 179 (1975).
284. G. Smolinsky and M. J. Vasile, *Int. J. Mass Spectrom. Ion Phys.* **22**, 171 (1976).
285. M. J. Vasile and G. Smolinsky, *Int. J. Mass Spectrom. Ion Phys.* **24**, 11 (1977); M. J. Vasile and G. Smolinsky, *J. Phys. Chem.* **81**, 2605 (1977).
286. A. R. Westwood, *Eur. Polym. J.* **7**, 363 (1971).
287. L. F. Thompson and K. G. Mayhan, *J. Appl. Polym. Sci.* **16**, 2317 (1972).
288. A. Inspektor, U. Carmi, R. Avni, and H. Nickel, *Proc. Int. Symp. Plasma Chem., 4th, Zurich, 1979*, 390 (1979).
289. G. K. Vinogradov, Yu. A. Ivanov, and L. Polak, *High Energy Chem.* **15**, 120 (1981).
290. E. Kay and A. Dilks, *ACS Symp. Ser.* **108**, 195 (1979); E. Kay and A. Dilks, *J. Vac. Sci. Technol.* **18**, 1 (1981).
291. M. M. Millard and E. Kay, *J. Electrochem. Soc.* **129**, 160 (1982).
292. E. Kay, A. Dilks, and U. Hetzler, *Macromol. Sci.-Chem. A* **12**, 1393 (1978).
293. F. Cramarossa, R. d'Agostino, and S. de Benedictis, *Proc. Int. Conf. Ion Plasma Assisted Techniques, 4th, Kyoto, 1983*, (1983).
294. R. d'Agostino, F. Cramarossa, and S. de Benedictis, *Plasma Chem. Plasma Process.* **2**, 213 (1982).
295. R. d'Agostino, F. Cramarossa, V. Colaprico, and R. s'Ettole, *J. Appl. Phys.* **54**, 1284 (1983); R. d'Agostino, F. Cramarossa, and S. de Benedictis, *Proc. Int. Symp. Plasma Chem., 6th, Montreal, 1983*, 394 (1983); R. d'Agostino, S. de Benedictis, and F. Cramarossa, *Plasma Chem. Plasma Process.* **4**, 1 (1984).
296. E. Kay and M. Hecq, *Proc. Int. Symp. Plasma Chem., 6th, Montreal, 1983*, 490 (1983).
297. R. Kirk, *in* "Techniques and Applications of Plasma Chemistry" (J. R. Hollahan and A. T. Bell, eds.), Chapter 9, Wiley, New York, 1974.
298. J. R. Hollahan and R. S. Rosler, *in* "Thin Film Processes" (J. L. Vossen and W. Kern, eds.), p. 354. Academic Press, New York, 1978.
299. M. J. Rand, *J. Vac. Sci. Technol.* **16**, 420 (1979).
300. A. R. Reinberg, *Ann. Rev. Mater. Sci.* **9**, 341; *J. Electron. Mater.* **8**, 345 (1979).
301. T. D. Bonifield, *in* "Deposition Technologies for Films and Coatings" (R. F. Bunshah, ed.), pp. 365–384, Noyes Publ., Park Ridge, New Jersey, 1982.
302. D. A. Anderson and W. E. Spear, *Philos. Mag.* **35**, 1 (1977).
303. H. F. Sterling and R. C. G. Swann, *Solid-State Electron.* **8**, 653 (1965).
304. Y. Catherine and G. Turban, *Int. Round Table Surf. Treat. Plasma Polym., IUPAC, Limoges, France* (1977).
305. M. LeContellec, J. Richard, A. Guivarc'h, E. Ligeon, and J. Fontenille, *Thin Solid Films* **58**, 407 (1979).
306. Y. Catherine and G. Turban, *Thin Solid Films* **60**, 193 (1979).
307. D. Engemann, R. Fischer, and J. Knecht, *Appl. Phys. Lett.* **32**, 567 (1978).
308. H. Yoshihara, H. Mori, and M. Kiuchi, *Thin Solid Films* **76**, 1 (1981).
309. M. Katz, A. Grill, D. Itzhak, and R. Avni, *Proc. Int. Symp. Plasma Chem., 4th, Zurich, 1979*, 444 (1979).
310. F. J. Hazelwood, *Int. Conf. Adv. Surf. Coat. Technol., London, 1978*, 29 (1978).
311. G. Lewicki and C. A. Mead, *Phys. Rev. Lett.* **16**, 266 (1971).
312. M. Sokolowski, A. Sokolowska, A. Michalski, B. Gokieli, Z. Romanowski, and A. Rusek, *J. Cryst. Growth* **42**, 507 (1977).
313. S. Veprek, C. Brendel, and H. Schafer, *J. Cryst. Growth* **9**, 266 (1971).
314. H. Arnold, L. Biste, D. Bolze, and G. Eichhorn, *Krist. Tech.* **11**, 17 (1976).
315. J. Bauer, L. Biste, and D. Bolze, *Phys. Status Solidi A* **39**, 173 (1977).
316. S. B. Hyder and T. O. Yep, *J. Electrochem. Soc.* **123**, 1721 (1976).
317. H. F. Sterling, J. H. Alexander, and R. J. Joyce, *Spec. Ceram.* **4**, 139 (1968).

318. S. Veprek and J. Roos, *J. Phys. Chem. Solids* **37**, 554 (1976).
319. Y. Kuwano, *Jpn. J. Appl. Phys.* **7**, 88 (1968); Y. Kuwano, *Jpn. J. Appl. Phys.* **8**, 876 (1969).
320. R. Gereth and W. Scherber, *J. Electrochem. Soc.* **119**, 1248 (1972).
321. A. R. Reinberg, *Int. Round Table Surf. Treat. Plasma Polym. IUPAC, Limoges, France* (1977).
322. H. Dunn, P. Pan, F. R. White, and W. Doufe, *J. Electrochem. Soc.* **128**, 1555 (1981).
323. R. C. G. Swann, R. R. Mehta, and T. P. Cauge, *J. Electrochem. Soc.* **114**, 713 (1967).
324. R. J. Joyce, H. F. Sterling, and J. H. Alexander, *Thin Solid Films* **1**, 481 (1968).
325. A. W. Horsley, *Electronics* **84**, 3 (1969).
326. E. Taft, *J. Electrochem. Soc.* **118**, 1341 (1971).
327. R. S. Rosler, W. C. Benzig, and J. Baldo, *Solid State Technol.* **19**(6), 45 (1976).
328. V. J. Kumav, *J. Electrochem. Soc.* **123**, 262 (1976).
329. W. Kern and R. S. Rosler, *J. Vac. Sci. Technol.* **14**, 1082 (1977).
330. Y. Ron, A. Raveh, U. Carmi, A. Inspektor, and R. Avni, *Thin Solid Films* **107**, 183 (1983).
331. H. Katto and Y. Koga, *J. Electrochem. Soc.* **118**, 1619 (1971).
332. D. R. Secrist and J. D. Mackenzie, *Bull. Am. Ceram. Soc.* **45**, 784 (1966); D. R. Secrist and J. D. Mackenzie, *J. Electrochem. Soc.* **113**, 914 (1966).
333. D. R. Secrist and J. D. Mackenzie, *Adv. Chem. Ser.* **80**, 242 (1969).
334. D. Kueppers, J. Koenings, and H. Wilson, *J. Electrochem. Soc.* **123**, 1079 (1976); D. Kueppers, J. Koenings, and H. Wilson, *J. Electrochem. Soc.* **125**, 1298 (1978).
335. D. Kueppers and H. Lydtin, *in* "Plasma Chemistry I" (S. Veprek and M. Venugopalan, eds.), pp. 107–131, Springer-Verlag, Berlin, 1980.
336. L. L. Alt, S. W. Ing, Jr., and K. W. Laendle, *J. Electrochem. Soc.* **110**, 465 (1963).
337. S. W. Ing and W. Davern, *J. Electrochem. Soc.* **111**, 120 (1964).
338. S. P. Mukherjee and P. E. Evans, *Thin Solid Films* **14**, 105 (1972).
339. J. R. Hollahan, *J. Electrochem. Soc.* **126**, 931 (1979).
340. M. Mashita and K. Matsushima, *Jpn. J. Appl. Phys., Suppl. 2*, Part 1, 761 (1974).
341. M. R. Haque, H. R. Oswald, and S. Veprek, *Proc. Int. Conf. Phenom. Ionized Gases, Part I, 12th, Eindhoven, 1975*, (1975).
342. G. V. Samsonov and I. M. Vinitskii, "Handbook of Refractory Compounds," IFI/Plenum, New York, 1980.
343. W. A. Brainard and D. R. Wheeler, *J. Vac. Sci. Technol.* **15**, 1800 (1978).
344. (a) R. Avni, Invited Lecture at *Int. Conf. Metall. Coatings, 11th, San Diego, April 1984*, (1984); (b) R. Avni, U. Carmi, A. Inspektor, and R. Rosenthal, *NASA Tech. Paper 2301*, April (1984).
345. D. M. Mattox, *in* "Deposition Technologies for Films and Coatings" (R. F. Bunshah, ed.), Noyes Publ., Park Ridge, New Jersey, 1982.
346. J. M. Tibbitt, R. Jensen, A. T. Bell, and M. Shen, *Macromol.* **10**, 647 (1977).
347. A. R. Denaro, P. A. Owens, and A. Crawshaw, *Eur. Polym. J.* **4**, 93 (1968).
348. A. R. Denaro, P. A. Owens, and A. Crawshaw, *Eur. Polym. J.* **5**, 471 (1969); A. R. Denaro, P. A. Owens, and A. Crawshaw, *Eur. Polym. J.* **6**, 487 (1970).
349. D. K. Lam, R. F. Baddour, and A. F. Stancell, *in* "Plasma Chemistry of Polymers," (M. Shen, ed.), pp. 53–82, Marcel Dekker, New York, 1976.
350. H. Chatham, D. Hils, R. Robertson, and A. C. Gallagher, *J. Chem. Phys.* **79**, 1301 (1983).
351. A. Raveh, A. Inspektor, U. Carmi, and R. Avni, *Thin Solid Films* **108**, 39 (1983).
352. K. Koizlik, KFA Report 1128RW, Kernforschungsanlage, Juelich (1974).
353. H. Luhleich, D. Seeberger, L. Suetterlin, and R, von Seggern, *High Temp. High Pressures* **9**, 283 (1977).

4 PREPARATION, STRUCTURE, AND PROPERTIES OF HARD COATINGS ON THE BASIS OF i-C AND i-BN

C. Weissmantel

Sektion Physik/EB
Technische Hochschule
Karl-Marx-Stadt, German Democratic Republic

I. INTRODUCTION

The fourfold coordinated crystalline phases of carbon and boron nitride, i.e., diamond and cubic BN, are distinguished by their unique hardness, which has led us to call them superhard materials [1, 2]. In current research, evidence has been accumulated that less-ordered metastable film phases of these materials can be prepared by condensation of energetic ionized particles, thus yielding layers that show resemblance to the superhard crystalline counterparts with respect to macroscopic properties. In the case of carbonaceous films these remarkable properties include great hardness, high electrical resistivity, low surface friction, high refractive index, and a more or less pronounced transparency in the visible and infrared region. These properties, either alone or through combined utilization, imply widespread applications as hard and protective coatings, tribological layers, or functional films in optics and electronics. Furthermore, the kinetics of metastable film formation from energetic free particles under conditions far apart from thermodynamic equilibrium raises principal questions concerning the feasibility of producing film materials of unusual microstructure [3].

This chapter deals with the ion-enhanced preparation, the structure, properties, and possible applications of hard films with C or BN as major constituents; certain aspects of this topic have been reviewed by several authors [4–8]. The interesting attempts to synthesize fine diamond powder or epitactic diamond layers by means of laser or plasma activation will not be discussed [9–12].

According to a proposition [13] that has been widely accepted, the hard films are designated as i-phases, e.g., i-C or i-BN. The prefix i refers to the ion enhancement that seems to be the essential common feature of all reported preparation techniques. The designation diamondlike carbon introduced by Aisenberg and Chabot [14] is still widely used, but care should be taken since the microstructure of this material has been found to differ greatly from that of diamond. With reference to the substantial hydrogen concentration detected in many, but not all, kinds of hard carbon film, other authors [8, 15, 16] prefer to use the term hydrogenated amorphous carbon or a-C : H in analogy to the related silicon material a-Si : H. However, since the hard films are clearly distinct from the soft, conducting, and opaque amorphous carbon obtained by thermal processes, hydrogenated hard layers may better be designated as i-C : H.

In recent investigations it has been demonstrated also that composite films of the type Me/i-C or B/i-BN, containing finely dispersed metal or excess boron, can be prepared [3, 17]. Examples of such metastables composites, which look promising for applications as hard and protective coatings of good adhesion and low stress, will be presented.

II. PREPARATION—MECHANISM AND TECHNIQUES

A. History and Basic Mechanisms

The deposition of inert and glossy carbonaceous films on the walls of gas discharge tubes containing hydrocarbons had already been noticed decades ago [18, 19]. In the 1950s Schmellenmeier studied the formation of such films via dc plasma decomposition of C_2H_2 and other hydrocarbons [20, 21]. As a conclusion of these studies, the presence of diamondlike structures in the layers was considered and applications were anticipated. Much later Aisenberg and Chabot [14, 22] initiated the present period of steadily increasing interest in the topic with reports on the deposition of very hard and transparent carbon films that were obtained by condensation of 50-eV carbon ions together with argon ions. These findings were confirmed by another group [23], and the preparation of apparently similar films from ionized hydrocarbon species generated in dc or rf plasmas has been revived [4, 24–26]. In addition, it has been demonstrated that i-C films can be grown as well by ion beam sputter deposition of carbon, in particular if the forming layers are bombarded by inert Ar^+ ions of some 10^2 eV [27, 28].

These concepts of ion-enhanced deposition, which have been developed in the earlier work, are being employed still in the numerous recent investigations on i-C and i-BN preparation. Thus, before going into details of the deposition techniques, we should ask for the basic mechanisms involved in film formation from energetic ionized or neutral particles [29–32].

1. Properties of the Film-Forming Species

In ion-enhanced film deposition a significant fraction of the particles arriving in vacuum on the substrate is generated in the form of ions that can be accelerated easily by a proper bias voltage. The incoming particle flux, in general, will be composed of energetic ions, fast neutrals, and thermal neutrals. Since by common means of ionization only some fraction of the free particles can be ionized, the thermal neutrals will normally be predominant if no ion/neutral separation is applied. Fast neutrals are generated by secondary processes, such as gas phase collisions, ion neutralization, or sputtering. Hence, the mean kinetic energy imparted per particle to the substrate or the growing film is given by

$$\bar{\varepsilon} = N^{-1}\left(N_{\mathrm{i}} \int \nu_{\mathrm{i}} E \, dE + N_{\mathrm{n}} \int \nu_{\mathrm{n}} E \, dE + N_{\mathrm{t}} \int \nu_{\mathrm{t}} E \, dE\right), \tag{1}$$

where N is the total number of particles arriving per unit of area and time; N_{i}, N_{n}, and N_{t} are the numbers of ions, energetic neutrals, and thermal neutrals, respectively; and ν_{i}, ν_{n}, and ν_{t} designate the corresponding normalized energy distribution functions.

For simplicity in Eq. (1) only a single species of each kind has been taken into account. In practice, even for elemental particles generated by electron beam evaporation or sputtering a small percentage of atomic clusters, such as C_2, and C_3, cluster ions, and multiple charged ions must be considered [33]. In work with molecular species quite a variety of neutral and ionized fragments, e.g., C_xH_y and $(C_xH_y)^+$, must be expected to participate in the film formation.

Typical values of $\bar{\varepsilon}$ in ion-enhanced deposition of i-C and i-BN cover the range from a few to several hundred electron volts per condensing atom. An upper limit is set when resputtering begins to dominate over deposition. Due to the low sputtering yields [34] of carbon and boron this threshold lies at fairly high energies for the materials of interest here; however, a preferred sputter removal of other components such as H and N is likely to occur at high ion energies. Since the temperature equivalent to a kinetic energy of 1 eV is about 1.16×10^4 K, it follows that ion-enhanced deposition offers access to a region of nonequilibrium processes far beyond the capabilities of thermal reactions [35, 36].

2. Ion and Fast Neutral Actions

Although we are far from being able to describe the intricate kinetics of film formation from energetic free particles quantitatively, a crude picture of the basic mechanisms can be outlined by analyzing the relevant elementary processes of ion–solid interaction [30–32].

Obviously the incident energetic particles impart their excess energy and momentum to the near-surface region of the substrate, i.e., the growing film. There is evidence that for the energies considered here, the energy transfer takes place through binary elastic collisions between the impinging particle and the struck atom, whereas electronic interactions play a significant but minor role. With the laws of energy and momentum conservation we find for the energy E_2 imparted to an initially resting atom of mass m_2 by an impinging particle of energy E_1 and mass m_1

$$E_2 = [4m_1m_2/(m_1 + m_2)^2]E_1 \cos^2 \Theta, \tag{2}$$

where Θ denotes the angle between the directions of the incident particle and the knock-on atom.

As for the incident particle, it can become reflected, trapped at the surface, or penetrate the bombarded solid. In the latter case it will be slowed down in a sequence of further collisions before it comes to rest as an implanted particle. If the energy E_2 acquired by the recoil exceeds the displacement energy of the solid, which is a few tens of electron volts, it can leave its position and initiate secondary collisions. Some of the knock-on atoms will escape through the surface barrier as sputtered particles.

With respect to film formation from energetic free particles the following consequences can be expected.

a. Sputter Cleaning. The removal of adsorbed impurities and weakly bound species prior to deposition is generally acknowledged as being important to achieve good adhesion and surface coverage [37]. Since the equipment used in i-C or i-BN deposition can be easily operated with inert gases, the film preparation is nearly always preceded by sputter cleaning with Ar^+ or Kr^+ bombardment at energies of 1–3 keV. It should be noted, however, that at such high energies the preferential sputter removal of weakly bound species is much less pronounced than at lower energies, and impurities may become knock-on implanted into the substrate as well [38]. According to the author's experience the best layers are formed if the operation is altered smoothly from etching to deposition by proper variation of the parameters. During film growth the continuous preferred removal of weakly bound species due to the impacting film-forming particles of relatively low energy may be of considerable importance [39].

b. Surface Migration. Particles that become trapped at the surface will still have sufficient excess energy to hop around the surface by enhanced diffusion before they finally find a favorable bonding site. Although this process is generally assumed to play a vital role, experimental verification has been obtained so far mainly with metal films deposited by cluster beam techniques [40].

c. Defect Creation. Because of the relatively low ion energies applied in i-C and i-BN preparation, the production of point defects will take place mainly in a shallow region of a few atomic layers adjacent to the bombarded surface. Nevertheless, the creation of vacancies and the reoccupation by penetrating primary particles or diffusing interstitials are considered as being the crucial steps in the formation of a cross-linked microstructure [41]. Defects created at the substrate surface can be assumed to act as nucleation sites, however, little is known at present about the first stage of i-C and i-BN film deposition.

d. Atomic Mixing. Even at particle energies as low as a few hundred electron volts a considerable material transport over several atomic distances will be caused by the penetration of fast ions and neutrals as well as by the recoil displacement of struck atoms [30, 42]. In addition, at sufficiently high particle energies and impact rates an enhanced diffusion of the thermally agitated atoms near the film surface can occur via the continuously produced vacancies [43]. In the formation of i-C, i-BN, and related composites these effects could help to improve film adhesion by broadening the interface and to promote chemical reactions leading to interlinked bonding. On the other hand, excessive defect production and the associated atomic displacements are likely to account for the detrimental effects to the film structure observed at higher ion energies [7, 44].

3. Spike Effects

The treatment of ion–solid interactions on the basis of isolated binary collisions is obviously not justified when collective agitations of all atoms contained in a certain region must be taken into account. This was realized already decades ago by Seitz and Koehler and by Brinkman [45, 46]. Such nonlinear effects, called energy spikes or high-density cascades, have attracted much attention in conjunction with work on ion implantation and ion-induced phase transitions [47, 48]. For the relatively low particle energies involved in i-C or i-BN preparation, two closely related types of energy spikes affecting the outermost atomic layers of the growing films should be considered.

a. Thermal Spikes. If the energy of impinging particles is less than about 100 eV, according to Eq. (2) no major displacements of the struck atoms can be caused; single recoils moving over a few interatomic distances would not make much difference. The trapped particles instead are expected to impart their excess energy to all the atoms adjacent to the site of impact. Eventually, through mutual interactions with the second- and third-nearest neighbors a short-living region of highly agitated atoms will emerge. Estimations based on different assumptions have shown that the maximum temperature within the spike region would be in the range from some 10^3 K up to more than 10^4 K, and for the lifetime of the rapidly collapsing spikes values on the order of 10^{-11} have been assessed [41, 45, 49, 50]. Furthermore, due to the strong agitation at the spike surface, a pressure increase on the order of 10 GPa has been estimated [41, 47, 50]. Although the application of such terms as temperature and pressure becomes questionable for events in which only about hundred atoms are involved, it follows that on an atomic scale in ion-enhanced film deposition the conditions can be equivalent to very high temperatures and pressures. According to Grant [48], the crucial question whether the extremely short period of agitation would suffice to promote atomic rearrangements and chemical reactions can be answered as affirmative; however, there is a great likelihood that metastable structures would form. This is exactly what seems to happen in the formation of i-phases. It should be noted that the probability of overlapping spike events due to consecutive or adjacent impacts is negligible; hence the formation of i-C and i-BN must be regarded to proceed in localized steps. This conjecture may explain the origin of the high compressive stress that is typically found for these films [7, 8].

b. Displacement Spikes. More extended collective agitations are supposed to emerge when consecutive displacements caused by penetrating energetic ions are separated only by one or few interatomic distances. Then the almost simultaneously displaced atoms will leave a multiple vacancy whereas a highly agitated region surrounding this core is formed by the recoils and the overlapping thermal spikes associated with them [46]. It has been demonstrated that the regime of displacement spikes depends on the mass and the energy of the projectile [47]. Particles of low mass number, as used in i-C and i-BN preparation, should create small spikes of this type only at rather low energies of a few hundred electron volts. However, a situation quite similar to the formation of displacement spikes can be expected to arise when energetic molecular species such as $(C_xH_y)^+$ impinge on the surface of a solid. In the case of overlapping spike events the previously discussed temperature and pressure effects are expected to multiply [48].

4. Further Effects

The role of ion charge and of the bonding state remains to be briefly discussed.

a. Charge Effects. There is evidence that charge exchange becomes effective as soon as an ion enters the surface potential well of a solid [51]. Therefore, the actions of energetic ions and neutrals can be regarded as similar, provided that the charge accumulating does not influence the deposition process. In the case of insulating i-C films it is necessary to care for an effective charge compensation in order to prevent serious disturbances. While in plasma deposition neutralizing electrons will often be delivered automatically, in ion beam work charge compensation is usually accomplished by electrons generated from a hot cathode filament. Despite this measure it has been observed that locally small i-C film segments become detached and re-embedded during the deposition process [7].

b. Dissociation of Molecular Species. Investigations on reactive ion beam etching with ions obtained from O_2, N_2, CF_4, and C_2F_6 gave evidence that molecular ions of energies higher than some hundred electron volts dissociate completely upon the first impact on the substrate [52]. For particles of lower energy this is less evident, and a partial dissociation depending on the energy seems to be more plausible. Of course, the thermal neutrals arriving on the surface of the growing film may condense in their original bonding state, but fragmentation or even preferred sputter removal of these species is likely to be caused by the bombarding energetic ions and neutrals.

B. Deposition Techniques

Table 4.1 summarizes available data on the techniques reported for the preparation of i-C, i-BN, and related composite films. According to the principal methods the list is divided into dc plasma deposition, rf plasma deposition, sputter techniques, ion plating techniques, and ion-beam deposition. The relatively few reports on i-BN preparation are included at the end.

Some remarkable common features should be emphasized. First, the most important parameters influencing the film properties are the ion energy, often

TABLE 4.1

Deposition Data for i-C, i-BN, and Related Composites

Technique	Reactants	p_d (Pa)[a]	U_b (V) or ε (eV)[b]	Comments	References
DC plasma deposition	C_2H_2	2–50	—	Glossy films	[20, 21]
	C_2H_4, (Ar)	1.6–2.8	2000–5000 V	Black hard films	[24, 53, 54]
	C_2H_2	120	300–400 V	Yellow or brown inert films	[55, 56]
RF plasma deposition	C_4H_{10}, (CH_4, C_3H_8)	1–60	100–1900 V	Hard insulating films, depending on parameters	[4, 25, 57–61]
	C_2H_2	50–110	—	Yellow-brownish insulating films	[26]
	C_2H_6, (C_2H_2, C_2H_4, C_3H_8)	0.1–20	500–1000 V	Hard insulating films, $\rho = 1.9–2.0$ g/cm^3	[5, 62, 63]
	C_2H_4	0.05–13	10–1400 V	Very low friction coefficients	[6, 64a]
	C_3H_8	20	400–700 V	Optical coatings	[64b]
	C_6H_6	1–6	400–1800 V	Optical and protective coatings	[65, 66]
	CH_4, C_2H_6	0.1–1	500–2000 V	Coatings with diamondlike properties	[67, 68]
	CH_4	1–20	540 V	Investigations on kinetics and structure	[69, 70]
	CH_4, C_2H_2, C_2H_4	1–100	—	Particle conglomerates or films containing microcrystals	[71]
Sputter deposition	C, (Ar, Kr)	0.02–0.1	—	Ion-beam sputtering	[72–77]
	C, H_2, (Ar)	0.02–0.1	—	Ion-beam sputtering with H^+ participation	[78, 79]
	C, CH_4, (Ar)	0.02–0.1	200–1000 eV	Dual-ion-beam sputtering	[7, 27, 28, 74, 80–82]
	C, (Ar)	0.6	0–300 V	RF bias sputtering	[83]
	C, C_2H_2, (Ar)	0.15–40	—	Hybrid process: dc magnetron sputtering and plasma decomposition	[84–89]
	C, C_4H_{10}	3–13	0–100 V	Hybrid process, low-stress films	[90]

Ion plating techniques	C_6H_6, (tetraline, anthracene)	0.01–0.05	20–5000 V	Deposition rates of 30–100 nm/min at low pressures	[7, 13, 41, 91, 92]
	C_6H_6, Al, Si, Ti, Cr	0.01–0.05	100–3000 V	Preparation of Me/i-C composites	[3, 7, 17, 93–96]
	C, CH_4, (Ar)	1–4	3000–5000 V	Columnar film morphology	
Ion-beam deposition	C, (Ar)	$(3–7) \times 10^{-3}$	20–100 eV	Very hard, inert, transparent films	[14, 23, 25, 39, 97–99a]
	C, CH_4	—	900 eV	Epitaxial layers on diamond	
	CH_4	10	20–800 V	Hard films containing microcrystals	[99b]
Other techniques	C, (Ar)	—	0–980 eV	Laser evaporation with ion bombardment	[75, 100]
	CH_4	10–100	—	Pulsed plasma process	[101]
	C, (Ar)	$10^2–10^4$	—	Ion generation from randomly moving arc spot, high deposition rates	[102]
Deposition of i-BN	BN, N_2, (Ar)	1	—	RF sputtering, stoichiometric films	[103, 104]
	BN, N_2, (Ar)	0.6	0–100 V	RF bias sputtering, inert, dense films	[105]
	B, N_2, NH_3, (Ar)	0.05–0.5	0–3500 V	Ion plating combined with electron beam evaporation	[7, 13, 17, 64b, 73]
	BH_3, B_2H_6, N_2, NH_3	—	—	Plasma decomposition	[106–108]
	$B_3N_3H_6$, (Ar)	0.01–0.06	200–1000 V	Plasma decomposition	[13, 73, 109]
	B, N_2	10^{-3}	40 keV	Electron beam evaporation of B with nitrogen ion bombardment	[110, 111]

[a] Working pressure during deposition.
[b] Substrate bias voltage U_b or ion energy ε.

expressed by the negative substrate bias voltage U_b, and the deposition pressure p_d or in beam techniques the ion energy ε and the flux density.

Second, typical i-C or i-BN films are formed on cool substrates; substrate heating to temperatures above a few hundred degrees Celsius tends to have detrimental effects on the hardness and other film properties.

Third, the substrate material seems to have little influence; apparently similar films have been grown on various substrates such as glass, fused silica, silicon, sodium chloride, diverse metals and alloys, diamond, and even on polymers. In the following, some aspects of preparation techniques are discussed, with emphasis given to the methods used in the author's laboratory.

1. Preparation of i-C by Hydrocarbon Decomposition

Most of the reported work on i-C has been performed using the deposition of ionized and neutral species obtained from hydrocarbons. In fact, it seems that any hydrocarbon can be used as starting material; for reasons of easy control of the flow rate and vapor pressure simple aliphatic or aromatic hydrocarbons, particularly CH_4, C_2H_2, C_2H_4, C_2H_6, C_3H_8, C_4H_{10}, and C_6H_6, are chosen. As hydrogen is provided together with the carbon, the resulting films inevitably can be expected to contain appreciable amounts of this element.

Deposition of the species formed in a dc plasma works reliably at not too low pressures with not too large substrate areas and growth rates [53–56]. To prevent troubles arising from space charge accumulation, deposition from rf plasma has been introduced and is now being used widely [57–71]. Capacitive or inductive coupling through matching networks is applied to connect the discharge system with the rf power supply. For obvious reasons the industrial frequencies 2.3 or 13.6 MHz are used. In the rf discharge, typically sustained at 300–1900 V and a gas pressure of 1–50 Pa, reactive species including ionized and neutral fragments are formed. Ion acceleration takes place in the plasma sheath formed between the discharge and the powered cathode on which the substrates are placed. It has been shown that due to the strong negative self-bias U_b of the substrate region the voltage fluctuation is highly asymmetric. Thus, the ions are accelerated by a voltage at least one order of magnitude higher than the electrons in the half-cycles of reversed polarity [8]. Despite their low contribution to the power consumption, the electrons arriving on the substrates or the growing films accomplish an effective space charge neutralization. A welcome side effect of this potential distribution is that dust particles, which are known to charge up negatively within the plasma, will be repelled from the films. With rf plasma deposition fairly high film growth rates up to 250 nm/min over substrate areas on the order of 100 cm^2 can be achieved. The main limitation is due to the relatively high deposition pressure that makes it impossible to prevent a major fraction of thermal neutrals to participate in the film formation, thus lowering the effective mean energy according to Eq. (1).

Closely related to dc and rf plasma deposition is the technique using magnetron sputter sources operated with hydrocarbon gas [84–87]. In this case some, presumably small, contribution of species sputtered from the carbon-coated electrode can be assumed. The low mass density and other properties of the films indicate, however, that the layers prepared by hydrocarbon decomposition in a magnetron are formed from particles having a lower mean energy $\bar{\varepsilon}$ than in rf plasma deposition at high bias voltage and low pressure.

As a general rule it is observed that diamondlike properties of i-C deposits become more pronounced with increasing mean energy of the film-forming species, while films grown under conditions of comparatively small activation tend to exhibit properties resembling polymers or ordinary amorphous carbon.

In order to study the role of ion actions more systematically the ion plating configuration shown in Fig. 4.1 has been devised [7, 91, 92]. Efficient ionization of hydrocarbon species at low deposition pressures (0.01–0.05 Pa) is achieved by impacting electrons generated from a thermionic tungsten cathode. The ionization efficiency per electron is greatly increased by forcing the electrons to move along oscillating trajectories between the upper and lower reflection electrode before they have a chance to find the wide-mesh anode grid. Since it was found that the floating plasma sheath formed near

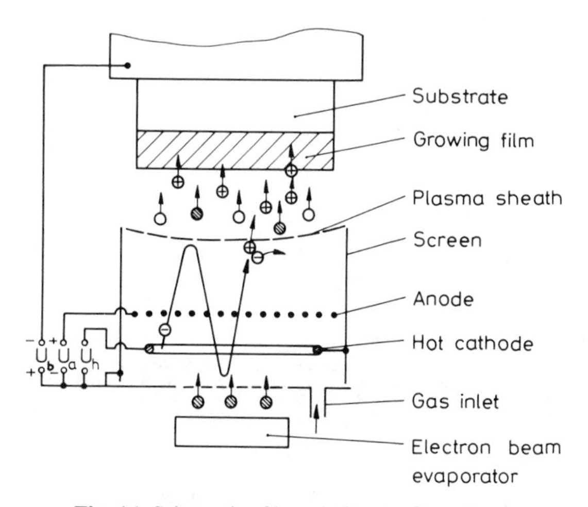

Fig. 4.1 Schematic of ion plating configuration.

the substrate region also acts as electron reflector, the upper reflection electrode grid is often omitted, which helps to reduce contamination by sputtered electrode material. A tantalum or steel cylinder held at the potential of the lower electrode is usually installed to confine the low-pressure discharge. Contamination of the films is low because all the inner parts soon become covered with i-C.

The bias voltage that determines the ion acceleration is applied between the lower electrode and the water-cooled substrate holder; normally U_b has been varied in the 20–3000-V range. As in other techniques of ion-enhanced film deposition the substrates are cleaned prior to film growth by operating the system with Ar or Kr. Then, the operation is changed smoothly from etching to deposition by reducing the inert gas feed-in and opening the inlet of hydrocarbon vapor.

Although the system has been shown to work reliably with quite a variety of hydrocarbons, including even condensed aromatic compounds like anthracene, benzene is preferred for two reasons: (1) the deposition pressure can be easily controlled by the temperature of the ampoule containing the liquid and (2) the aromatic rings of benzene are likely to withstand ionizing impacts without fragmentation [112, 113]. Figure 4.2 illustrates the abundance of different ionized species formed under conditions equivalent to those in the ion plating system. With about 70% $(C_6H_x)^+$ ions, the rest being fragments, the kinetic energy acquired by acceleration is for the majority of the ionized species shared by six carbon atoms. Consequently, at high energies a particularly strong activation and eventually the formation of displacement

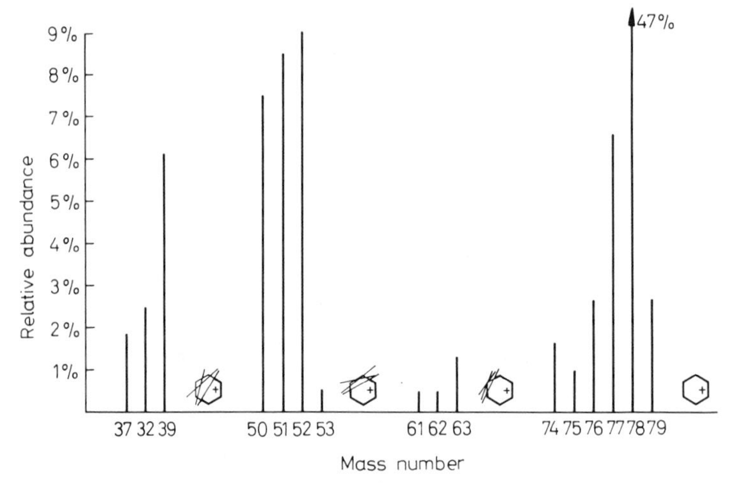

Fig. 4.2. Mass distribution and relative abundance of ionized species obtained from benzene (50–100-eV electron energy).

spikes, as considered in Subsection II.A.3.b, can be expected when these ions strike the surface of the growing film.

Depending on the bias voltage and the pressure, film growth rates in the range of 30 to 100 nm/min have been measured. By comparing the measured ion current and growth rate with reasonable corrections with regard to the secondary electron current, it has been estimated that typical i-C films in this technique are predominantly formed from the condensing energetic ions, whereas the contribution of embedded thermal neutrals was found to be less than 35%. This is much less than the thermal collision rates for the corresponding gas pressures would suggest. This can be explained by a low sticking probability or, as indicated in Subsection II.A.2.a., by the preferred sputter removal of weakly bound species. Since the intense ion flux extracted from the system forms a broad beam with little scattering by gas phase collisions, this deposition technique may also be characterized as ion beam plating [36, 73].

Several of these ion plating systems have been operated for years with reliable performance, and units for coating areas up to about 1 m^2 were developed for industrial application [114].

Furthermore, by means of an attached evaporation unit, shown in Fig. 4.1, composite coatings of i-C with metals or Si can be prepared in a controlled single process [7, 93]. Since both the ion current of carbonaceous species and the evaporated metal flux can be varied independently within wide limits, composite films covering a considerable range of composition or with a gradient along the surface normal can be deposited.

2. Preparation of i-C by Ion-Beam Deposition and Sputtering

Methods based on the deposition of carbon ions or neutrals obtained by sputtering from carbon targets are distinguished in that they allow us to study the formation of i-C films with little hydrogen content. To produce films free of hydrogen would indeed be very difficult because available carbon materials, such as graphite or vitreous carbon and even diamond, are known to contain various amounts of this element.

Aisenberg and Chabot, in their pioneering work [14, 22], condensed carbon ions of 40–100 eV that were generated by argon ion sputtering in a special source. Some simultaneous argon ion bombardment of the growing layers can be assumed. The resulting films were reported to be transparent, insulating, chemically inert, and very hard. On the other hand, the low growth rate ($\sim$5 nm/min) was a handicap for applications. Attempts to produce similar films by rf bias sputtering [83] or by magnetron sputtering [39, 115] were only partially successful since the deposits proved to be less "diamondlike."

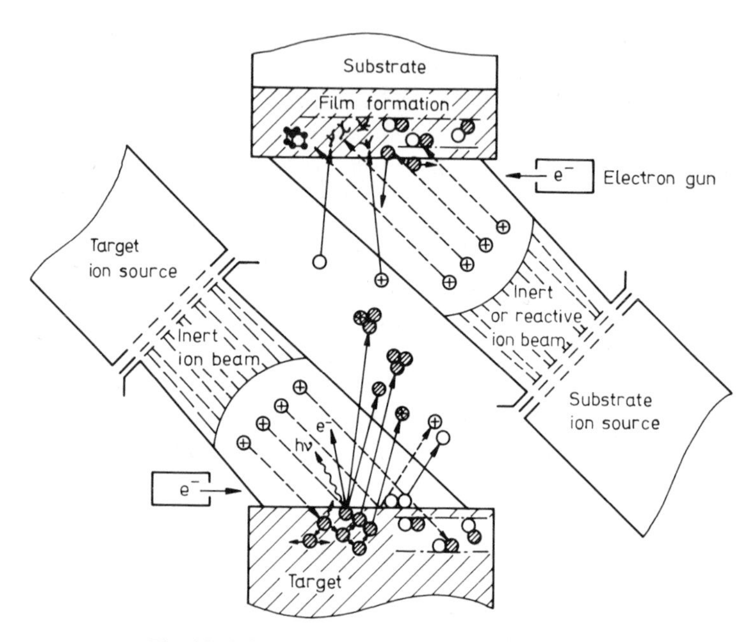

Fig. 4.3. Schematic of dual ion-beam sputter system.

The capabilities of ion beam sputtering have been investigated in our laboratory using the device shown schematically in Fig. 4.3 [17, 72–74]. This apparatus, which is equipped with an oil-free turbomolecular pumping system, can be operated in different modes: (1) In single-beam sputtering an argon ion beam ($\sim$10 mA, 4–10 keV) delivered from the target ion source is used to sputter a graphite target; (2) dual-beam sputtering is performed with simultaneous operation of the substrate ion gun (3–5 mA or 150–350 $\mu A/cm^2$, 0.2–1.0 keV); and (3) sometimes additionally a laser has been attached to study the simultaneous action of intense light [7].

Experiments with normal ion beam sputtering using only the beam of the substrate source gave evidence that the as-deposited films exhibit at least more pronounced "diamondlike properties" than those obtained by rf or magnetron sputtering. Meanwhile, these findings have been confirmed by publications of other groups [75–78]. The ion action can be enhanced if oblique angles of ion incidence and substrate position are chosen to produce a high intensity of reflected primary ions that impinge upon the growing film, as was proposed by Wehner [116]. Based on similar experience obtained with ion beam preparation of a-Si : H [17], first investigations to prepare hydrogenated i-C films by sputtering with a mixed beam of argon and hydrogen have begun [79].

As for the dual beam technique, colorless and apparently very hard films corresponding to those described by Aisenberg and Chabot [14] could be prepared, provided that careful space charge compensation by flooding the substrate region with electrons was applied (see Subsection II.A.4.a.). For reasons that are not yet understood, such films could be grown only on insulating substrates such as glass or NaCl. Simultaneous laser irradiation has been observed to modify the film structure and to reduce intrinsic stress. These effects are the object of work in progress. It is noteworthy that ion beam bombardment of laser-evaporated carbon films after their formation yields only conducting layers ($\sim 10^{-1}$ Ω cm) of moderate hardness [75].

Because of the low sputtering yield of carbon and the special geometry of ion beam configurations the achievable growth rate is only about 5 nm/min. A considerable improvement by a factor 3–5 has been observed if instead of pure argon a mixture of argon and methane is applied to feed the sputter ion source [74]. Another promising way to increase the growth rate is offered by the construction of coaxial ion sources with cylindrical carbon targets that are capable of delivering a much stronger flux of carbon ions and sputtered neutrals [7, 39, 74]. Several reports on the deposition of partially ionized beams obtained from methane agree in that hard deposits containing fractions of diamond microcrystals are formed [68, 96, 99a, 99b].

In a different approach the long-known phenomenon of the randomly moving arc spot on a cathode has been utilized to generate a particle flux containing a high percentage of ions [102]. For arc current densities of 10^5–10^6 A/cm^2, a flashlike evaporation without melting and an exceptionally high ionization degree in the dense plasma are obtained. Based on that mechanism new techniques of i-C preparation from carbon electrodes are expected to emerge.

3. Preparation of i-BN Films

Both scientific and technological motivations have stimulated research on the eventual formation of metastable i-BN films, i.e., hard deposits with resemblance to cubic boron nitride. The hitherto reported techniques are basically similar to those applied to i-C preparation.

Stoichiometric boron nitride films with properties similar to the soft hexagonal phase have been deposited by rf sputtering in Ar/N_2 discharges [103, 104], and additional application of a negative bias voltage was found to increase the film hardness [105]. With plasma chemical vapor deposition (CVD) from boranes, nitrogen, and/or ammonia, soft coatings of potential use as insulating films in microelectronic devices have been prepared [106, 107], but the formation of deposits containing cubic BN has been reported if strong activation is applied [108].

In our laboratory, efforts have been focused on the development of an efficient and versatile ion plating technique to prepare hard i-BN films [7, 117]. To that purpose electron-beam evaporators of 3 to 6-kW maximum power were combined with postionization systems. A smaller apparatus has been equipped with a unit much like that shown in Fig. 4.1, whereas in the more powerful plant a separate electron gun was installed in the ionization chamber. Films of the composition B_xN, with x in the range from 2 to 1, are obtained using mixtures of N_2 or NH_3 with Ar as working gas and simultaneously generated boron vapor. Growth rates up to 0.3 μm/min have been realized with the large plant, and typically a bias voltage of a few kilovolts was applied to deposit hard coatings. Since free boron particles are known to be extremely reactive [33], much care had to be taken to avoid oxygen contamination originating from the boron material and from the residual gas. Composite films such as Si/i-BN or Me/i-BN can be synthesized by means of a second evaporator.

In conjunction with our earlier reports on the formation of hard i-BN films by plasma decomposition of borazine ($B_3N_3H_6$) vapor [13, 73], Shanfield and Wolfson have reported the preparation of films containing the cubic phase that were obtained by condensation of an ion beam extracted from a borazine plasma [109].

Finally, Satou *et al.* have published interesting work in which i-BN was prepared by electron-beam evaporation of boron and simultaneous ion bombardment of the growing films with 40-keV N_2^+ ions [110, 111]. Since ions of such a high energy can penetrate over a considerable distance, the reaction must be assumed to occur well below the film surface between activated boron atoms and the implanted nitrogen atoms formed after dissociation of the molecular ions.

III. PROPERTIES AND STRUCTURE OF i-C FILMS

A. General Behavior and Macroscopic Properties

A careful analysis of all available properties and structural data is required if we are to understand the nature of a highly disordered material such as i-C. Before going into details, the general behavior of i-C films prepared at different degrees of ion enhancement is outlined. To avoid confusion we deal at first with the behavior of films obtained by ion plating from benzene and ask then to what extent the findings can be generalized.

Films deposited at low bias voltage of several tens of volts proved to be polymerlike; actually their IR absorption spectra revealed a striking similarity with polystyrene [118]. In the bias voltage range from about 100 to 800 V a gradual and dramatic change of nearly all properties takes place. This is

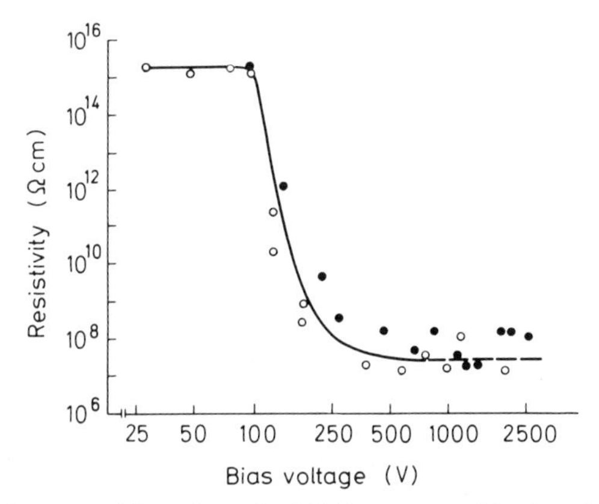

Fig. 4.4. Resistivity versus bias voltage for i-C films prepared by ion plating from benzene. Silicon substrate precleaned with Kr^+ (○) ($\rho = 10^{-2}$ Ω cm) and precleaned with Ar^+ (●) ($\rho = 10$ Ω cm).

illustrated in Figs. 4.4 and 4.5 for the electrical resistivity and the mass density, respectively. Further alterations include a drastically rising hardness and an increasing refractive index. This broad transition region is followed by a relatively narrow region of optimum hardness in the range from about 800 to 1100 V, as will be discussed later. With further increased bias voltage up to 3 kV the film structure is found to degrade, as can be concluded from the decreasing hardness and the vanishing optical transparency. For still higher bias voltages predominating sputter removal prevents the formation of well-defined films. All these statements refer to films deposited at room temperature on various substrate materials such as fused silica, steel, hard alloys, and single-crystalline Si or NaCl. Substrate heating up to about 300°C seems to have little influence, however, at higher temperatures (300–400°C) the films, while still retaining some of the hardness, become graphitic with electrical conductivities on the order of 10^2–10^{-1} Ω cm. On first sight

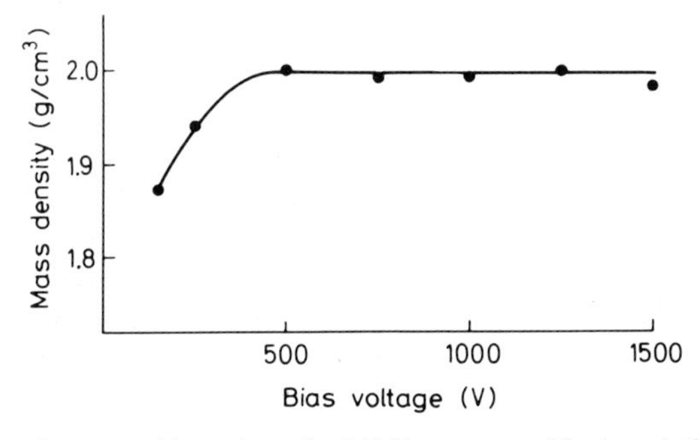

Fig. 4.5. Mass density versus bias voltage for i-C films prepared by ion plating from benzene.

this strong influence of relatively small temperature changes on deposition processes, involving particles of energies corresponding to very high temperatures, appears to be strange. It turns out that the effects caused by random thermal motion in the film–substrate system as a whole must be distinguished from the strictly localized and rapidly collapsing thermal spikes created by impacting particles (see Subsection II.A.3.a.) [32, 94].

A comparison with the reported behavior of i-C films prepared by plasma decomposition of hydrocarbons reveals satisfying agreement as far as the general influence of the bias voltage and the substrate temperature is concerned [6, 8, 16, 78, 87, 89]. The results obtained for the mass density, electrical resistivity, and refractive index show even quantitative accordance, if we account for the different mean ion mass. However, it follows that most of the i-C films deposited under plasma conditions correspond to the lower or middle transition region. Apparently in plasma deposition the fraction of energetic ions contained in the flux of film-forming species cannot easily be raised to the level that is required to obtain films of optimum hardness and density.

On the other hand, i-C films prepared by ion beam techniques seem to belong mostly to the regions of optimum hardness or beginning degradation; their mass density, as shown in Fig. 4.5, is slightly higher than that of films obtained form hydrocarbons [74, 83].

1. Film Composition

The occurrence of characteristic bands in the infrared absorption spectra that can be assigned to different states of C–H bonds gives evidence that i-C films prepared from hydrocarbons contain substantial amounts of hydrogen [4, 8, 66]. Quantitative determinations of the hydrogen concentration in dependence on preparation and annealing conditions have been performed by pyrolysis [87] and by nuclear reaction analysis [78, 94]. Figure 4.6 shows data obtained for ion-plated i-C films using the resonant reaction [119]

$$^{1}\mathrm{H}(^{15}\mathrm{N};\alpha,\gamma)^{12}\mathrm{C}, \qquad W_{\mathrm{res}} = 6.385 \quad \mathrm{MeV}. \tag{3}$$

The measured hydrogen concentration for the lowest bias voltage comes close to the H/C ratio 1 : 1 that would correspond to polystyrene; after a steady decline with increasing bias voltage a constant concentration of about 23 at. % is reached near $U_b \simeq 800$ V. While the irradiation of films prepared at low bias voltage causes a significant time-dependent depletion in hydrogen, only a small decrease was observed at higher bias voltages.

Nyaiesh and Nowak [69] detected by differential scanning calorimetry a broad, asymmetric, exothermic peak at about 550°C and a larger energy

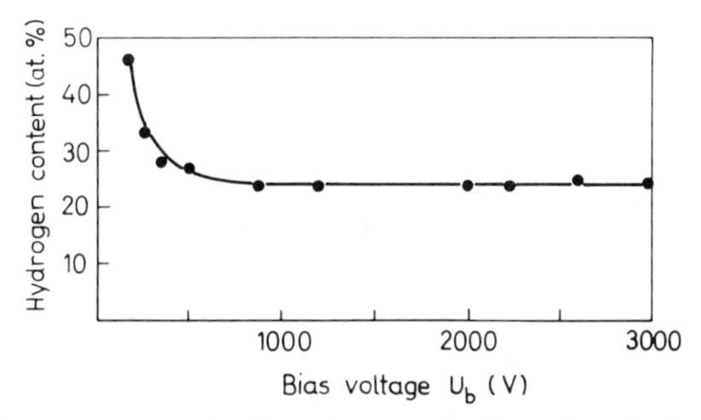

Fig. 4.6. Hydrogen concentration dependence on the bias voltage for ion-plated i-C films.

release near 750°C for i-C films prepared by rf plasma deposition from methane. The latter peak is attributed to graphitization whereas the peak near 550°C is interpreted as an indication of release of chemisorbed hydrogen. The total amount of chemisorbed hydrogen in the films was found to be about 6 at. % as compared with an estimated total hydrogen concentration of 25 at. %.

Craig and Harding [87] found, by analysis of the pyrolysis products obtained from i-C samples, hydrogen concentrations of 43 (resp. 34) at. % and additionally 6.7 at. % oxygen. Since these specimens, which were prepared by C_2H_2 decomposition in a cylindrical (resp. planar) magnetron, clearly correspond to the lower transition region, their hydrogen content would be consistent with the data of Fig. 4.6. As for the oxygen, the situation seems to be less clear. Although Berg and Andersson [120] also detected significant amounts of oxygen by ESCA in i-C films prepared by rf deposition from butane, in other work the absence of C–O bonds in the IR spectra is emphasized [121]. According to our preliminary measurements, the oxygen incorporation in most i-C films can be estimated to not exceed 1 at. %, as follows from x-ray microanalysis and ESCA.

Some contamination of the films originating from the sputtering of wall, electrode, and crucible materials or from thermionic tungsten cathodes is likely to occur in most of the applied preparation systems [122]. Some catalytic action of metallic contaminants to the formation of diamondlike crystallites cannot be excluded.

2. Mechanical Properties

a. Mass Density. Film densities are usually measured by weighing and thickness determination or by flotation techniques using suitable liquids such as bromoform mixed with benzene or methanol [7]. The reported values range from 1.2 to 2.0 g/cm^3 for films made from hydrocarbons and up to 2.25 g/cm^3 for sputtered films [5, 7, 8, 16, 63, 65, 74, 83]. Recent fractional studies with fine film particles floating in the region of density gradient formed between partially mixed bromoform and methanol indicates that the mass density of i-C is not uniform but exhibits a distribution of considerable width [74].

b. Hardness. The microhardness is determined from the traces left after applying a Vickers or Knoop indenter to the film surface. Particularly for hard films, difficulties can arise due to the influence of the underlying substrate [6]. Therefore, the data plotted in Fig. 4.7 were obtained from specimens consisting of 5–10-μm i-C films deposited on substrates of the WC/Co hard alloy HG 10 [17]. The upper part of this figure shows the measured Vickers hardness of i-C films prepared by ion plating from benzene in dependence on the bias voltage. Since the "cushionlike" shape of the indents betrays a significant influence of elastic recovery, the data presented in the lower part were calculated from the measured values by applying the correction term

$$VH_k = 1854L_{\mathrm{VH}}/(d_{\mathrm{VH}} + k_{\mathrm{VH}})^2, \tag{4}$$

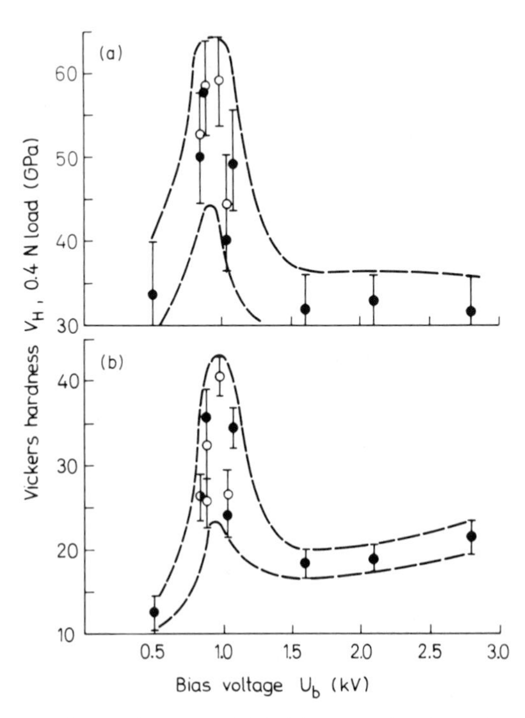

Fig. 4.7. Vickers hardness of ion-plated i-C films as a function of the bias voltage. (a) Measured (uncorrected) values and (b) corrected data.

where L_{VH} is the load in newtons, d_{VH} the corresponding diagonal length of the indent in micrometers, and k_{VH} a correction determined as the abscissa intersection of the plot $L_{VH}^{1/2}$ versus d_{VH}. These corrected data seem to be more realistic than the measured hardness values. Noteworthy is the pronounced maximum in the 800–1100-V range, which could be well established in repeated sets of experiments. Other reported hardness data support the opinion that plasma-deposited i-C films correspond to the transition region [6]; indications of a maximum were found in one publication [99b].

It should be noted that the maximum hardness of i-C films is higher than that of all other hard materials except diamond or cubic BN. In accordance with this result it has been found that the Mohs hardness, determined by interpolation from scratch-test results, for coatings of maximum hardness is about 9.4. Dynamic recording of the indent depth as a function of the load confirmed that i-C films as opposed to crystalline hard materials, are distinguished by a high elastic recovery [123].

c. Intrinsic Stress. The evolution of high compressive stress in i-C layers causes serious problems because buckling and lift-off of the films from the substrates may result due to stress relief, which limits many potential applications [4, 6–8]. Characteristic stress relief patterns observed for sputtered and ion-plated i-C films are displayed in Fig. 4.8. The buckling often starts after days when the coated samples are exposed to air, and the process may be provoked by applying a diamond indenter [7, 91]. Similar observations were reported by Nir [124], who also treated the propagation of buckling waves theoretically.

From substrate bending intrinsic stresses of several gigapascals have been measured for plasma-deposited i-C films. Our measurements with ion-plated films of 0.25–0.75 μm thickness yielded 4–6 GPa and 3–4 GPa for films deposited at bias voltages of 1 kV and 3 kV, respectively [125]. With increasing thickness of the coatings lower stresses were found (for example, 0.4–0.6 GPa for 5-μm coatings).

Zelez [90] demonstrated that i-C films of low or even tensile stress can be prepared by means of a hybrid deposition process involving rf sputtering and butane decomposition. Other promising ways to overcome the stress problem are offered by laser irradiation during deposition [32] and, as will be discussed later, by the preparation of Me/i-C composite coatings with a gradient of composition.

d. Strength and Friction Properties. Table 4.2 summarizes further mechanical properties that were measured or calculated for ion-plated i-C films. The hardness data of the top row are statistically weighed results. Employing the method proposed by Matuda *et al.* [126], the modulus of elasticity was calculated from the delamination pattern of buckled films.

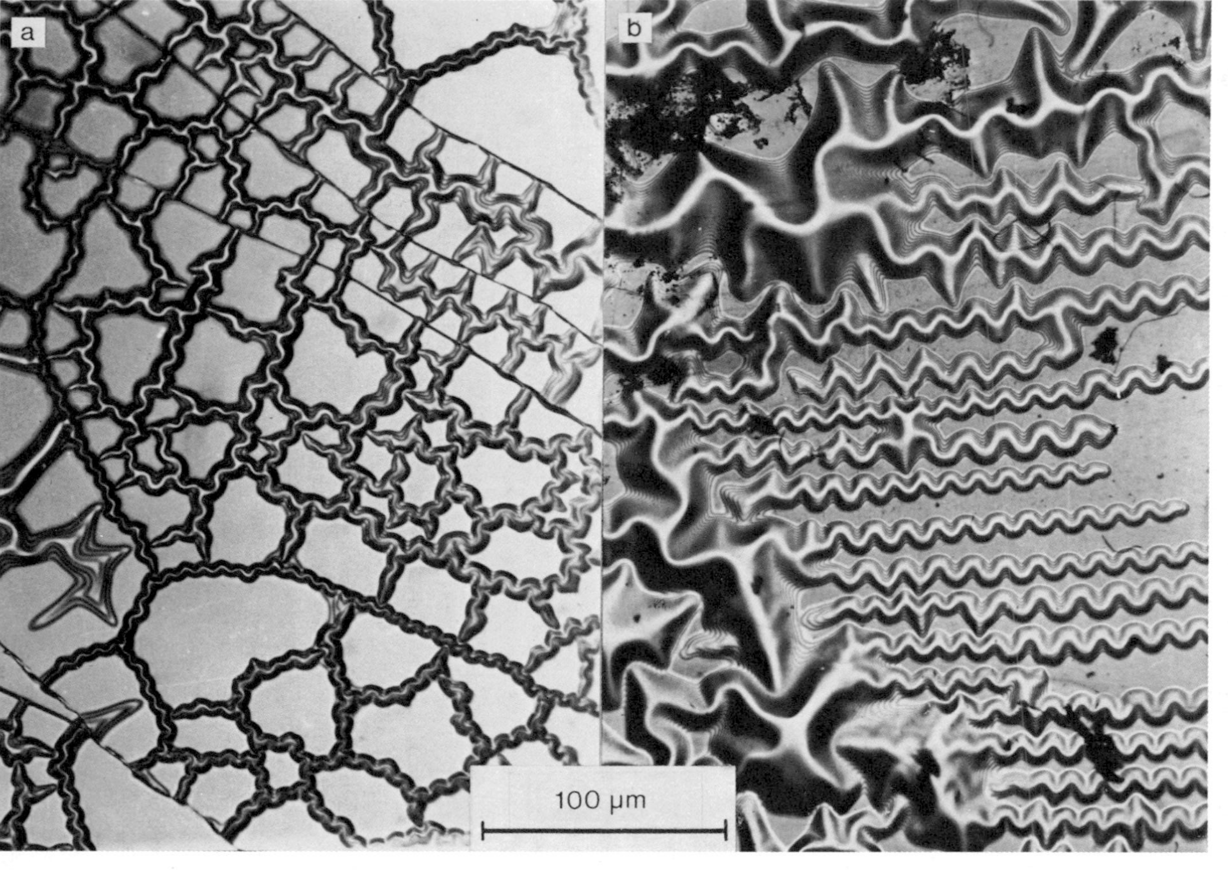

Fig. 4.8. Stress relief patterns observed for ion-beam-sputtered i-C films (a) on glass and (b) ion-plated i-C films on silicon.

TABLE 4.2

Mechanical Properties of i-C Films Prepared by Ion Plating from C_6H_6

Ion energy ε (eV)	800–1000	1500–3000
Film thickness d (μm)	5–10	5–10
Vickers microhardness VH (GPa)	58 ± 7 (max)[a]	32 ± 4
Corrected hardness VH_K (GPa)	40 ± 3	20 ± 2
Modulus of elasticity E (GPa)	190–290	120–160
Interface energy Φ (N/m)		
i-C/Si, $d = 0.5$ μm	4.9–7.0	4.2–5.6
i-C/steel, $d = 5$ μm	—	1.9–2.5[b]
Yield strength σ_y (GPa)[c]	23–35	13–18
Friction coefficient μ	0.04 (min)[d]	0.08–0.11

[a] Maximum values in the plot of VH versus bias voltage U_b.
[b] Estimated from "weak spots" in the coatings.
[c] Calculated from crack patterns under bend loading with VH_K.
[d] Minimum values measured in air with steel against i-C/steel.

Reasonable estimations of the interface energy were obtained from the equidistant cracks that appear, as shown in Fig. 4.9, in coatings subjected to bend loading. The yield strength was calculated from the E-modulus and the corrected microhardness; the results agree well with the cracking behavior of the coatings.

Of particular interest in view of applications is the tribological performance of i-C-coated materials. Using a steel ball sliding against an i-C-coated silicon disk, Enke measured a friction coefficient μ of about 0.04 that increased up to about 0.3 with rising humidity in the nitrogen environment; by means of an ultra-high-vacuum tester, exceptionally low μ values in the range from 0.005 to 0.02 were found [6, 127]. Here again a resemblance to diamond is evident; graphitization under the load pressure is thought to account for the low friction. The same author noticed a smooth wear of i-C coatings in endurance tests, but a sharp wear increase was observed for loads exceeding 30 N.

Our measurements made by using ball-to-disk and ring-to-cylinder devices operated in air revealed minimum friction coefficients for ion-plated i-C films corresponding to the region of optimum hardness (see Table 4.2). The friction between a diamond needle and i-C-coated steel was found to increase with the load when the region of plastic deformation is entered, as Fig. 4.10 indicates. Endurance tests performed with steel against i-C/steel in the ring-to-cylinder device yielded the curves shown in Fig. 4.11, in which the

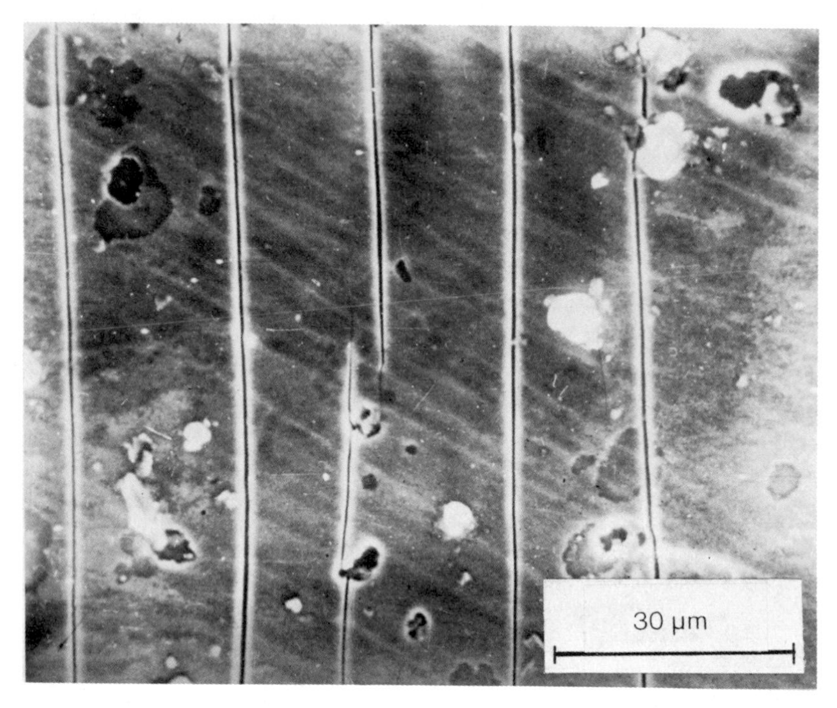

Fig. 4.9. Cracks in i-C coatings on steel that were caused by bend loading of the specimens.

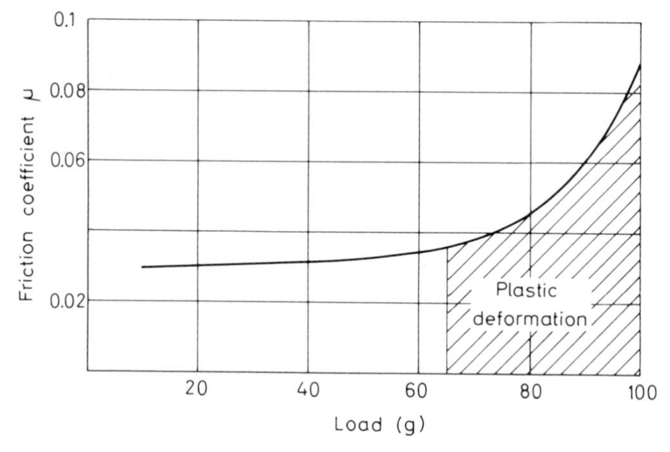

Fig. 4.10. Friction coefficient versus load for diamond against i-C on steel.

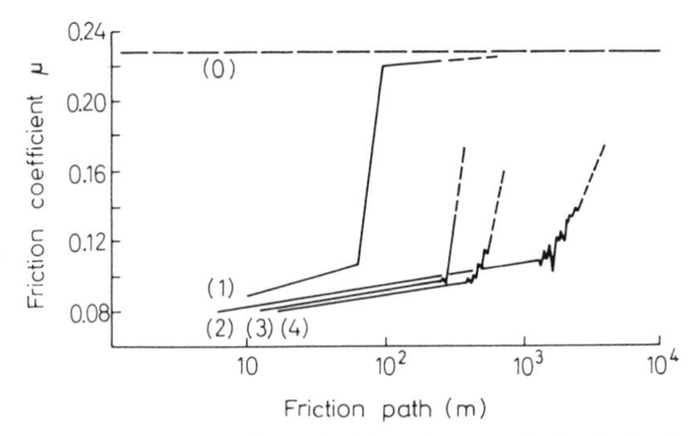

Fig. 4.11. Results of wear tests performed with a ring-to-cylinder device for i-C-coated steel against steel.

sudden increase of μ is a sign that the coatings are worn through [114, 125]. Examination of the specimens subjected to wear tests by scanning electron microscopy (SEM) revealed that the wear proceeds very smoothly.

3. Electrical and Optical Properties

a. Electrical Resistivity. The conductivity of i-C films has been measured either in transverse direction or parallel to the surface normal, apparently without indications of anisotropy. The reported results cover a wide range, extending from 10^{12}–10^7 Ω cm for typical hard films prepared by hydrocarbon decomposition or ion beam deposition to 10^{-1} Ω cm for graphitic layers formed by deposition on heated substrates or by annealing of the as-deposited i-C films [4–8, 16, 87–89]. McKenzie *et al.* [89] found that the conductivity as a function of temperature within the range of 77 to 500 K obeys Mott's $T^{-1/4}$ law of hopping conduction, and they obtained for the state density at the Fermi level $N(E_F) = 1.4 \times 10^{18}/\text{eV cm}^3$. Substantial doping effects have not yet been reported.

b. Optical Properties. There is considerable interest in the optical behavior of i-C coatings since their high and adjustable refractive index implies applications in infrared optics and solar energy devices. The fact that the transmittance of germanium in the wavelength region up to about 10 μm can be improved significantly by i-C coatings was demonstrated by Holland and Ojha [4]. By using spectrophotometric techniques and computer-aided fitting, the optical constants n, k, ε_1, and ε_2 have been determined over the quantum energy range from 0.5 to 7 eV (i.e., in the visible and parts of the UV and IR region) for i-C : H films prepared under various conditions [8, 16, 89, 128]. A selection of data is listed in Table 4.3.

Summarizing it can be stated that the absorption coefficient α decreases with increasing wavelength by orders of magnitude until a minimum is

TABLE 4.3

Optical Constants for an i-C:H Film Prepared at 250°C by C_2H_2 Plasma Decomposition[a]

Photon energy (eV)	Refractive index n	Extinction coefficient k
1.5	1.74	0.0030
2.0	1.74	0.0065
2.5	1.75	0.026
3.0	1.76	0.061
3.5	1.77	0.119
4.0	1.77	0.168
4.5	1.75	0.220

[a] Selection of data published by Smith [16].

reached near 2.8 μm or 0.45 eV. The typical yellow–brownish appearance of i-C films results from the stronger absorption in the short-wavelength region of the visible spectrum; nearly colorless films have been obtained so far only by the sophisticated ion-beam deposition and dual-beam techniques [7, 14, 39]. In the IR region characteristic absorption lines arise in the 3–3.6-μm range due to C–H stretch bands and in the range above 6 μm due to C–H deformation and C–C stretch vibrations; the lowest absorption coefficient ($\alpha = 121\ \mathrm{cm}^{-1}$) was measured near 2.8 μm [66].

The spectral dependence of the optical absorption in i-C films, which is distinctly different from that of diamond, can be described by a similar relation as was found to hold for a-Si : H [129, 130], namely,

$$(\alpha E)^{1/2} = G(E - E_{\mathrm{opt}}). \tag{5}$$

With the constant factor $G = 280/\mathrm{eV\ cm}$ the optical gap E_{opt} is obtained as the abscissa intersection of Eq. (5). Typical values for E_{opt} fall in the range of 0.8 to 2.6 eV, i.e., relatively small gaps that are interpreted by the presence of broad regions of tail states near the band edges of i-C [16, 66, 121].

The refractive index has been found to vary negligibly with the wavelength, indicating that there is no dispersion [16, 131], yet the finding that it can be adjusted by proper choice of the deposition conditions within the range of 1.8 to 2.3 is of considerable importance for applications [8, 131]. Oscillations of n in certain wavelength regions have been ascribed to film inhomogeneities [16]. Photoluminescence spectra for i-C films have been measured at 81 K [121].

All of the optical properties of i-C are reported to change considerably if the films are deposited at elevated substrate temperatures above 200°C [121], and similar changes result from annealing at temperatures above

450°C [16]. The striking increase of the extinction coefficient k with rising substrate or annealing temperature indicates a degradation of the i-C structure due to hydrogen release and graphitization.

4. Chemical Behavior

In many studies the excellent chemical inertness of i-C films against the attack of acids, bases, and organic solvents has been noticed [5–8]. In a series of inertness tests ion-plated i-C films prepared at a bias voltage of 1 kV were lifted off from glass substrates and then treated with various reagents. Over a period of 100 h the films proved to be resistant with respect to HF, HCl, H_2SO_4, HPO_3, HNO_3, $HClO_4$, $HNO_3 + H_2SO_4$, $CrO_3 + H_2SO_4$, NaOH, KOH, and various organic solvents [41].

In order to learn more about the protective action of i-C coatings on iron and steel, electrochemical measurements using the potentiostatic technique were carried out in dilute H_2SO_4 (1%) [41, 132]. The current density j was recorded with the samples held at the constant anode potential $\varphi_a = -0.170$ V relative to the saturated calomel electrode (SCE). After a transient period of about 5 s the current density was found to assume a constant value j_s. In evaluation of the results obtained for numerous i-C-coated iron specimens the dependance of j_s on the film thickness d was shown to follow approximately the relation

$$j_s = (9.03 - 1.32 \ln d) \quad \text{mA/cm}^2 \tag{6}$$

with d expressed in nanometers. According to this relation j_s reaches zero at $d = 950$ nm. Indeed a major fraction of i-C coatings corresponding to a thickness of 1.0–1.5 μm proved to be dense as could also be concluded from the high positive rest potentials, comparable with noble metals, that were measured in the currentless state. The relative porosity of coatings thinner than 0.95 μm could be determined from the ratio $f_p = j_s/j_{Fe}$, where j_{Fe} is the equivalent current density of uncoated iron samples. On the other hand, thick i-C coatings ($d > 2$ μm) and films deposited on rough steel surfaces proved to exhibit a certain porosity, which probably originates from the formation of microscopic cracks.

B. Structure of i-C Films

With the final goal to elucidate the atomic structure of i-C in relation to its unusual properties, the full arsenal of modern structural investigation methods has been employed to probe the films. As will turn out from the following discussion, much has been learned about the nature of i-C; however, essential details remain to be eluciated in further investigations.

1. Electron Diffraction and Microscopy

Electron diffraction patterns taken from thin i-C films prepared from hydrocarbons generally show a few diffuse rings that could arise either from an amorphous structure or from extremely small crystallites [61, 91, 98]. More accurate measurements over a wider range of the diffraction parameter $s = 4\pi(\sin \Theta)/\lambda$, where 2Θ is the diffraction angle and λ the electron wavelength, enable us to determine the reduced intensity function defined [133] by

$$i(s) = [I(s)/f^2(s)] - 1. \tag{7}$$

Here $I(s)$ denotes the normalized intensity, and $f(s)$ is the atomic scattering factor. From $i(s)$ the radial distribution function (RDF) can be calculated through Fourier transformation.

For i-C films such investigations have been carried out by McKenzie *et al.* [134] and by Bewilogua and Holzhüter [135, 136]. Although the examined films were prepared by distinctly different techniques, i.e., by C_2H_2 decomposition in a dc magnetron and by ion plating from C_6H_6, respectively, the results agree in that a predominating threefold coordination equivalent to sp^2 bonding has been unambiguously verified. Moreover, Fig. 4.12 shows instructively the amazing similarity of the intensity functions obtained for a very hard i-C film and for a thermally evaporated film of black amorphous carbon. An analysis of the small differences indicates that the degree of near

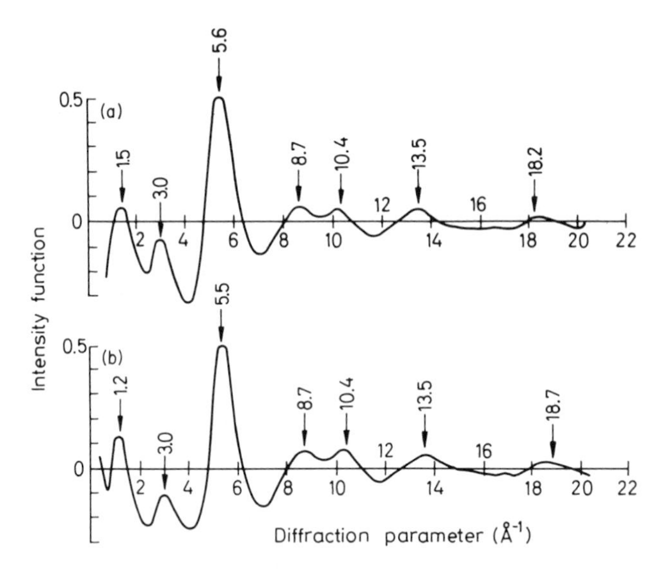

Fig. 4.12. Intensity functions of electron diffraction (a) ion-plated i-C ($U_b = 950$ V) and (b) evaporated a-C.

order is more pronounced in i-C than in a-C. It must be emphasized that these results do not exclude the presence of a certain fraction of sp^3-bonded carbon, especially if its contribution to the mean coordination number would be compensated by twofold coordinated atoms.

In electron micrographs of moderate magnification, some authors observed a columnar topography, but in hydrogenated i-C films normally no crystals could be discerned [87, 89]. Figure 4.13 shows a high-resolution micrograph (TEM) taken at optimum defocusing from an i-C film prepared by ion plating at a bias voltage of 250 V [137]. The image reveals a distinct topological disorder; only in certain regions can structure elements be identified. The enlarged section, which corresponds to the indicated region of the original micrograph, could be interpreted as showing tilted graphite planes. Similar elements with tilt angles in the range from 100 to 140° have been detected by Iijima in evaporated and sputtered carbon films [138, 139].

Small crystallites have been found in i-C films prepared by ion-beam deposition or ion-beam sputtering [23, 73, 98, 99b]. Provided that some

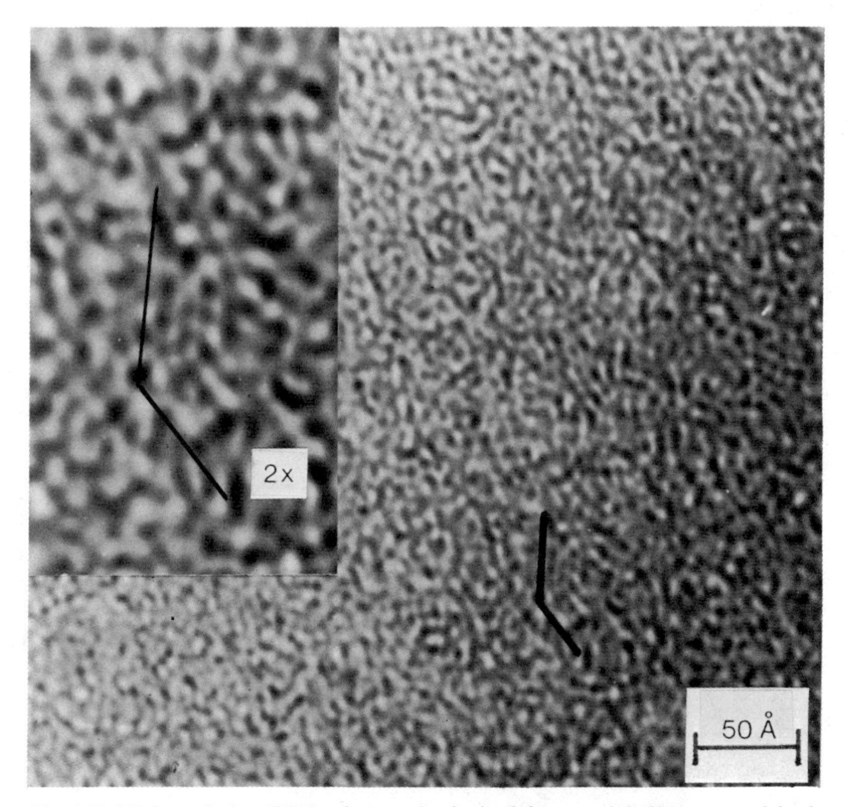

Fig. 4.13. High-resolution TEM micrograph obtained from an i-C film prepared by ion plating from benzene.

forbidden reflexes can be explained by multiple scattering or ordered incorporation of noble gas atoms [23, 61], the measured diffraction patterns are essentially consistent with the assumption that these microcrystals represent small cubic diamonds. Nonfitting reflexes have been ascribed to other carbon phases or to impurities [23, 98]. The diffraction pattern as well as the bright- and dark-field TEM micrographs of an ion-beam-sputtered i-C film are shown in Fig. 4.14. Since these crystallites constitute only a small fraction of the film material, local high stress and the catalytic action of metal contamination may play a role in their formation [74]. There are indications that crystallites belonging to the hexagonal diamond phase (lonsdaelite) or to unknown phases may be formed under conditions of particularly strong ion enhancement, as in dual-beam sputter or pulsed plasma techniques [73, 101].

For the normally amorphous i-C films prepared by ion plating from benzene, in addition to graphitization, the growth of numerous microcrystals up to about 80 nm in diameter has been observed when the freshly deposited layers were annealed at a temperature of 800–1000 K inside the lens chamber of the electron microscope [7, 13, 137]. The TEM micrographs and the diffraction pattern are displayed in Fig. 4.15. Convincing evidence that these crystals indeed consist of carbon was obtained, as shown in Fig. 4.16, by

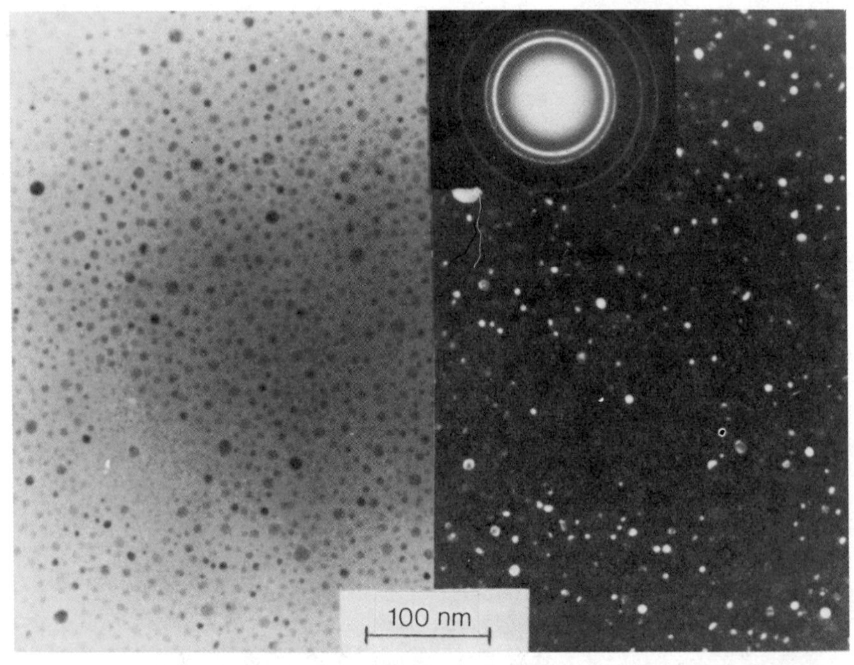

Fig. 4.14. Diffraction pattern as well as bright- and dark-field TEM micrographs indicating the presence of microcrystals in ion-beam-sputtered i-C films.

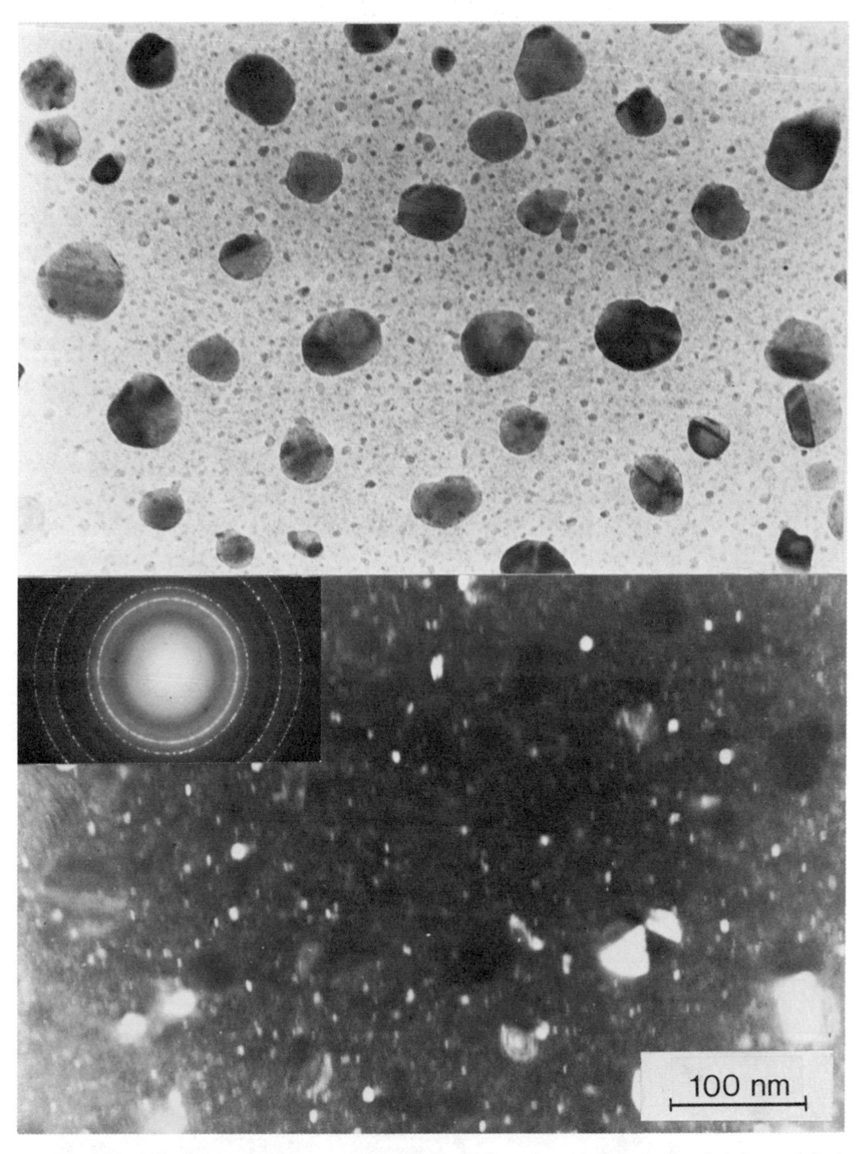

Fig. 4.15. Microcrystals grown upon annealing in ion-plated i-C, including bright- and dark-field TEM micrographs as well as diffraction pattern.

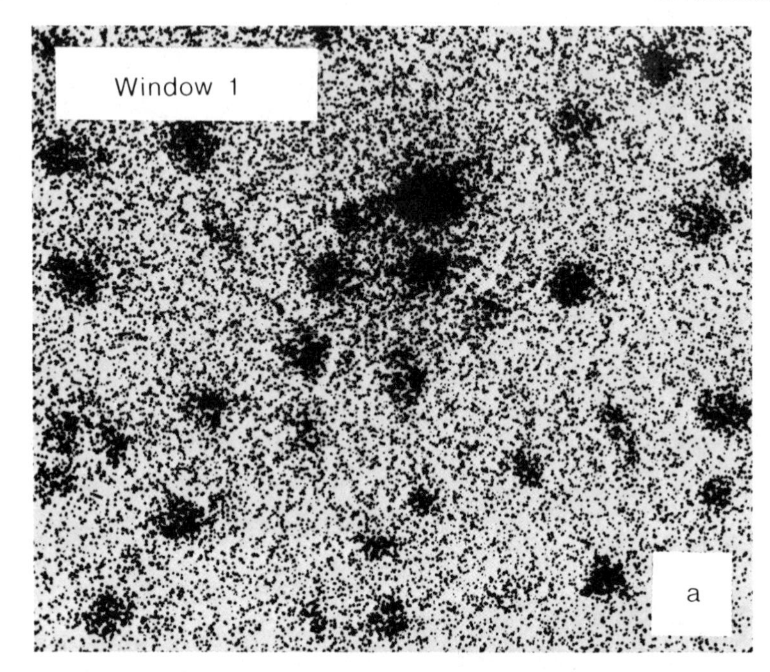
Window 1
a

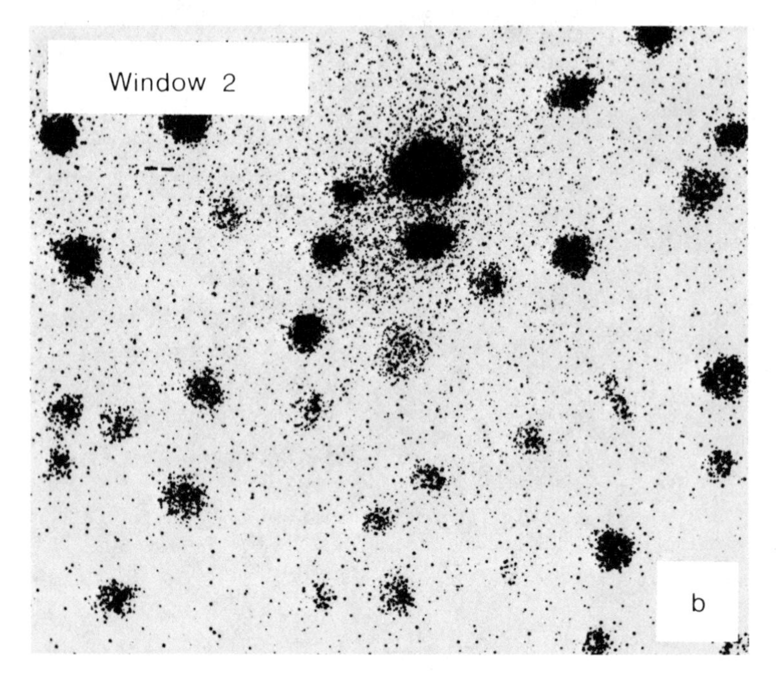
Window 2
b

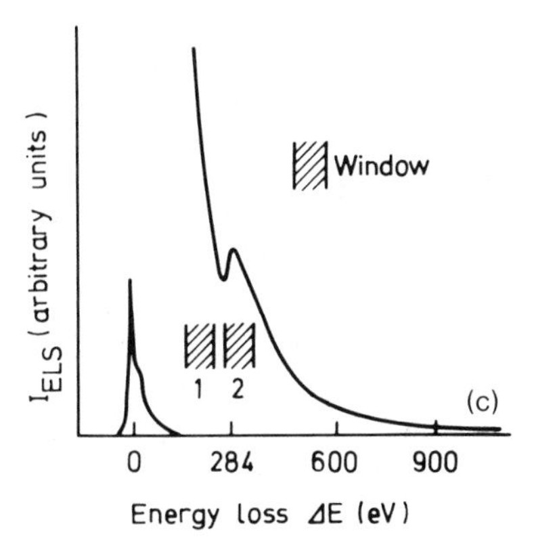

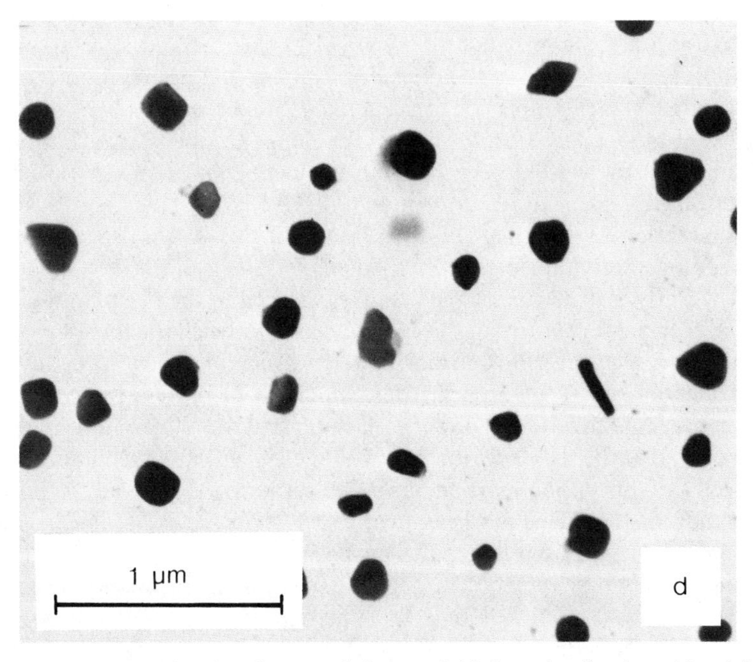

Fig. 4.16. Evidence that the microcrystals in annealed i-C consist of carbon: (a) and (b) are electron micrographs obtained with energy-loss electrons corresponding to the windows indicated in the graph (c); (d) shows a STEM micrograph of the same film region.

comparison of a scanning TEM micrograph with images formed by electrons that had undergone energy losses corresponding to the K-shell excitation of carbon atoms [137]. Presumably due to high intrinsic stress, the larger crystallites were found to disintegrate spontaneously into smaller fragments during storage of the samples in atmospheric environment for more than a year. From the viewpoint of thermodynamics, the growth of small diamonds at normal pressure would be plausible if we assume that the i-C films correspond to states that are more apart from equilibrium than the equally metastable diamond phase [3]. Then, the transition upon annealing would mean just a jump from one metastable position to a more favorable one.

2. Electron Spectroscopy and EXAFS

Diagnostics by use of Auger electron spectroscopy (AES) and by photoelectron spectroscopy (ESCA) confirm that i-C films, with regard to their atomic and electronic structure, must be placed much closer to amorphous carbon than to graphite or diamond [62, 63, 98]. For some films, however, similarities with the ESCA spectra of diamond were observed, and characteristic details were noticed [120].

Electron energy loss spectrometry (EELS) has proved to provide valuable information on the film structure [7, 98], and this method is being used in our laboratory for checking the reproducibility in film preparation.

The apparatus in use, a reconstructed electron microscope with electrostatic lenses, allows us to probe the films in the forward direction (transmission) by a 40-keV beam in three energy loss regions: (1) photon interaction equivalent to IR absorption (0.01–0.5 eV), (2) plasmon excitation with interaction by interband transitions (2–40 eV), and (3) core excitation into band or interband states (100–1000 eV).

Particularly instructive are the results measured in the plasmon region (2) because even small features in the spectra prove to be characteristic and well reproducible. We have introduced the term plasmonometry to denote film diagnostics based on the accurate determination of plasmon positions and intensities [7]. Figure 4.17 shows the plasmon spectrum for an ion-plated i-C film prepared at the relatively low bias voltage $U_b = 250$ V, in comparison to the spectra of amorphous carbon, graphite, and diamond. From the position of the main $(\pi + \sigma)$ plasmon peak near 22 eV, evidence is obtained that the i-C films are more closely related to evaporated amorphous carbon than to graphite or diamond. A shoulder near the position of the diamond plasmon peak at 34 eV was detected only in films deposited at relatively low bias voltages corresponding to the transition region; sometimes this bulge disappeared within a few minutes under the electron irradiation ($\sim$0.1 mA/cm^2) in the spectrometer. The occurrence of a small π-plasmon peak at

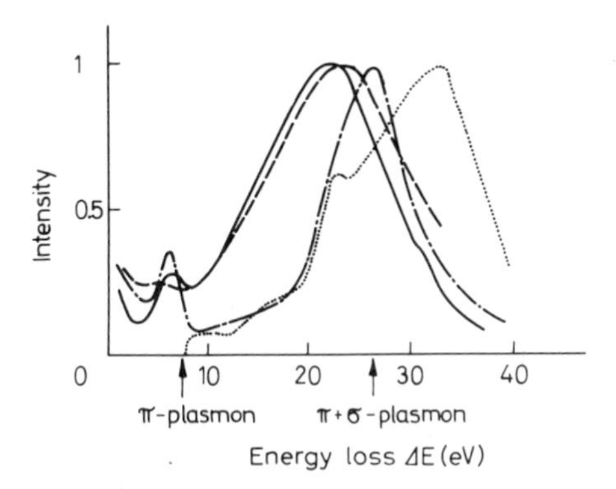

Fig. 4.17. Electron energy loss spectra (EELS) of i-C and of other carbon phases (the i-C spectrum corresponds to a film prepared by ion plating from benzene at a bias voltage of 250 V). (——) i-carbon, (– – –) a-carbon, (— – —) graphite, and (·····) diamond.

6.8 eV is interpreted as a sign for the presence of a polymeric component containing aromatic rings. In this context it merits to mention that retained benzene molecules as well as high-molecular-mass aggregates have been detected in benzene subjected to shock waves up to 210 GPa [140]. Further information concerning this polymeric component was obtained from plots of the peak intensity versus bias voltage. It turns out that the height of the π-plasmon peak decreases linearly with increasing bias voltage until it reaches zero at about 1000 V. It could be concluded that the ion-plated i-C films of optimum hardness do not contain this component [13]. This behavior may be associated with a dissociation of energetic molecular species upon impact (see Subsection II.A.4.b.).

Perhaps the most interesting result of plasmonometry is the detection of a characteristic depression in the plot of the $(\pi + \sigma)$ plasmon position as a function of the bias voltage, shown in Fig. 4.18. This depression, which was

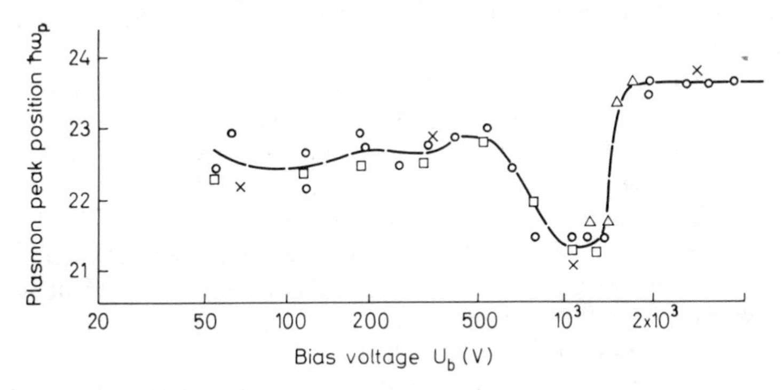

Fig. 4.18. Position of the $(\pi + \sigma)$ peak in i-C EELS spectra as a function of the bias voltage applied in ion plating deposition.

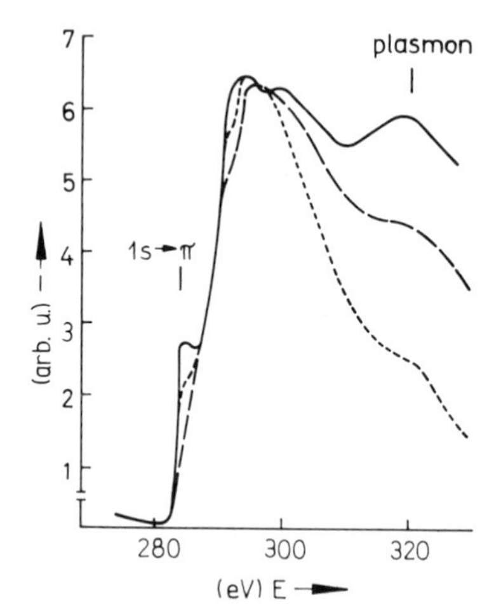

Fig. 4.19. Electron energy loss spectra corresponding to the core excitation region for three ion-plated i-C films prepared at different bias voltages: (——) $U_{b_1} = 250$ V; (– – –) $U_{b_2} = 600$ V; and (······) $U_{b_3} = 950$ V.

observed in reproduced series of experiments and even for samples stored at air for three years, coincides almost exactly with the region of maximum hardness that was discussed in Subsection III.B.2.b. [7, 17, 94]. Since such a striking coincidence can hardly be considered to be fortuitous, this observation implies some structural peculiarity within the relatively narrow U_b region from 800 to 1100 V, but no satisfying explanation has been found as yet.

Electron energy loss spectra taken in the region of core electron excitation can also provide valuable information, as is demonstrated by Fig. 4.19. In particular the slopes following the absorption peaks were found to vary sensitively with the bias voltage, structural changes during film aging, and other influences [7].

Equivalent yet more precise and detailed studies on the structure of i-C films have become feasible by using synchrotron radiation to probe the layers via absorption (EXAFS) and photoemission measurements. The first published results have shown the plasma-deposited films to be amorphous, with an essentially featureless valence band density of electron states [141, 142]. A considerable amount of π-bonding was deduced from the near-edge structure of the carbon K absorption, and the optical constants were calculated from the loss functions for energies less than 40 eV [142]. The estimated contribution of about two-third tetrahedrally bonded carbon atoms and only about one-third trigonally bonded particles to the i-C structure is, however, in contradiction to the results of electron diffraction analysis that were discussed in the preceding section. Certainly further work is required to elucidate the bonding structure.

3. SIMS and ESR Diagnostics

Analysis by secondary ion mass spectrometry (SIMS) reveals that i-C films made by rf plasma deposition from C_3H_8 contain mainly carbon and hydrogen, while oxygen and OH are present near the surface, probably originating from background gas. No impurities greater than several parts per million could be detected [61].

The density of unsaturated spins has been determined by the ESR technique. For i-C films prepared by rf plasma decomposition of C_2H_4 at 200–250°C, the spin density was about $3 \times 10^{17}/cm^3$, and an increase to more than $10^{19}/cm^3$ was found when the deposition temperature was raised to 350°C [121].

4. IR Spectroscopy

Since the optical properties were discussed already in Subsection III.A.3.b., only some results dealing with the structure of i-C films are added here.

The C–H stretch bands observed at 3300 cm^{-1}, 3000 cm^{-1}, 2920 cm^{-1}, and 2850 cm^{-1} can be assigned to monohydride bonded to sp^1-, sp^2-, and sp^3-type carbon and to dihydride (CH_2) bonded to sp^3-type carbon, respectively [66]. From the intensities, a predominant monohydride bonding state has been verified for hard i-C films made from benzene, with the abundance of the adjacent type of carbon decreasing in the order sp^3, sp^2, and sp^1. Evidently these important statements refer only to those carbon atoms that are linked with hydrogen and tell nothing about the bonding state of carbon atoms solely bonded to each other.

Smith [116] has investigated in detail the alteration of the optical constants of i-C films deposited at 250°C by dc plasma decomposition of C_2H_2 upon annealing at temperatures up to 750°C. The behavior could be described by an effective medium approximation (EMA), assuming that i-C : H can be considered to be a composite of the following components: (1) amorphous diamond, (2) amorphous graphite, (3) a polymeric component, and (4) a void component. With reasonable assumptions, it has been concluded that the as-prepared films consist of volume fractions amounting to about 0.14, 0.11, 0.75, and 0.00 for components (1)–(4). Upon annealing, particularly in the temperature range of 350–550°C, the polymeric and diamondlike components were calculated to decrease, reaching zero at about

650°C, whereas the graphitic component grows and a void component develops to a maximum volume fraction of nearly 0.3 at 550°C.

From these considerations two conclusions have been suggested. First, the diamondlike component is supposed to serve as a "glue" that binds together the other components in the hard films. Second, hydrogen release becomes significant at 450°C, and at 650°C (where the optical gap was observed to fall to zero), only negligible amounts of this element are left.

5. Hypothetical Structural Model

What can be told at present about the atomic structure of i-C? Earlier conjectures, including ours, that this material would represent some kind of amorphous or extremely fine-crystalline diamond had to be revised in view of the overwhelming evidence discussed in the foregoing sections. Successful simulations of the electron diffraction data by trial-and-error calculations considering contributions from different structure elements have encouraged us to propose a structure model for i-C that seems to be consistent with most reported properties and structural investigations [17, 136]. The basic carbon matrix of hard films, especially for the ion-plated films of optimum hardness, is thought to consist of puckered and tilted ring structures that are strongly interlinked through tetrahedral cross bonds. In particular for films prepared at higher bias voltages some contribution of n-fold carbon rings, where n equals 3, 4, 5, 7, or 8, is taken into account. Figure 4.20 shows some plausible models of the atomic arrangement [137]. The number of broken bonds in such structures is in principal agreement with the hydrogen concentration detected in i-C films of optimum hardness. For i-C films corresponding to the transition region, i.e., for most of the films obtained by plasma techniques, additional polymeric inclusions are assumed to be incorporated and "glued" to the carbon matrix via cross links. Such a picture seems to be very close to the EMA model suggested by Smith [16]. Less understandable is why hard i-C films produced by ion beam techniques despite the lack of bond-saturating hydrogen are found to be relatively stable; a larger fraction of C–C cross links may account for the slightly higher mass density of these layers. Local accumulation of sp^3 carbon can be expected to result in the formation of diamondlike nuclei, which due to strong ion activation and presumably supported by high stress and catalytic action of metal impurities may grow to real microcrystals. As an alternative, upon annealing the observed graphitization of hydrogenated i-C films will arise from hydrogen release that leaves unsaturated bonds.

The model sketched here should be regarded as a hypothetical basis for further discussion; its final confirmation, including improvements, or rejection will depend on more detailed experimental evidence.

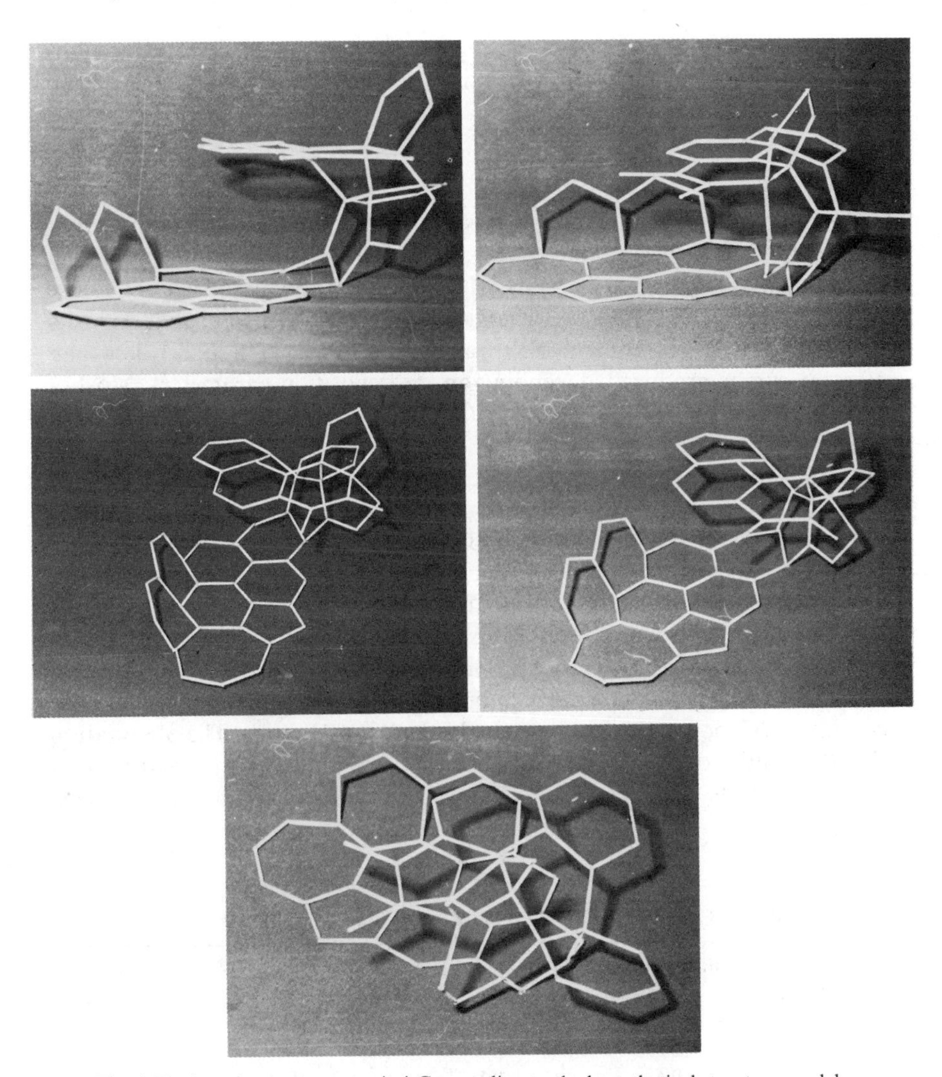

Fig. 4.20. Atomic arrangements in i-C according to the hypothetical structure model.

IV. PROPERTIES AND STRUCTURE OF i-BN AND COMPOSITE FILMS

A. Hard i-BN Films

The investigation of i-BN films prepared by the techniques described in Subsection II.B.3. is still in a relatively early stage. Nevertheless, the results obtained so far suggest that for i-BN coatings synthesized from evaporated boron and NH_3 or N_2 by ion plating a structure similarity with i-C films exists. In the following brief discussion some of the established facts are presented and assessed.

1. Film Composition and Properties

In comparative studies the composition and properties of films prepared from B and NH_3 have been investigated [17, 103]. The B/N ratio was mainly determined by x-ray microanalysis, with additional checks carried out by use of EELS, AES, ESCA, and wet-chemical analysis. It turns out that deposits of virtually any boron content in the range from 100 at. % to slightly less than 50 at. % can be produced, depending on the deposition parameters. Most of the measurements were performed with films corresponding to B/N ratios of about 3 : 2 or 1 : 1. Nearly stoichiometric films were even obtained by the "neutral process" without operation of the ionization system. The hydrogen concentration of the films was measured by nuclear reaction analysis (see Subsection III.A.1.). In Table 4.4 results obtained for films of two B/N ratios, prepared by the neutral process using different degrees of ion enhancement, are listed. From the data it follows that ion enhancement causes a distinct decrease of the hydrogen concentration, as compared with the hydrogen-rich films formed in the neutral process. This effect is clearly more pronounced for films of B/N ~ 1 : 1, however, approximately the same minimum concentration of about 6.5 at. % is reached for both B/N ratios. The observation that the relative decrease of the hydrogen content under the bombardment of the probing $^{15}N^+$ beam (~5 W/cm^2 power) was found to rise strongly with

TABLE 4.4

Hydrogen Content[a] of i-BN Films Prepared by Ion Plating from B and NH_3

B/N ratio	Bias voltage U_b (kV)	Substrate current J_s (mA)	H-concentration c_H (at. %)	Decrease[b] Δc_H (%)
1 : 1	0[c]	0	30	−1
	0.5	0.1	5	−4
	0.5	1	6.5	−4
	2.0	1	7.0	−36
3 : 2	0[c]	0	26.5	Not measured
	0.5	0.1	16.0	
	0.5	1	12.0	
	2.0	1	6.5	

[a] Measured by nuclear reaction analysis $^1H(^{15}N; \alpha, \gamma)^{12}C$.
[b] Decrease of the hydrogen concentration c_H after 400 s ^{15}N irradiation.
[c] Neutral CVD process without ion enhancement.

increasing ion enhancement during film preparation deserves attention. Hence, it seems that the residual hydrogen incorporated in films deposited at $U_b = 2$ kV is more loosely bonded than in the other layers.

The as-prepared films ($U_b = 2$ kV) proved to be partially transparent in the visible region, well-adherent on glass, silicon, and hard alloy substrates, inert against acids and bases, and stable in air up to temperatures of at least 1400 K. The latter property seems to be particularly important in view of applications.

Other macroscopic properties have been measured for relatively thick coatings prepared by means of a powerful ion-plating plant from B and NH_3 or N_2 in the presence of Ar. The mass density is in the range of 1.8 to 2.4 g/cm^3; the higher values were measured for films made with N_2 under strong ion enhancement. The Vickers microhardness at loads $\leqslant 0.4$ N falls in the range of 20 to 30 GPa; maximum achievable values have probably not yet been realized. The electrical resistivity is in the range of 5×10^{13} to 8×10^{14} Ω cm [105].

2. Structure of i-BN Films

Examination of i-BN films by means of electron diffraction, x-ray diffraction (for thick films), and transmission electron microscopy reveal an amorphous structure with some indication of microcrystalline fractions. The diffraction patterns obtained with pure boron films and boron-rich deposits resemble that reported by Matuda *et al.* [143]. The presence of boron icosahedrons seems to be likely [144]. With growing nitrogen content a broadening of the diffuse diffraction rings is observed, and for B/N ratios smaller than 3 contributions arising from B–N bonds begin to dominate. The essential features of the patterns obtained with typical ion-plated i-BN films could be simulated by a superposition of icosahedral and hexagonal i-BN fractions; some finer details may be explained by tetrahedrally coordinated structure elements. As for the crystalline fractions, the reflexes may both be assigned to hexagonal, including turbostratic, BN, or to cubic BN. The formation of some cubic BN, as has been reported by other groups (64b, 106–111], cannot be excluded.

In the EEL spectra a main plasmon peak near 25 eV and a very small one at about 9 eV was detected. The ESCA spectrum of i-BN, shown in Fig. 4.21, was found to be intermediate between a spectrum for a film prepared by the neutral process and that of a rf-sputtered turbostratic BN layer. The position of the N 1s level is in accordance with data reported for hexagonal and cubic BN [145].

In the IR spectra several characteristic absorption bands could be identified, as is designated in Fig. 4.22. Whereas some of the hydrogen

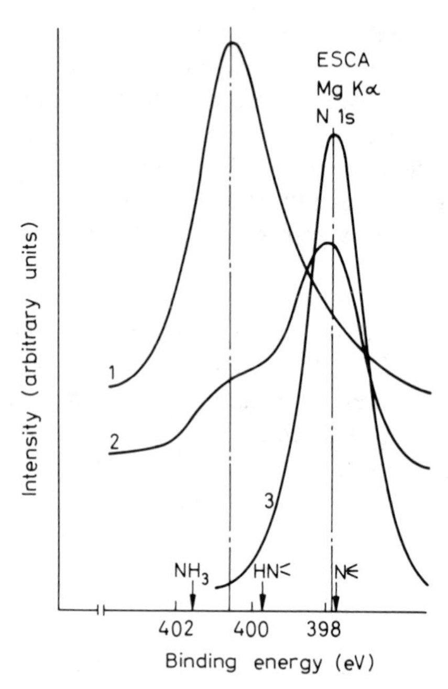

Fig. 4.21. ESCA spectra of (1) BN prepared from B and NH_3 without ion enhancement, (2) i-BN prepared by ion plating from B and NH_3, and (3) BN deposited by rf sputtering.

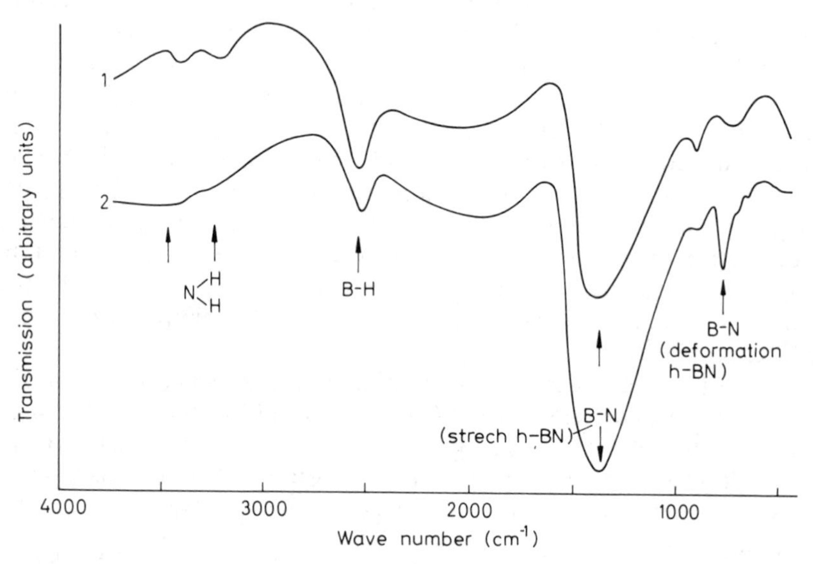

Fig. 4.22. IR spectrum of an ion-plated i-BN film (2) in comparison to that of a BN film prepared without ion enhancement (1).

incorporated in the soft films prepared by the neutral process is bonded as NH_2, in ion-plated hard coatings only B–H bonding could be detected.

Summarizing, it follows that i-BN films exhibit the same structural resemblance to soft, amorphous boron nitride deposits as i-C films to evaporated, amorphous carbon. Therefore, it is possible to interpret their particular properties in terms of ion-induced cross linkage in which tetrahedral bonding is involved, but this is still to be proved.

B. Composites on the Basis of i-C and i-BN

The relatively loosely packed atomic arrangement in i-C and i-BN should be capable of accommodating foreign atoms up to a considerable concentration. In micropores of 1–4 nm size, which according to calorimetric and electrochemical investigations may form easily in i-C films [69, 132], even clusters of foreign atoms would find place. These considerations suggested looking for an eventual formation of composite films, which might be found to retain some of the properties of i-C or i-BN.

So far, the composite system Cr/i-C has been investigated in some detail [93, 137], and tentative experiments with Al/i-C, Si/i-C, and Ti/i-C were carried out. As mentioned in Subsection II.B.1., the films were prepared by ion plating from benzene and simultaneous evaporation of the other component. All results indicate that at low deposition temperatures (20–100°C) metastable composites can indeed be produced within a wide range of composition. Moreover, there seems to be little tendency to form carbides, as has been also found for $Si_xC_yH_z$ films prepared by glow discharge or reactive sputtering techniques [146, 147].

For Cr/i-C coatings the Vickers hardness was found to increase almost linearly with rising atomic concentration of carbon [3, 7, 17]. Figure 4.23 shows the photometric records of the electron diffraction patterns measured

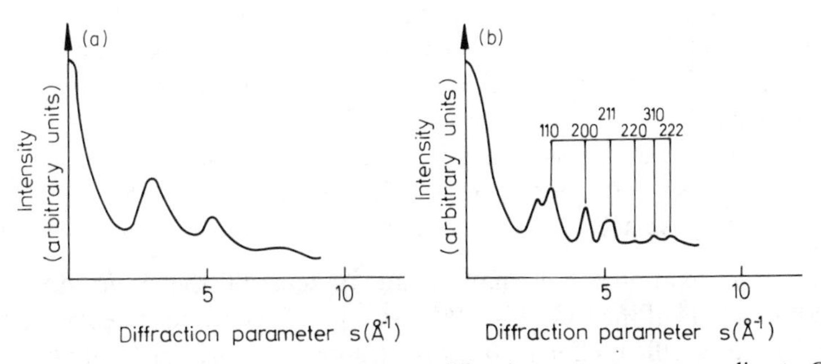

Fig. 4.23. Photometric records of the electron diffraction patterns corresponding to Cr/i-C composite films deposited (a) on a water-cooled substrate and (b) at ambient substrate temperature.

for a composite film deposited with water cooling of the substrate and for a similar film deposited at ambient temperature. In the first case, the structure is clearly amorphous, whereas chromium microcrystallites of about 2 nm in diameter can be identified in the second film.

In accordance with the assumption of amorphous atomic mixtures, the plasmon spectrum of all these composites reveals only a single main peak, the position of which is found to be unequivocally correlated to the composition and the plasmon peaks of the components [149]. Upon annealing, however, the dissociation of the mixture, which for Al/i-C sometimes occurred rather vehemently, could be detected by separate emerging peaks [3, 17]. Furthermore, Al/i-C composites, which are golden-yellowish, proved to be stable in contact with water over long time periods as opposed to aluminium carbide, which is known to react forming Al_2O_3 and CH_4.

The boron-rich i-BN films may also be regarded as composites of the B/i-BN type, and tentative experiments have demonstrated that i-BN composites including other elements can be made as well. An interesting question is whether the incorporation of certain elements such as Si could increase the fraction of tetrahedral bonds in i-C or i--BN films [16].

V. OUTLOOK ON APPLICATIONS

We will resist the temptation to list here all the applications that have been proposed or already practiced, at least in tests or small scale production, in conjunction with coatings based on i-C or i-BN, for reviews we refer to some of the numerous publications in the field [4–8, 39, 87, 150]. According to the desired effects, such applications can be categorized as follows:

(1) hard coatings on tools and machine parts,
(2) tribological coatings on various components,
(3) coatings for corrosion protection in nearly all kinds of environment,
(4) functional optical coatings for the infrared region and protective coatings on solar cells or glass fibers,
(5) insulating and passivating coatings in electronics,
(6) coatings of low sputter erosion for fusion reactors [151], and
(7) coatings to achieve decoration effects.

Certainly more applications may be anticipated, and the circumstance that the components of the i-C and i-BN-based coatings are relatively cheap, will stimulate intensified efforts for a widespread utilization. Final success, however, will depend on the optimization of film properties and on the efficiency and controllability of the deposition technology.

ACKNOWLEDGMENTS

The author wants to express his gratitude to all his co-workers as well as to colleagues of many countries for contributions and stimulating discussions. Their names are found in the references. Thanks are given to Professor G. Leonhardt, Academy of Sciences, Berlin, who made the ESCA analysis, and to Dr. W. Rudolph and Dr. C. Heiser, Academy of Sciences, Dresden, who carried out the nuclear reaction analysis. Dr. B. Rau and Dipl.-Phys. B. Rother are thanked for help in preparing the manuscript and Mrs. I. Grösel for drawing the figures.

REFERENCES

1. J. E. Field, ed., "The Properties of Diamond", Academic Press, New York, 1979.
2. N. V. Novikov and P. S. Kisly, *Thin Solid Films* **64**, 205 (1979).
3. C. Weissmantel, K. Breuer, and B. Winde, *Thin Solid Films* **100**, 383 (1983).
4. L. Holland and S. M. Ojha, *Thin Solid Films* **58**, 107 (1979).
5. L. P. Andersson, *Thin Solid Films* **86**, 183 (1981).
6. K. Enke, *Thin Solid Films* **80**, 227 (1981).
7. C. Weissmantel, K. Bewilogua, K. Breuer, D. Dietrich, U. Ebersbach, H.-J. Erler, B. Rau, and G. Reisse, *Thin Solid Films* **96**, 31 (1982).
8. A. Bubenzer, B. Dischler, G. Brandt, and P. Koidl, *J. Appl. Phys.* **54**, 4590 (1983).
9. B. V. Derjaguin, D. V. Fedoseev, and V. P. Varnin, *Zh. Eksp. Teor. Fiz.* **62**, 1250 (1975).
10. V. I. Spitsyn, L. L. Bonilov, and B. V. Derjaguin, *J. Cryst. Growth* **52**, 219 (1981).
11. D. V. Fedoseev, B. V. Derjaguin, I. G. Varshavskaya, and A. V. Lavrentev, *Carbon* **21**, 243 (1983).
12. D. V. Fedoseev, V. L. Bukhovets, I. G. Varshavskaya, A. V. Lavrentev, and B. V. Derjaguin, *Carbon* **21**, 237 (1983).
13. C. Weissmantel, K. Bewilogua, D. Dietrich, H.-J. Erler, H.-J. Hinneberg, S. Klose, W. Nowick, and G. Reisse, *Thin Solid Films* **72**, 19 (1980).
14. S. Aisenberg and R. W. Chabot, *J. Appl. Phys.* **42**, 2953 (1971).
15. D. R. McKenzie, L. C. Botten, and R. C. McPhedran, *Phys. Rev.* **51**, 280 (1983).
16. F. W. Smith, *J. Appl. Phys.* **55**, 764 (1984).
17. C. Weissmantel, *Proc. Int. Vac. Congr., 9th, Madrid, 1983*, 299 (1983).
18. R. L. Stewart, *Phys. Rev.* **45**, 488 (1934).
19. H. König and G. Helwig, *Z. Phys.* **129**, 491 (1951).
20. H. Schmellenmeier, *Exp. Tech. Phys.* **1**, 49 (1953).
21. H. Schmellenmeier, *Z. Phys. Chem.* **205**, 349 (1955–1956).
22. S. Aisenberg and R. W. Chabot, *J. Vac. Sci. Technol.* **10**, 104 (1973).
23. E. G. Spencer, P. H. Schmidt, D. J. Joy, and F. J. Sanssalone, *Appl. Phys. Lett.* **29**, 118 (1976).
24. D. S. Whitmell and R. Williamson, *Thin Solid Films* **35**, 255 (1976).
25. L. Holland and S. M. Ojha, *Thin Solid Films* **38**, L17 (1976).
26. D. A. Anderson, *Philos. Mag.* **35**, 17 (1977).
27. C. Weissmantel, *Proc. Int. Vac. Congr., 7th, Vienna, 1977*, 1533 (1977).
28. C. Weissmantel, *Vide* **183**, 107 (1976).
29. L. Pranevičius and J. Dudonis, "Modifikazia Twerdych Tel Ionnymi putshkami" ("Ion Beam Modification of Solids"), Mosklas, Vilnius, 1980.
30. G. Carter and D. G. Armour, *Thin Solid Films* **80**, 13 (1981).
31. T. Takagi, *J. Vac. Sci. Technol. A* **2**, 382 (1984).

32. C. Weissmantel, *in* "Surface Modification of Metals" (C. R. Clayton and J. Lumsden, eds.), Academic Press, New York, 1984.
33. K. J. Klabunde, "Chemistry of Free Atoms and Particles," Academic Press, New York, 1980.
34. H. H. Andersen and H. Bay, *in* "Sputtering by Particle Bombardment I" (R. Behrisch, ed.), Vol. I, p. 146, Springer-Verlag, Berlin and New York, 1981.
35. S. Aisenberg and M. Stein, *in* "Laser Induced Damage in Optical Materials," p. 313, U.S. Gov. Printing Office, Washington D.C., 1980.
36. C. Weissmantel, *Thin Solid Films* **92**, 55 (1982).
37. D. M. Mattox, *Thin Solid Films* **53**, 81 (1978).
38. D. M. Mattox, *Vide* **212**, 175 (1982).
39. S. Aisenberg, *J. Vac. Sci. Technol. A* **2**, 369 (1984).
40. T. Takagi, I. Yamada, and A. Sasaki, *Thin Solid Films* **45**, 569 (1977).
41. G. Reisse, C. Schürer, U. Ebersbach, K. Bewilogua, K. Breuer, H.-J. Erler, and C. Weissmantel, *Wiss. Z. Tech. Hochsch. Karl-Marx-Stadt* **22**, 653 (1980).
42. W. A. Grant and J. S. Colligon, *Vacuum* **32**, 675 (1982).
43. S. Matteson and M.-A. Nicolet, *in* "Metastable Materials Formation by Ion Implantation" (S. T. Picraux and W. J. Choyke, eds.), p. 3, North-Holland Publ., New York and Amsterdam, 1982.
44. C. Weissmantel, *Thin Solid Films* **58**, 101 (1979).
45. F. Seitz and J. S. Koehler, *Solid State Phys.* **2**, 305 (1956).
46. J. A. Brinkman, *J. Appl. Phys.* **25**, 961 (1954).
47. D. A. Thompson, *Radiat. Eff.* **56**, 105 (1981).
48. W. A. Grant, *Proc. Int. Ion Eng. Congr. Kyoto, 1983*, 675, IEEJ, Tokyo (1983).
49. G. H. Vineyard, *Radiat. Eff.* **29**, 245 (1976).
50. R. Kelly, *Radiat. Eff.* **32**, 91 (1977).
51. I. A. Sellin and R. Laubert, *in* "Inelastic Particle-Surface Collisions" (E. Taglauer and W. Heiland, eds.), pp. 120–137, Springer-Verlag, Berlin and New York, 1981.
52. C. Steinbrüchel, *J. Vac. Sci. Technol. B* **2**, 38 (1984).
53. M. Pickering, N. R. S. Tait, and D. W. L. Tolfree, *Philos. Mag.* **A42**, 257 (1980).
54. N. R. S. Tait and D. W. L. Tolfree, *Phys. Status Solidi A* **69**, 329 (1981).
55. B. Meyerson and F. W. Smith, *J. Non-Crystal. Solids* **35/36**, 435 (1980).
56. B. Meyerson and F. W. Smith, *Solid State Comm.* **34**, 531 (1980).
57. L. Holland and S. M. Ojha, *Thin Solid Films* **48**, L21 (1978).
58. L. Holland, *Surface Technol.* **11**, 145 (1980).
59. S. M. Ojha and L. Holland, *Proc. Int. Vac. Congr., 7th, Vienna, 1977*, 1667 (1977).
60. S. M. Ojha, H. Norström, and D. McCulluch, *Thin Solid Films* **60**, 213 (1979).
61. H. Vora and T. J. Moravec, *J. Appl. Phys.* **52**, 6151 (1981).
62. L. P. Andersson and S. Berg, *Vacuum* **28**, 449 (1978).
63. L. P. Andersson, S. Berg, H. Norström, and S. Towta, *Thin Solid Films* **63**, 155 (1979).
64. (a) R. J. Gambino and J. A. Thompson, *Solid State Commun.* **34**, 15 (1980); (b) H. A. Beale, *Ind. Res./Dev.* (June), 143 (1979).
65. A. Bubenzer, B. Dischler, and A. Nyaish, *Thin Solid Films* **91**, 81 (1982).
66. B. Dischler, A. Bubenzer, and P. Koidl, *Appl. Phys. Lett.* **42**, 636 (1983).
67. N. Fujimori, A. Doi, and T. Yoshioka, *Proc. Conf. Japan Society of Applied Physics, 30th, Tokyo, 1983*, 214 (1983).
68. A. Doi, N. Fujimori, T. Yoshioka, and Y. Doi, *Proc. Int. Ion Eng. Cong., Kyoto, 1983*, IEEJ, Tokyo, 1137 (1983).
69. A. R. Nyaiesh and W. B. Nowak, *J. Vac. Sci. Technol. A* **1**, 308 (1983).
70. K. Hashimoto, T. Sakamoto, N. Matuda, S. Baba, and A. Kinbara, *Proc. Int. Ion Eng. Cong., Kyoto, 1983*, IEEJ, Tokyo, 1125 (1983).

71. Z. Haś, St. Mitura, and B. Wendler, *Proc. Int. Ion Eng. Cong., Kyoto, 1983*, IEEJ, Tokyo, 1143 (1983).
72. G. Gautherin and C. Weissmantel, *Thin Solid Films* **60**, 135 (1978).
73. C. Weissmantel, *J. Vac. Sci. Technol.* **18**, 179 (1981).
74. J. Preusse, thesis, Technische Hochschule, Karl-Marx-Stadt, German Democratic Republic, 1984.
75. S. Fujimori, T. Kasai, and T. Inamura, *Thin Solid Films* **92**, 71 (1982).
76. D. Mathine, R. O. Dillon, A. A. Khan, G. Bu-Abbud, J. A. Woollam, D. C. Liu, B. Banks, and S. Domitz, *J. Vac. Sci. Technol. A* **2**, 365 (1984).
77. V. E. Strelnizki, V. G. Padalka, and S. I. Vakula, *Zh. Teck. Fiz.* **48**, 377 (1978).
78. J. M. Mackowski, M. Berton, P. Ganau, and Y. Touze, *Vide* **212**, 99 (1982).
79. W. Nowick, Technische Hochschule, Karl-Marx-Stadt, German Democratic Republic, unpublished.
80. C. Weissmantel, H.-J. Erler, and G. Reisse, *Surf. Sci.* **86**, 207 (1979).
81. N. N. Nikiforova, M. B. Guseva, and V. G. Babaev, *Proc. USSR Conf. Interactions Atomic Particles Solids, 6th, Minsk, 1982*, 112 (1982).
82. B. A. Banks and S. K. Rutledge, *J. Vac. Sci. Technol.* **21**, 807 (1982).
83. A. Sathyamoorthy and W. Weisweiler, *Thin Solid Films* **87**, 33 (1982).
84. G. L. Harding, B. Window, D. R. McKenzie, A. R. Collins, and C. M. Horwitz, *J. Vac. Sci. Technol.* **16**, 2105 (1979).
85. G. L. Harding, *Sol. Energy Mater.* **7**, 101 (1980).
86. G. L. Harding, S. Craig, and B. Window, *Appl. Surf. Sci.* **11/12**, 315 (1982).
87. S. Craig and G. L. Harding, *Thin Solid Films* **97**, 345 (1982).
88. D. R. McKenzie and L. M. Briggs, *Sol. Energy Mater.* **6**, 79 (1981).
89. D. R. McKenzie, R. C. McPhedran, N. Savvides, and L. C. Botten, *Philos. Mag. B* **48**, 341 (1983).
90. J. Zelez, *J. Vac. Sci. Technol. A* **1**, 305 (1983).
91. C. Weissmantel, G. Reisse, H.-J. Erler, F. Henny, K. Bewilogua, U. Ebersbach, and C. Schürer, *Thin Solid Films* **3**, 315 (1979).
92. C. Weissmantel, K. Bewilogua, H.-J. Erler, H.-J. Hinneberg, S. Klose, W. Nowick, and G. Reisse, *in* "Low Energy Ion Beams" (I. H. Wilson and K. G. Stephens, eds.), p. 188, Institute of Physics Conf. Series 54, Institute of Physics, Bristol and London, 1980.
93. K. Bewilogua, E. Bugiel, B. Rau, C. Schürer, and C. Weissmantel, *Krist. und Tech.* **15**, 1205 (1980).
94. C. Weissmantel, *Proc. Int. Ion Eng. Cong., Kyoto, 1983*, IEEJ, Tokyo, 1257 (1983).
95. D. G. Teer and M. Salama, *Thin Solid Films* **45**, 553 (1977).
96. M. S. Salama, *Proc. Int. Ion Eng. Cong., Kyoto, 1983*, IEEJ, Tokyo, 1131 (1983).
97. T. Furuse, T. Suzuki, S. Matsumoto, K. Nishida, and Y. Nannichi, *Appl. Phys. Lett.* **33**, 317 (1978).
98. T. J. Moravec and T. W. Orent, *J. Vac. Sci. Technol.* **18**, 226 (1981).
99. (a) J. H. Freeman, W. Temple, D. Beanland, and G. A. Gard, *Nucl. Instrum. Methods* **135**, 1 (1976); (b) T. Mōri and Y. Namba, *J. Vac. Sci. Technol. A* **1**, 23 (1983).
100. S. Fujimori and K. Nagai, *Jpn. J. Appl. Phys.* **20**, L194 (1981).
101. M. Sokolowski and A. Sokolowska, *J. Crystal Growth* **57**, 185 (1982).
102. A. M. Dorodnov, *Zh. Tekh. Fiz.* **9**, 48 (1978).
103. C. R. Aita and Ngoc C. Tran, *J. Appl. Phys.* **54**, 6051 (1983).
104. M. D. Wiggins, C. R. Aita, and F. S. Hickernell, *J. Vac. Sci. Technol. A* **2**, 322 (1983).
105. D. Roth, H.-J. Erler, B. Rother, L. Richter, K. Peter, and B. Winde, *Proc. Conf. Hochvakuum, Grenzflächen, Dünne Schichten, 8th, Dresden, 1984*, Vol. 1, p. 265, Phys. Soc. of GDR, Berlin, (1984).
106. O. Gafri, A. Grill, D. Itzhak, A. Inspektor, and R. Avni, *Thin Solid Films* **72**, 523 (1980).

107. H. Miyamoto, M. Hirose, and Y. Osaka, *Jpn. J. Appl. Phys.* **22**, L216 (1983).
108. M. Sokolowski, A. Sokolowska, A. Michalski, Z. Romanowski, A. Rusek-Mazurek, and M. Wronikowski, *Thin Solid Films* **80**, 249 (1981).
109. S. Shanfield and R. Wolfson, *J. Vac. Sci. Technol. A* **1**, 323 (1983).
110. M. Satou, K. Matuda, and F. Fujimoto, *Proc. Symp. Ion Sources and Ion-assisted Technology, 6th, Tokyo, 1982*, 425 (1982).
111. M. Satou and F. Fujimoto, *Jpn. J. Appl. Phys.* **22**, L171 (1983).
112. W. Benz, "Massenspektrometrie organischer Verbindungen", Akad. Verlagsges., Leipzig, 1969.
113. V. Yu. Orlov, L. A. Taranenko, and M. V. Gurev, *Khim. Vys. Energ.* **3**, 145 (1969).
114. T. Lunow and R. Wilberg, *Proc. Conf. Hochvakuum, Grenzflächen, Dünne Schichten, 8th, Dresden, 1984*, appendix, Vol. 33, Phys. Soc. of GDR, Berlin, (1984).
115. P. G. Turner, R. P. Howson, and C. A. Bishop, *in* "Low Energy Ion Beams" (I. H. Wilson and K. G. Stephens, eds.), Chapter 6, p. 229, Institute of Physics Conf. Series 54, Bristol and London, 1980.
116. G. K. Wehner, University of Minnesota, Minneapolis, personal communication, 1980.
117. D. Roth, H.-J. Erler, B. Rother, I. Richter, K. Peter, and B. Winde, *Proc. Conf. Hochvakuum, Grenzflächen, Dünne Schichten, 8th, Dresden, 1984*, Vol. II, p. 265, Phys. Soc. of GDR, Berlin (1984).
118. D. Dietrich, P. Reinhardt, and C. Weissmantel, *Proc. Conf. Hochvakuum, Grenzflächen, Dünne Schichten, 7th, Dresden, 1981*, Vol. II, P. 416, Phys. Soc. of GDR, Berlin (1981).
119. M. H. Brodsky, M. A. Frisch, J. F. Ziegler, W. A. Ziegler, and W. A. Lanford, *Appl. Phys. Lett.* **30**, 561 (1977).
120. S. Berg and L. P. Andersson, *Thin Solid Films* **58**, 117 (1979).
121. I. Watanabe, S. Hasegawa, and Y. Kurata, *Jpn. J. Appl. Phys.* **21**, 856 (1982).
122. O. Fiedler, G. Reisse, B. Schöneich, and C. Weissmantel, *Proc. Int. Vacuum Congr., 4th, Manchester, 1968*, 569 (1968).
123. C. Weissmantel, C. Schürer, F. Fröhlich, P. Grau, and H. Lehmann, *Thin Solid Films* **61**, L5 (1979).
124. D. Nir, *Thin Solid Films* **112**, 41 (1984).
125. B. Rau, thesis, Technische Hochschule, Karl-Marx-Stadt, German Democratic Republic, 1984.
126. N. Matuda, S. Baba, and A. Kinbara, *Thin Solid Films* **81**, 301 (1981).
127. K. Enke, H. Dimigen, and H. Hübsch, *Appl. Phys. Lett.* **36**, 291 (1980).
128. R. C. McPhedran, L. C. Botten, M. S. Craig, M. Nevière, and D. Maystre, *Opt. Acta.* **29**, 289 (1982).
129. E. C. Freeman and W. Paul, *Phys. Rev.* **20**, 716 (1979).
130. G. D. Cody, T. Tiedje, B. Abeles, B. Brooks, and Y. Goldstein, *Phys. Rev. Lett.* **47**, 1480 (1981).
131. T. J. Moravec and J. C. Lee, *J. Vac. Sci. Technol.* **20**, 338 (1982).
132. U. Ebersbach, C. Schürer, and C. Weissmantel, *Z. Phys. Chem.* **260**, 938 (1979)
133. S. C. Moss and J. F. Graczyk, *Phys. Rev. Lett.* **23**, 1167 (1969).
134. D. R. McKenzie, L. C. Botten, and R. C. McPhedran, *Phys. Rev. Lett.* **51**, 280 (1983).
135. K. Bewilogua, D. Dietrich, L. Pagel, C. Schürer, and C. Weissmantel, *Surf. Sci.* **86**, 308 (1979).
136. K. Bewilogua, D. Dietrich, G. Holzhüter, and C. Weissmantel, *Phys. Status Solidi A* **71**, K57–59 (1982).
137. K. Bewilogua, thesis (B), Technische Hochschule, Karl-Marx-Stadt, German Democratic Republic, 1983.
138. S. Iijima, *J. Microsc.* **119**, 99 (1980).

139. S. Iijima, *J. Cryst. Growth* **50**, 675 (1980).
140. W. J. Nellis, F. H. Ree, R. J. Trainor, A. C. Mitchell, and M. B. Boslough, *J. Chem. Phys.* **80**, 2789 (1984).
141. D. Wesner, S. Krummacher, R. Carr, T. K. Sham, M. Strongin, W. Eberhardt, S. L. Weng, G. Williams, M. Howells, F. Kampas, S. Heald, and F. W. Smith, *Phys. Rev. B* **28**, 2152 (1983).
142. J. Fink, T. Müller-Heinzerling, J. Pflüger, A. Bubenzer, P. Koidl, and G. Crecelius, *Solid State Commun.* **47**, 687 (1983).
143. N. Matuda, S. Baba, and A. Kinbara, *Thin Solid Films* **89**, 139 (1982).
144. K. Katada, *Jpn. J. Appl. Phys.* **5**, 582 (1966).
145. K. Hamrin, G. Johansson, U. Gelius, C. Nordling, and K. Siegbahn, *Phys. Scr.* **1**, 277 (1970).
146. D. A. Anderson and W. E. Spear, *Philos. Mag.* **35**, 1 (1977).
147. T. Shimada, Y. Katayama, and K. F. Komatsubara, *J. Appl. Phys.* **50**, 5530 (1979).
148. Y. Katayama, K. Usami, and T. Shimada, *Philos. Mag. B* **43**, 283 (1981).
149. B. Breuer and R. Henning, Technische Hochschule, Karl-Marx-Stadt, German Democratic Republic, personal communication, 1984.
150. B. Winde, *Proc. Conf. Hochvakuum, Grenzflächen, Dünne Schichten, 8th, Dresden, 1984*, Vol. I, p. 138, Phys. Soc. of GDR, Berlin (1984).
151. S. Das, M. Kaminsky, L. H. Rovner, J. Chin, and K. Y, Chen, *Thin Solid Films* **63**, 227 (1979).

5 HIGH-VACUUM DEPOSITION METHODS INVOLVING SUPERTHERMAL FREE PARTICLES

G. Gautherin, D. Bouchier, and C. Schwebel

Institut d'Electronique Fondamentale
Université Paris-Sud
Orsay, France

I. INTRODUCTION

The aim of this chapter is to present some recently developed deposition techniques. These techniques involve the condensation of superthermal free atoms in high vacuum.

The successful thin film deposition of III–V compounds by molecular beam epitaxial (MBE) techniques [1, 2] has demonstrated the importance of

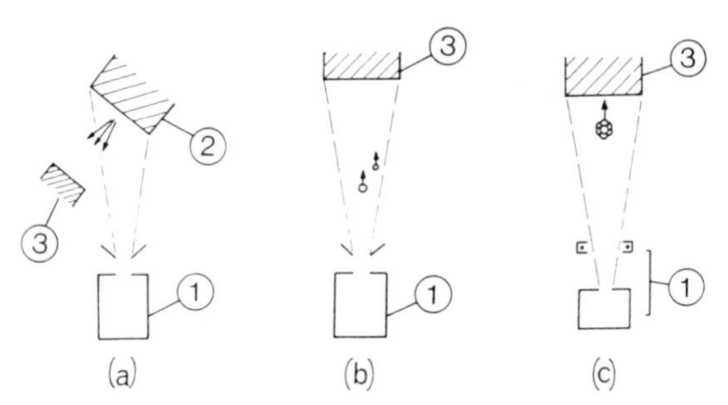

Fig. 5.1. Deposition configurations. (a) Ion-beam sputtering deposition (IBSD), (b) ion-beam deposition (IBD), and (c) ionized cluster beam deposition (ICBD). In the case of IBSD the ion-beam generated by the source [1] is directed onto the target [2], sputtered material is deposited on the substrate [3]. For IBD or ICBD, ionized material is deposited directly onto the substrate [3].

ultra-high vacuum (UHV) techniques for obtaining high-purity films in well-defined conditions, monitored by *in situ* analysis (AES, LEED, mass spectrometry, etc.). Unfortunately, if thermal evaporation is the most useful high-vacuum deposition process, some materials of particular interest (e.g., very hard coatings or nitrides) are beyond the scope of this application.

This limitation, in addition to the problems created by the poor adhesion of certain evaporated layers, may generally be overcome in the case of methods utilizing superthermal energies. The most classical of these superthermal methods, i.e., glow discharge sputter deposition and ion plating, has been widely exploited in the majority of technological processes [3]. Unfortunately, the presence of a plasma in the deposition chamber increases the difficulty of using *in situ* analysis techniques, and, furthermore, the growth of the layers may be disturbed by the impact of various particles and radiations emitted from the plasma [4]. Consequently, a method allowing greater isolation of the layer growth from the ion generation process would be quite attractive. This goal is achieved in the framework of ion-beam sputtering deposition (IBSD), ion-beam deposition (IBD), and a variation of the latter, ionized cluster beam deposition (ICBD). The principles of these techniques are schematically illustrated in Fig. 5.1. In these new methods, superthermal energy is supplied to the atoms of a solid target by momentum transfer from an energetic ion beam (IBSD) or by accelerating ionized species in an electric field (IBD and ICBD).

II. ION–SURFACE INTERACTIONS

In this section, the sputtering from a solid target and related phenomena are analyzed in order to determine the rate of arrival and the energy of the species from which ion-beam-sputtered thin films are constructed. In the case of IBD and ICBD, these mechanisms occur during the condensation of films and determine their growth kinetics. For more complete information, the reader is referred to review articles [5, 6] and to a pedagogical overview [7].

A. Basic Considerations

The interaction of an energetic particle of mass M_1 and atomic number Z_1 with a solid can be roughly characterized by the more efficient energy-loss process, which is characteristic of either electronic or nuclear collisions. The interaction of the ion with the lattice electrons results in secondary electron emission due to kinetic ejection [8] or de-excitation by Auger processes. The latter mechanism is involved in the potential ejection process, which leads to the neutralization [9] of the ion when it approaches the solid surface. Let us emphasize that the related neutralization probability is nearly equal to one [9] and that the difference between potential and kinetic ejection has been clearly established by experiments [10].

To a first approximation, the rate of energy loss to the lattice electrons is proportional to the ion velocity [11]. Furthermore, this rate is much smaller than energy loss by momentum transfer to the target nuclei if the ion energy does not reach $Z_1^{5/3}$ keV, i.e., a velocity less than $0.1 \times (Z_1^{2/3}e^2/\hbar)$ [11]. The latter condition is valid for ions at the energies that are most useful in ion-beam sputtering or ion-beam deposition processes. Consequently, the electronic energy loss will be neglected in the remainder of this chapter.

The calculation of the nuclear energy loss and theories relating to phenomena induced by momentum transfer (e.g., penetration of ions in a solid [12], back scattering [13], generation of lattice damage [14], and sputtering [15]) are based on the assumption that an incident ion interacts with only one target atom at a time. Thus, the path of the ion is determined by a series of binary collisions described by conservation of energy and momentum. This hypothesis is assumed to be suitable for collisions between target atoms [15].

In the context of these assumptions, the specific energy loss (i.e., the average energy loss per unit path length) is defined by

$$dE/dx = NS_n(E), \tag{1}$$

where N is the number of scattering centers per unit volume and $S_n(E)$ is the nuclear stopping cross section per center or nuclear stopping power.

In order to describe a binary collision and determine the related differential cross section, a realistic evaluation of the interatomic potential is required. Firsov [16] showed that a Thomas–Fermi potential is appropriate for interatomic distances less than 10^{-8} cm. For larger values, a Born–Mayer potential can be used [15]. A calculation of the nuclear stopping

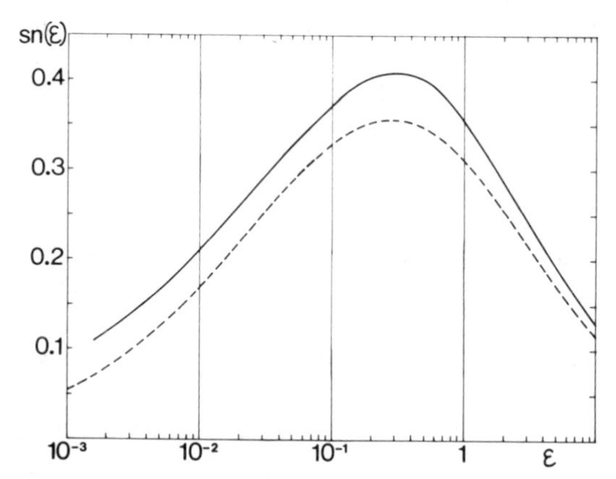

Fig. 5.2. Reduced nuclear stopping power $s_n(\varepsilon)$ as a function of ε. Solid line represents data from Lindhard *et al.* theory (T-F potential) [12]; dashed line represents graph of the Wilson's formula [17]. $s_n(\varepsilon) = \frac{1}{2}\ln(1+\varepsilon)/[\varepsilon + (\varepsilon/385)^{3/8}]$.

power $S_n(E)$ has been performed by Lindhard *et al.* [12] using a Thomas–Fermi potential. Their result is expressed in terms of a universal function $s_n(\varepsilon)$, which represents the reduced stopping power as a function of the reduced energy ε (Fig. 5.2). Here ε is defined by

$$\varepsilon = 32.5 M_2 E/[(M_1 + M_2) Z_1 Z_2 (Z_1^{2/3} + Z_2^{2/3})^{1/2}], \tag{2}$$

where E is expressed in kilo-electronvolts, and $S_n(E)$, expressed in units of electron volt-square Ångstroms, is given by

$$S_n(E) = \{84.75 M_1 Z_1 Z_2/[(M_1 + M_2)(Z_1^{2/3} + Z_2^{2/3})^{1/2}]\} s_n(\varepsilon), \tag{3}$$

where M_2 and Z_2 are the mass and the atomic number of the target atoms, respectively.

B. Reflection and Trapping

Multiple collision processes as encountered during the thermalization of energetic ions [12] or sputtering [15] can be conveniently treated within the framework of transport theory. Cascades of binary collisions are assumed to occur in an infinite medium composed of randomly located atoms. The surface of the target is defined by an arbitrary plane $x = 0$ from which the ions with energy E start at $t = 0$. The probability that an ion will come to rest at a distance x from the starting point is given by the range distribution $F_R(x, E, \cos\theta)$, where θ is the initial angle between the ion beam and the x axis. Here $F_R(x)$ has been observed to have a nearly Gaussian line shape with a maximum at $x = \hat{x}$. Within a reasonable approximation, the projected range $\hat{x}$ can be assumed to be proportional to the ion beam energy and is given by

$$\hat{x} \approx C_1(M_2/M_1)M_2[(Z_1^{2/3} + Z_2^{2/3})^{1/2}/Z_1Z_2]E\,, \tag{4}$$

where E is expressed in kilo-electronvolts, C_1 is a coefficient ranging from 0.5 to 0.8 for useful M_2/M_1 ratios (0.2–10) [18], and $\hat{x}$ is expressed in units of micrograms per square centimeter.

The total reflection coefficient is given by

$$R_e(E, \cos\theta) = \int_{-\infty}^{0} F_R(x, E, \cos\theta)\,dx. \tag{5}$$

When the electronic energy loss can be neglected, R_e is found to depend principally on the M_2/M_1 ratio and on $\cos\theta$ [13]. This evaluation of R_e is shown in Fig. 5.3; it gives reasonable results for light ions or for heavy ions in the kiloelectron-volt energy range, but appears to be inadequate for heavy ions in the electron-volt energy range. This breakdown of the theoretical predictions is illustrated by experimental results deduced from measurements of the trapping probabilities T_r of various ions on tungsten [19] ($T_r = 1 - R_e$). Nevertheless, the large values of R_e obtained for heavy ions at energies less than 1 keV can be explained by the possibility that subsequent ion bombardment could release previously trapped atoms. This hypothesis is supported by the fact that the implanted layer does not exceed 0.5 nm for 1 keV [(see calculation of $\hat{x}$ in Eq. (4)]. It may, therefore, be assumed that the theory is suitable for back-scattered ions that have retained a large fraction of their initial energy. In fact, few experiments have been reported about the energy distribution of reflected species (ions and neutrals). Their energy

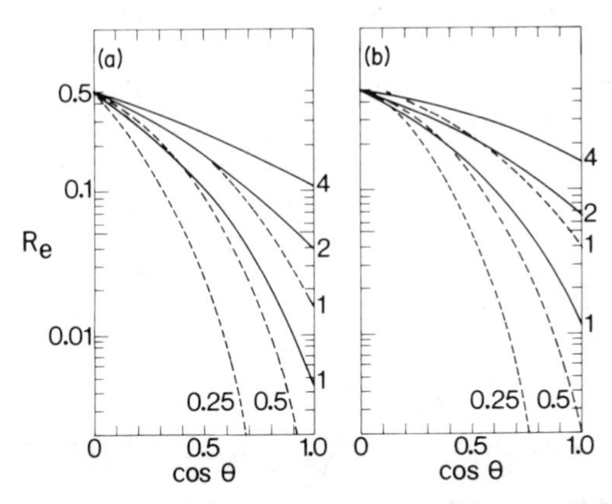

Fig. 5.3. Calculated reflection coefficients R_e as a function of the angle of incidence with the mass ratio M_2/M_1 as a parameter in two domains of reduced energy (a) $10^{-2} < \varepsilon < 2$ and (b) $10^{-3} < \varepsilon < 10^{-1}$. For mass ratios $M_2/M_1 \leqslant 1$, upper bounds are shown as dashed lines. (From results of Bottiger *et al.* [13].)

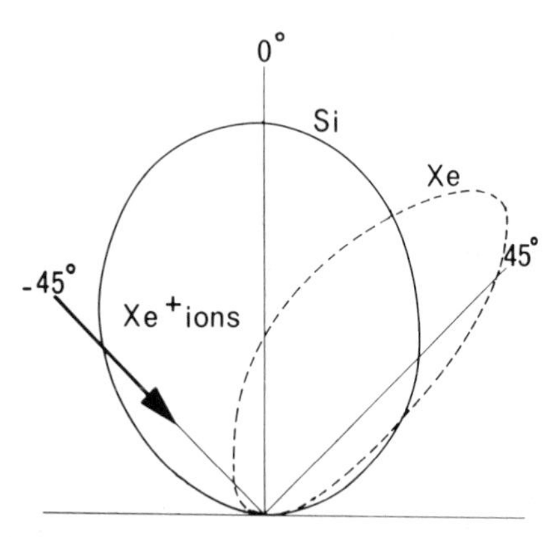

Fig. 5.4. Angular distributions of Si and Xe emitted from a silicon target sputtered with a 20-keV Xe^+ ion beam. (From Pellet *et al.* [23].)

spectra exhibit a large peak at high energy [20] corresponding to binary collisions with surface atoms [21]. The corresponding angular distribution is not well known, but some experiments [22] seem to indicate that specular reflection is more likely. This possibility is supported by the results shown in Fig. 5.4 in which it has been assumed that noble gas incorporation in ion-beam sputtered silicon layers is due to back-scattering processes [23].

C. Sputtering

Successful theories on the sputtering of amorphous and polycrystalline solids have been published in recent years [15, 24]. Sigmund's theory, however, is the most general but is only strictly valid at low fluence and at low beam current on homogeneous targets. The calculation of sputtering yield (i.e., the number of target atoms leaving the solid per impinging ion) is based on a solution of the Boltzmann's equation, which describes the motion of the recoil atoms. The density of energy deposited per bombarding ion near the surface $F_D(E, \theta, x = 0)$, is converted into the density of recoil atoms at $x = 0$. Finally, the fraction of these atoms able to overcome the surface binding forces is selected.

The sputtering yield is then given by

$$S(E, \theta) = (0.042/U_2)\alpha(M_2/M_1, \theta)S_n(E), \tag{6}$$

where U_2 is the binding energy for a surface atom, or the cohesive energy, evaluated by correcting experimentally measured latent heats of sublimation to absolute zero. These corrected values can be found elsewhere [25]. The calculation of binding energies for cubic metal [26] is in good agreement with this evaluation. In Eq. (5), α can be approximated by an energy-independent function that relates $S_n(E)$ to $F_D(E, \theta, x)$ at $x = 0$ [14, 15]. An empirical formula has been proposed [27] to represent the variation of α with M_2/M_1, i.e.,

$$\alpha(M_2/M_1, 0) = 0.15 + 0.13M_2/M_1. \tag{7}$$

Here α increases as a function of the M_2/M_1 ratio as the effect of back-scattered ions becomes more pronounced.

The variation of α with θ is given by $\alpha(M_2/M_1, \theta) = \alpha(M_2/M_1, 0)(\cos\theta)^{-f}$, where f may be taken equal to $\frac{5}{3}$ for nearly equal masses M_2 and M_1 and to 1 for $M_2 \gg M_1$.

For a comparison between theory and experiment, the reader is referred to the compilation of Andersen and Bay [6]. The region of validity of the linear analytic sputtering theory has been discussed in detail elsewhere [28], but, briefly, this model fails when the collective motion of target atoms can no longer be neglected. This thermal effect (or spikes effect) leads to anomalously large sputtering yields for heavy ions on heavy targets (see, for example, the self sputtering of gold [29]) The model also fails at low energy where the cascades of collisions are no longer isotropic. In this energy range, various corrections to Eq. (6) have been suggested [15, 30–32].

For ion-beam deposition techniques, the deposition rate is, of course, maximized when no sputtering event can occur; this is true for ion energies less than the threshold value E_{th}. For self sputtering, $E_{th} \approx 8U_2$ seems to be a reasonable approximation [33].

The energy and angular spectra of sputtered atoms can be deduced from the basic considerations of Thomson [24] and Sigmund [15]. The energy distribution should exhibit a maximum at nearly $U_2/2$ [15] and decrease as E^{-2} [24]. This behavior is confirmed by the experimental results reported by Stuart and Wehner [34] and later by Stuart [35]. These authors showed that the energy spectra of sputtered atoms do not depend on the primary-ion energy (Fig. 5.5). Other measurements prove that excited [36] or ionized [37] sputtered atoms give one important contribution in the high-energy end of the energy spectra. The angular distribution is generally assumed to vary as $\cos\theta$ (see Fig. 5.4); nevertheless the specular direction tends to predominate when the ion energy decreases (Fig. 5.6). This tendency is in agreement with the increasing effect of knock-on sputtering.

The sputtering of a binary alloy target has been discussed by Andersen and Sigmund [39] and Sigmund [6, 40]. For roughly equal masses M_2 and M_3 of the constituents, the approximation of Haff and Switkowski [41] can be used [6]. The sputtering yield S_i of the ith element is then given by

$$S_i(c_i, E) \approx [c_i/U_i(c_i)][c_2U_2(1)S_2(1, E) + c_3U_3(1)S_3(1, E)], \tag{8}$$

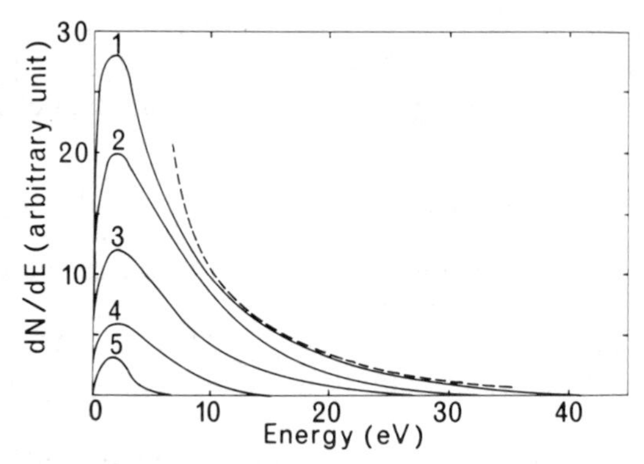

Fig. 5.5. Energy distributions of Cu atoms ejected in the normal direction from the (110) surface for primary ion-beam energies; (1) 1200 eV, (2) 600 eV, (3) 300 eV, (4) 150 eV, and (5) 80 eV. The distribution function obtained at 1200 eV is compared with the E^{-2} law (dashed line). (Data from Stuart and Wehner [34] and Stuart [35].)

where $U_i(c_i)$ is the binding energy of the ith element, which has the concentration c_i in the target. Thus $U_i(1)$ and $S_i(1, E)$ are the binding energies and the sputtering yields of the corresponding pure materials, respectively. An evaluation of $U_i(c_i)$ can be made by using a linear combination of the interatomic binding energies U_{ij} [42].

In the most general case, the bombardment of a virgin alloy, following Eq. (8), results in the nonstoichiometric ejection of the elements. Consequently, this process results in a surface enrichment of the more tightly bonded element. The evolution of an alloy target to the steady state has been extensively studied in the case of a PtSi target [43]. The conclusion of this study is presented in the following section.

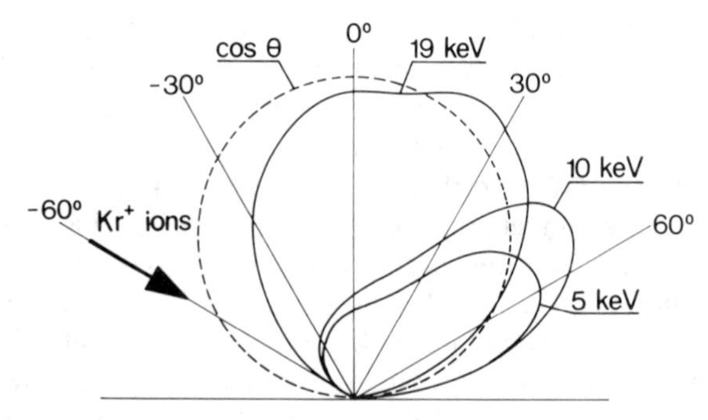

Fig. 5.6. Angular distribution of sputtered material from polycrystalline tungsten bombarded with oblique incident Kr^+ ions of various energies E. These results are compared with the cosine law (dashed line). (Data from Gurmin [38].)

D. Study of a Bombarded Target in the Steady State

For an alloy target, it has been clearly established that the surface concentration develops around values leading to a stoichiometric sputtering of the target in the absence of diffusion. The ratio of concentrations is then given by

$$(c_2/c_3)_{x=0} = (U_2/U_3)(c_2/c_3)_{\text{bulk}}. \tag{9}$$

Through the implantation of the bombarding species, prolonged bombardment of a pure target leads to the formation of a surface compound. The stationary distribution of implanted atoms has been studied as a function of sputter etching and diffusion [44, 45]. The total amount N of implanted ions is found not to depend on diffusion, and N is given by

$$N = [(1 - R_e)j\hat{x}]/v \sim [(1 - R_e)\hat{x}]/S_2(E), \tag{10}$$

where j is the current density of bombarding ions; then $(1 - R_e)j$ is the flux of ions that is effectively trapped in the target and v is the speed of surface erosion. Let us recall that the projected range $\hat{x}$ may be reasonably considered to be proportional to the ion energy E. Therefore, N is (in a first approximation) proportional to $E/S_2(E)$. In a range of beam energy values in which $S_2(E)$ is nearly constant, we have measured N for N^+ and N_2^+ ions bombarding a Si target. The results, shown in Fig. 5.7, allow us to conclude that a molecular nitrogen ion can be considered as two N^+ ions with the same initial velocity; i.e., each has one-half the initial energy. This behavior (the dissociation of the molecular ions or "shrapnel effect") seems to be characteristic of molecular ions in general [47, 48] if we exclude the heavy ions who cause the spikes effect [49]. Let us emphasize that, in the framework of ionized cluster deposition, the average energy of the deposited atoms is determined by this shrapnel effect.

In the case where diffusion is absent, the stationary profile $c_1(x)$ is simply proportional to the primitive of the range distribution $F_R(x)$. Concentration profiles that have this characteristic form are obtained for the case of N_2^+ ions bombarding silicon (Fig. 5.8).

The surface concentration of the implanted atoms can be deduced from preceding considerations about the sputtering of a binary alloy [Eq. (9)]. Assuming that the $(1 - R_e)$ implanted atom is sputtered per impinging atomic ion in the steady state, we can write the concentration ratio at $x = 0$ of the implanted target constituents as

$$(c_1/c_2) = [(1 - R_e)/S_2(c_2, E)][U_1(c_1)/U_2(c_2)], \tag{11}$$

where $S_2(c_2, E)$ is the sputtering yield value obtained when the stationary state has been established. For metal ions, the variation of S_2 with fluence may be very important [51]. For heavy noble gas ions in a light target (i.e.,

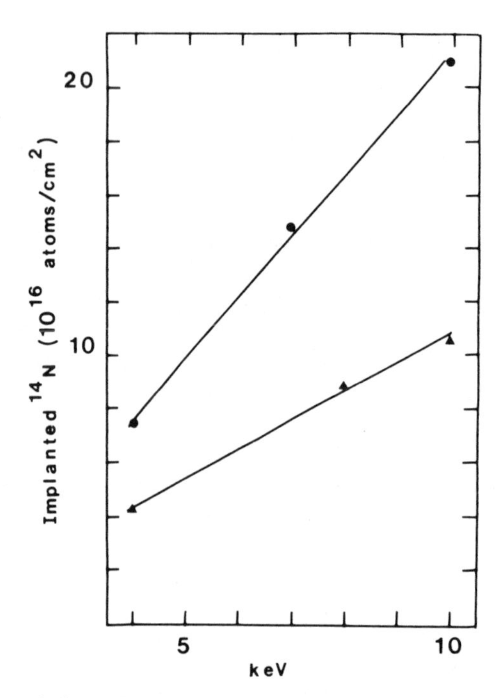

Fig. 5.7. Amount of nitrogen in (111) silicon target implanted with N^+ (●) and N_2^+ (▲) beams as a function of the primary energy. Dose used is 10^{18} ions/cm^2. (From Bouchier *et al.* [46]. Reprinted by permission of the publisher, The Electrochemical Society, Inc.)

Kr^+ on Si), the increase of the stopping power following the implantation of heavy atoms brings about an appreciable increase in the sputtering yield [52]. Reactive gas ions have a strong influence on the binding energy of the target atoms (for oxygen or nitrogen ions on metals, see Steinbrüchel and Gruen [53]).

Reactive molecules from the gas phase produce a similar effect via adsorption on the target. The resulting compound may be volatile at room temperature (this effect is involved in chemical sputtering [7, 54]). On the other hand, for the case of molecular oxygen the sputtering yield decreases with adsorption [55]. In addition, the oxygen atoms are incorporated into the superficial region of the target under the effect of ion-induced diffusion processes [56]. In this case, the steady state is obtained when an equilibrium has been established between the adsorption and the processes of sputtering, ion-induced desorption, and incorporation [56].

Let us emphasize that a study [57] of ion-induced desorption has shown that direct knock-on collisions and knock-on by reflected ions leads to

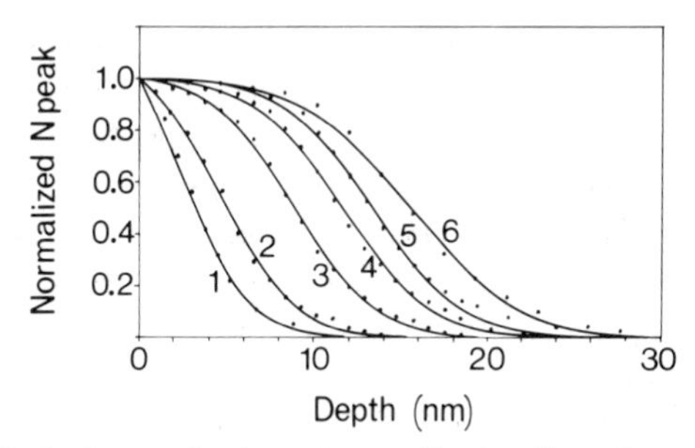

Fig. 5.8. Normalized nitrogen implantation profiles in silicon for various primary beam energies; (1) 0.25 keV, (2) 0.50 keV, (3) 1.0 keV, (4) 1.5 keV, (5) 2.0 keV, and (6) 2.5 keV. The projected range $\hat{x}$ has been located for 2.5 keV. The measurement has been performed by Auger electron spectroscopy. (Data from Malherbe [50].)

surprisingly large desorption yields at low energy on heavy targets, e.g., N atoms adsorbed on a W target.

In the limit of very weak sputtering yields and in the absence of diffusion, the stationary state is difficult to achieve. For a large projected range (light ions at high energy), the phenomena of blistering is able to arise (for example, see [58]). Alternatively, for the metallic ions of weak energy, we are able to obtain the growth of a layer. The beginning of the film growth is subject to the condition that the self-sputtering yield $S_1(c_1, E)$ is never able to reach unity regardless of the surface concentration c_1 of the implanted ion.

III. EXPERIMENTAL—PRINCIPLE AND DESCRIPTION

In Section I, we have seen that the methods studied in this chapter utilize charged particles in a manner that permits superthermal species to be obtained directly (ionized cluster or ion beam deposition) or indirectly (ion-beam sputtering). Consequently, a typical installation contains (1) an ion source, (2) an electrodes configuration for the extraction and acceleration of ions at the desired energy, and (3) a system of transport for the beam up to the substrate or target. In the following discussion, we examine these three aspects with the understanding that the creation of ionized clusters will be presented separately.

A. Production of Ions

For the applications of deposition contemplated here, the most useful ion sources almost exclusively use the process of ionization by electronic bombardment. There exists a wide variety of discharge configurations using either a hot cathode, a magnetic confinement, an electrostatic trapping of ions, etc. It is not possible to treat such a diversity in these few pages and we refer the reader to several review articles that are well documented [59, 60]. However, we may summarize these references, and we attempt to do this in our discussion. It is always possible to produce ion beams with a sufficient intensity for the applications of thin film deposition. This may be accomplished with any gas although it should be noted that the indications of optical quality (emittance) and of ionization efficiency are not always furnished by the authors.

For the less-volatile substances, the situation is less satisfactory, and two solutions may be attempted. For example, the ionization may be performed from the saturated vapor, but the functioning of the ion source at the required high temperature generates technological complications that are sometimes insurmountable. Alternatively, we may introduce a gaseous molecule containing the element under consideration (e.g., organometallic) in the ion source. In the latter case, the transport system is complicated by the required presence of a mass filter.

We wish to attract the attention of the reader to a notion that is not always taken into account, i.e., the ionization efficiency defined as the ionized fraction η of the whole flux (i.e., ions and neutrals) emitted from the ion source. This efficiency will determine the working pressure in all apparatus in which the deposition chamber is placed directly beside the ion source (Fig. 5.1a). This pressure P will be given by the expression

$$P = (2.6 \times 10^{-2} I^{+})/\eta D_{\mathrm{v}} \qquad \text{SI units,} \tag{12}$$

where I^{+} represents the beam intensity and D_{v} is the pumping speed of the pumping system for the gas or the vapor under consideration. With the selection of the type of source and of its working regime η can vary from about 10^{-4} to nearly 1.

B. Extraction of Ions

In the most general case, the ions can be extracted from the plasma in the manner shown schematically in Fig. 5.9a. It is essential to realize that the emissive surface of the plasma can be modeled as a deformable membrane whose equilibrium is established by two important experimental conditions. These conditions are the electric field generated by the extractor and the diffusion of the plasma particles from the bulk into the boundary region.

The emitted current is governed by the law of Child–Langmuir (see Fig. 5.9a), which is valid for monocharged ions and a planar geometry. Figure 5.10 represents the work of Aubert [61] at our laboratory and shows the deformation of the emissive surface under the influence of the extraction voltage (note that the ionic density in the plasma remains constant).

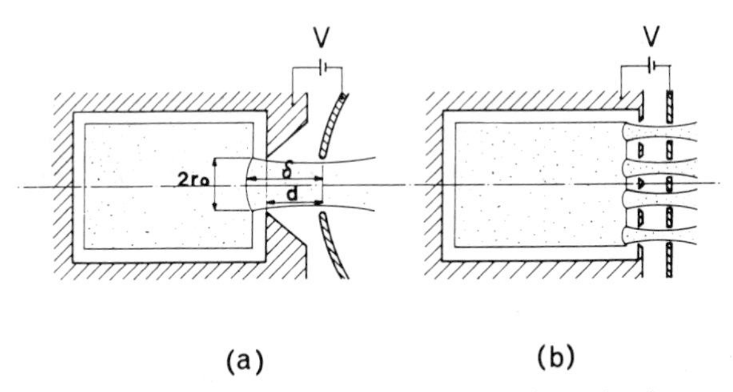

Fig. 5.9. Schematic illustration of ion extraction and beam formation from a plasma for (a) a single aperture and (b) multiple apertures. $I^+ = (4\pi\varepsilon_0/9)(2qZM^{-1})^{1/2}V^{3/2}\delta^{-2}r_0^2$.

Any analysis, even a rough one, of the extraction mechanisms must take into account the divergent lens effect due to the necessary opening in the extraction electrode. This effect causes the beam to have a divergence of around $r_0/3\delta$ (as defined in Fig. 5.9a). Thus r_0/δ must be minimized and experiments show that a r_0/δ ratio less than 0.5 is a good compromise. Under these conditions, the simple application of the formula of Child–Langmuir gives, in the case of argon, a current limit of 0.2 and 6 mA for extraction voltages of 1 and 10 kV, respectively.

This rapid analysis readily shows the difficulty of the extraction of intense beams at a low energy. On the other hand, this problem may be countered in two ways:

(1) extraction under high voltage and subsequent deceleration at the desired energy or

(2) multiplication of the number of extraction apertures, as shown schematically in Fig. 5.9b.

The latter solution restricts the beam to large diameters in which each elementary beam is governed by the law of Child–Langmuir. This type of extraction, initially studied for space applications (ionic propulsion), has been generalized for use in all applications requiring beams of low energy and high

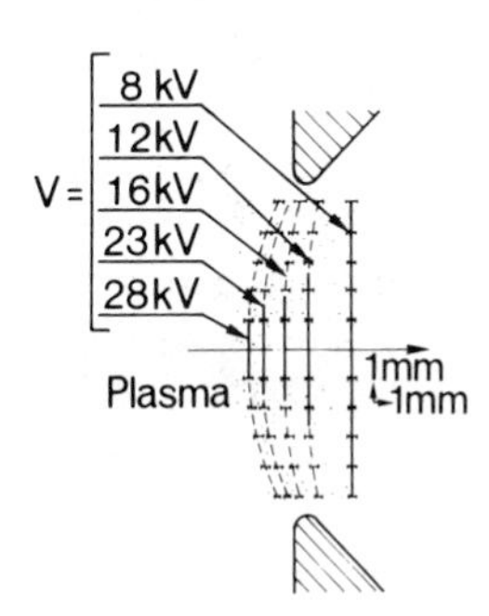

Fig. 5.10. Plasma meniscus for different values of the extraction voltage; $I^+ = 2.5$ mA, $r_0/d = 0.33$. (Data from Aubert [61].)

intensity. A review article by Kaufman *et al.* [59] exhaustively treats the technological problems underlying this type of extraction that permits us to obtain beams in a continuous range of intensities from a few tens to a few hundreds of milliamperes in a range of energy from 0.5 to 10 keV.

C. Transport of Beam

The ions must, of course, travel the distance between the ion source and the target or the substrate (in the case of IBD); this distance can vary from a few centimeters to tens of centimeters as a function of the experimental setup under consideration. The transport of the beam is thus governed by the laws of classical mechanics. In order to take into account all the electrostatic fields exactly, it is necessary to consider the following:

(1) the external electric fields imposed by the electrodes that are located along the path of the beam and that modify the trajectories of the ions and
(2) the internal field due to the coulombic repulsion in the ion beam as given by Gauss's theorem.

For the case in which the internal field is neglected, the trajectories have been calculated by numerous workers [62, 63], and this domain is commonly called electronic optics. At this point, we should recall that the trajectories in an electrostatic configuration are independent of the q/m ratio as long as all other parameters are held constant. Therefore, the interested reader will find specialized literature that describes the characteristics (focal distance, geometric and chromatic aberrations, etc.) of numerous electrostatic lens configurations.

When the electric field generated by the particles in the beam is no longer negligible when compared with the external field imposed by the electrodes of the focusing system, the analysis becomes significantly more complex since space charge phenomenon must be taken into account. The radial electric field due to space charge gives a radial velocity to the ions, and a beam with an initial radius r_0 will see its radius r increase as a function of the path z. Figure 5.11 represents the variation of r/r_0 as a function of z for various conditions of intensity, energy, and mass.

When the divergence has reached a value such that the radial velocity may no longer be neglected with respect to the axial velocity, we must take into account the potential difference that is created between the axis and the periphery of the beam. Therefore, the electric potential at the beam axis may increase to a value that is close to the ion-source potential. When this potential is achieved, a partial reflection of the beam occurs, and the transmission coefficient α is lowered. Figure 5.12 shows this effect for a calculation of α in which the flow is assumed to be unidimensional.

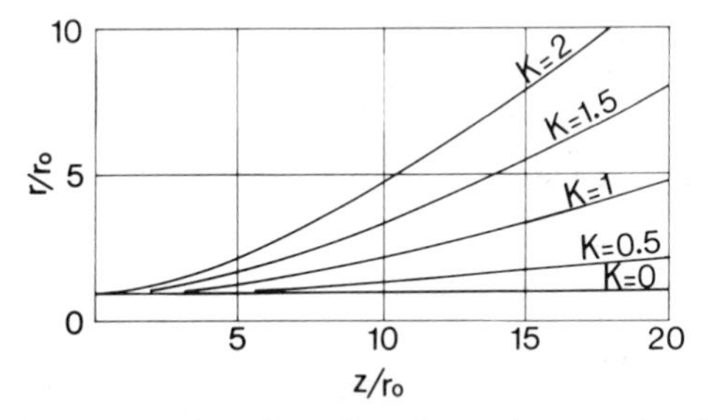

Fig. 5.11. Space charge expansion of a uniform beam of one species of particles in laminar flow as a function of the normalized path z/r_0, where r_0 is the initial radius of the beam: $K = 4.74 \times 10^3 M^{1/4} I^{1/2} V^{-3/4}$ where I is expressed in amperes, V in volts, and M in atomic mass units.

The difficulty of the transport of intense beams at low energy over large distances has been illustrated in a variety of ways in the preceding two figures (Figs. 5.11 and 5.12). We shall study in Section V solutions that may answer these questions of space-charge transport. This problem constitutes the most important limitation of the IBD method. One possible solution, which is rather useful, is the neutralization of charge-space in volume or in intensity. This method, described in Section V, is also applied in IBS deposition in particular for the case of an insulating material target.

D. Principles of Ionized Cluster Beam Deposition

This technique was developed by Takagi *et al.* [65] at Kyoto University, and its principle is represented in Fig. 5.13. Clusters are created by condensation of supersaturated vapor atoms produced by adiabatic expansion through a small nozzle into a high-vacuum region. These clusters

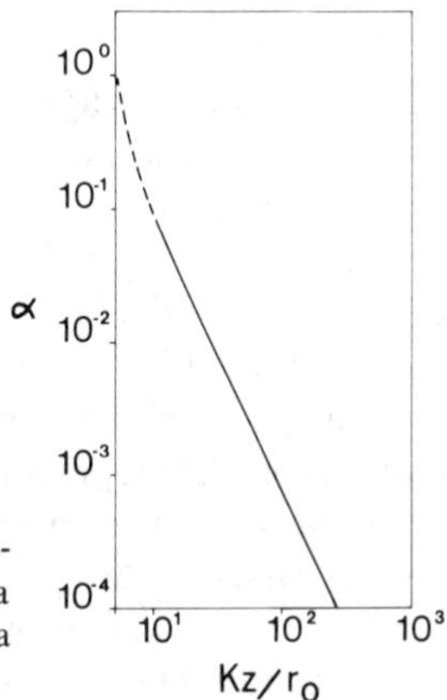

Fig. 5.12. Transmission coefficient α calculated for large-diameter beam assuming the flow to be unidimensional as a function of Kz/r_0. The coefficient K is defined in Fig. 11. (Data from Spiess [64].)

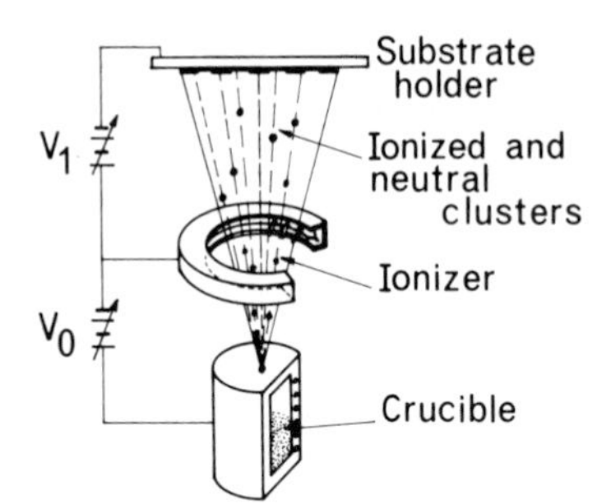

Fig. 5.13. Basic configuration of the ionized cluster source.

comprising N_1 atoms are then ionized with the help of electronic bombardment and accelerated by the voltage V applied between the region of ionization and the substrate. When a cluster dissolves into individual atoms upon its arrival on the substrate (see Section II), an average energy equal to qV/N_1 is transferred to each atom of the cluster.

Consequently, a 1-μA beam that is composed of single ionized clusters, where each cluster contains 10^3 atoms and is accelerated at 10 keV is equivalent to a 1-mA beam of monoatomic ions with an energy equal to 10 eV. We can evaluate the coefficient K for these two different beams (see Fig. 5.11). In the case of silicon

$$K_{\text{cluster beam}} = 6.13 \times 10^{-2},$$

$$K_{\text{monoatomic ion beam}} = 61.3$$

From this brief comparison, using Fig. 5.11, the transport of the monoatomic ion beam is not possible in practice.

The average energy of the individual atoms incident on the substrate is given by

$$\langle E \rangle = (1 - \eta)kT + \eta(qV/N_1), \tag{13}$$

where η is the ionization efficiency of the ionizer.

An examination of the formula shows that η, V, and N_1 are permitted to modify $\langle E \rangle$. We examine N_1 first. The leading work on silver, which is represented in Fig. 5.14, shows the role of the vapor pressure in the crucible on the size of the clusters. These results are only in qualitative agreement with a theoretical analysis based on the Gibbs–Thomas equation. The resulting discrepancy arises from the determination of the surface tension and further development of the theory is necessary. The ionization efficiency η depends on the intensity of the electronic bombardment and on the configuration of the ionizer. Thus η may be relatively important, from several percent to several tens of percent arising from the high value of the ionization cross section. (The factor of multiplication relative to a single atom is 10^2–

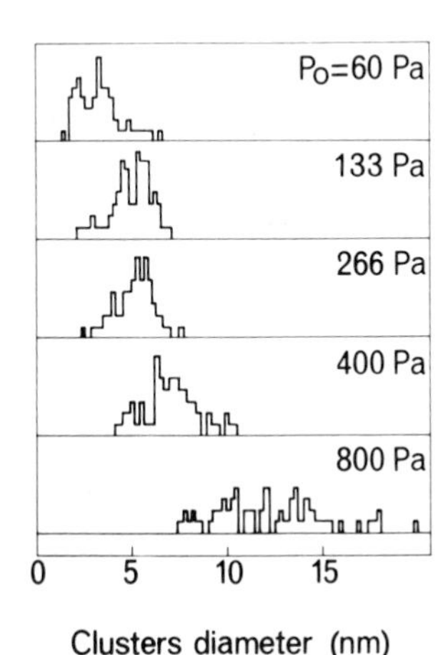

Fig. 5.14. Histograms of the cluster diameter distribution obtained by TEM observations. The pressure P_0 in the crucible has been varied from 60 to 800 Pa. (Data from Yamada [66].)

10^3 for clusters composed of 10^2–10^4 atoms.) It is important to note that the clusters are indeed the carriers of one single charge since a double ionization causes the dissociation of the cluster by coulombic repulsion.

In conclusion, the energies of several tenths of electron volts to several tens of electronvolts may, therefore, be obtained with deposition rates that are able to attain 1 μm/mn on large surfaces (multiple nozzles) and in a large range of materials. Other materials of interest may possibly be grown by the use of reactive gas (O_2, N_2, C_xH_y, ...) in the deposition chamber.

E. Experimental Setup Examples

Before examining the properties of thin layer materials elaborated by ion processes, it seems interesting to give for each method an example of each experimental apparatus. We have chosen to present the installations in which the *in situ* analysis methods are numerous and that permit us at best to reach the deposition mechanisms.

1. Ion-Beam Sputtering

Figure 5.15 illustrates the design of apparatus developed at the authors' laboratory. The vacuum system comprises two major parts: the ion beam generation unit and the deposition chamber. The generation unit consists of the duoplasmatron ion source that delivers a beam of 0.1–10 mA at an energy of 10–25 keV. By means of three Einzel lens units, the beam may be focused on the target with a diameter of between 10 and 50 mm. A great deal of caution is given to the materials in contact with the plasma or the beam and these elements are, in the more critical zones, realized in the same

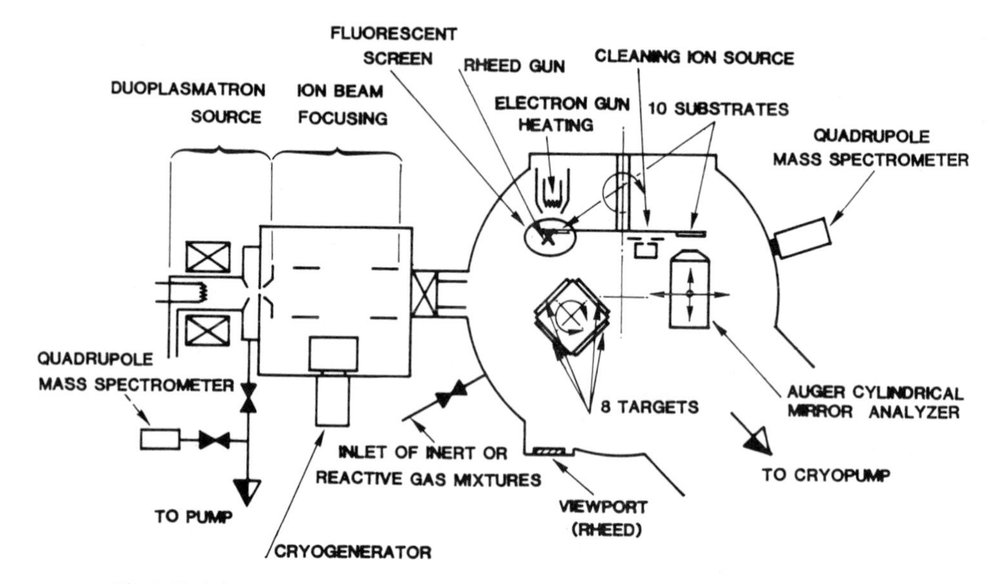

Fig. 5.15. Schematic diagram of the ion-beam sputtering system of Gautherin *et al.* [67].

material as the target. The deposition chamber, of large diameter (600 mm), receives a target holder at 8 positions and a heated substrate holder at 10 positions. It is pumped by a cryopump (5000 liters/s for Ar). It is equipped with one reflection high-energy electron diffraction (RHEED) apparatus, that permits *in situ* determination of the structure of the thin films. The chemical composition of the films can be analyzed by Auger electron spectroscopy during intervals of the deposition process. The cleaning of the substrates may be performed *in situ* by thermal treatment or by ion cleaning with a small source of ions. In this equipment, the thin films are deposited under a pressure of the order of 3×10^{-5} Pa, which consists primarily of the noble gas of the ion beam (99%).

2. Ion-Beam Deposition

Figure 5.16 represents an installation realized at Philips Research Laboratories for beams of Ag^+ and Ge^+. Briefly, the apparatus consists of three differentially pumped vacuum chambers. The first, the source chamber, contains a hot cathode ion source, a combined extracting/focusing ion lens, and a Wien type in line mass filter. The second chamber contains a second ion lens and serves as a differential stage. In the third chamber, or the UHV deposition chamber, facilities are present for Auger electron spectroscopy and reflection high-energy electron diffraction.

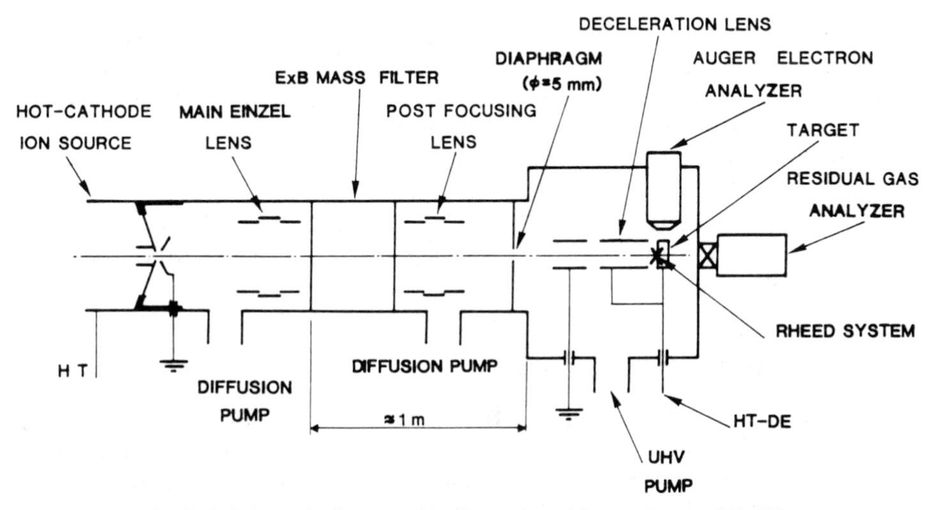

Fig. 5.16. Schematic diagram of ion-beam deposition equipment [68, 69].

A beam of 100 μA is extracted at a 15-keV energy. It may be decelerated at 20 eV, maintaining an intensity of 20 μA. The pressure in the deposition chamber is of the order of 6×10^{-8} Pa (turbomolecular pump supplemented by a liquid nitrogen cooled titanium sublimation pump). The standard *in situ* substrate (Si) preparation involves ion-bombardment cleaning at 2 keV with Ar^+ or Ne^+. This step is followed by annealing between 800–850°C.

3. Ionized Cluster Beam Deposition

The installation chosen and represented in Fig. 5.17 is that developed by Futaba Corporation for heteroepitaxial growth of GaP films on Si and Ge substrates. Gallium and phosphorus are heated at about 1500 and 500°C,

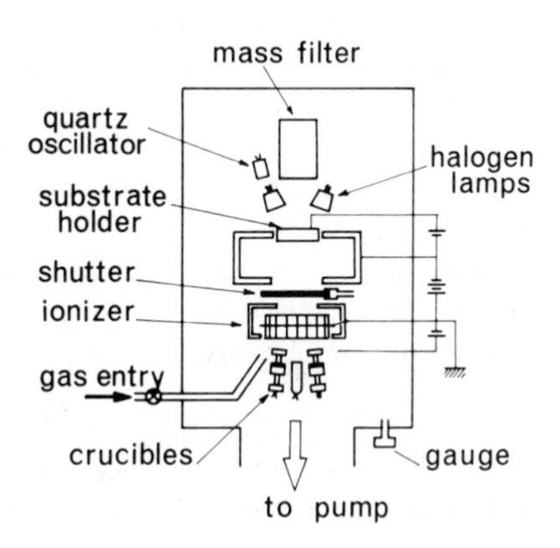

Fig. 5.17. Schematic diagram of the ionized cluster beam deposition equipment [70].

respectively, in separate crucibles with small nozzles (1 mm in diameter). The pressure in the bell jar is kept at about 10^{-5} Pa under no load condition and at 2×10^{-3} Pa during the deposition. The vapor pressure of phosphorus is measured with a quadruple mass spectrometer.

Prior to deposition the substrate is etched by argon ion sputtering and baked at 600°C. Good heteroepitaxial films of GaP/Si could be obtained with a deposition rate $\simeq 200$ nm/min under the conditions of the accelerating voltage $V_a = 2$ at 4 kV, the electron current for ionization $I_e \simeq 100$–200 mA and the substrate temperature $T_s = 550$°C.

IV. ION-BEAM SPUTTERING DEPOSITION

In this case, the free superthermal atoms are produced by ion-beam sputtering of a solid target (Fig. 5.1a). The films that are obtained are, therefore, a result of the respective contributions of all the emitted species from the bombarded target and from the gaseous phase. The first part of this section establishes the relative flux evaluation of the various sample constituents in conjunction with the discussion presented in Section II. The experimental results presented in Sections IV.B and IV.C are discussed in the context of these points.

A. Flux of Species Incident on the Growing Film

For a given target and substrate disposition, the flow rate ϕ_i of the ith component ejected from the target is given by

$$\phi_i = A_i \int_s j(x, y) f_i(\theta) \Omega(x, y)\, dx\, dy \tag{14}$$

where s is the surface of the target, $j(x, y)$ the ionic current density at point (x, y) of the target, $f_i(\theta)$ the angular distribution of the ith constituent, and $\Omega(x, y)$ the solid angle under which the substrate is seen from point (x, y). For the sputtered atoms, A_i must be equal to the corresponding sputtering yield S_i (see Section II.C), where $f_i(\theta)$ may be taken as equal to $(\cos\theta)/\pi$. For the reflected ions, $A_i = R_e$, where R_e is the total coefficient of reflection (see Section II.B); in this case, we possess little information on the distribution function $f_i(\theta)$.

If the profile of the current density can be approximated by an analytic expression (e.g., a Gaussian), the preceding equation may be cast into a simplified form, namely

$$\phi_i = A_i K I, \tag{15}$$

where I is the total ionic current expressed in units of ions per second, and K a geometric coefficient. For a distance of 4 cm between the target and the

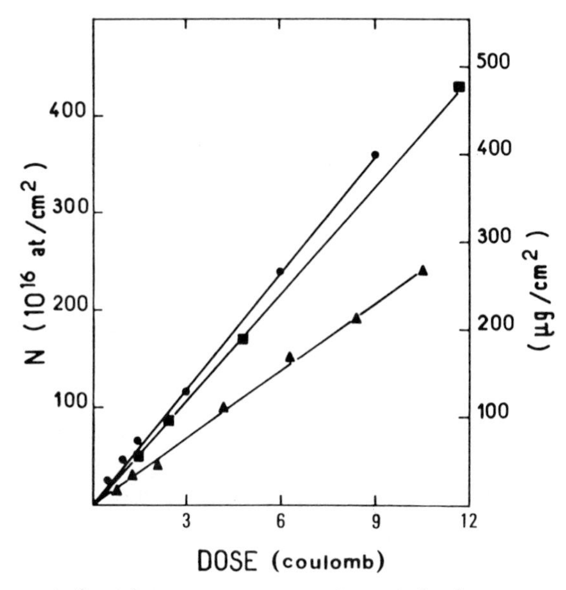

Fig. 5.18. Amount of metal atoms per square centimeter in ion-beam sputtered Nb–Ti films, as a function of the bombardment dose. This result has been obtained for 20 keV Ar^+ beam on a Nb–Ti target. The deposition yield, defined as the deposited amount per square centimeter and per coulomb, decreases with the beam current; 1 mA (●), 4 mA (■), and 7 mA (▲). (From Bouchier *et al.* [71].)

substrate, K varies from 1 to 2×10^{-2} cm^{-2} as a function of the ion beam focusing. This is illustrated in Fig. 5.18 in which we see that the effect of defocusing due to the space charge phenomena (see Section III) decreases the deposition yield.

In the case in which the angular distribution of the back-scattered and resputtered ions is not very different [a limit that is not always verified (see Fig. 5.4)], the flux ϕ_1 may be evaluated by satisfying the expression $A_1 = R_e + S_1 = 1$. The ratio ϕ_1/ϕ_2 is then equal to $1/S_2$. The impurities due to the desorbed or sputtered species in the ion source and contained in the ion beam may be incorporated in the samples with the same ratio (an impurity concentration β in the beam results in a contamination of the layers equal to β/S_2).

The effect arising from intentionally introduced gaseous molecules or from residual atmosphere is determined by two processes:

(1) Direct adsorption on the layer that is being grown: The corresponding flux ϕ_k is given by the kinetic theory of gases; for example, at room temperature, we obtain

$$\phi_k' = 1.54 \times 10^{19} M_k^{-1/2} P_k \tag{16}$$

where M_k is the molecular mass of the kth molecule and P_k the corresponding partial pressure is expressed in pascals.

(2) Through the adsorption on the target and subsequent bombardment of the resulting adsorbate.

For a weak value of the partial pressure ($\phi_k' \ll j$), the corresponding contribution is given by the following expression, which is strictly valid for a uniform current density, namely,

$$\phi_k'' = a_k'' \phi_k' \quad \text{K s} \tag{17}$$

where a_k'' is the sticking probability of the kth molecule on the bombarded target. We note that ϕ'' depends only on the focusing of the ion beam.

In the opposite case of a relatively high pressure ($\phi_k' \gtrsim j$), ϕ_k'' reaches a value of saturation. This occurs, following simple arguments, at a pressure value that is proportional to the current density j.

Finally, the sample composition may be evaluated by assigning each incident species a sticking probability a_i. The concentration ratios are given by

$$c_i/c_j = (a_i \phi_i / a_j \phi_j). \tag{18}$$

B. Ion-Beam Sputtering Deposition

Important initial work has already been performed for determining the potential for using IBSD techniques to develop metallic, dielectric, and semiconductor films. A summary of the corresponding published works is given in Table 5.1.

The sputtered films are found to be more adherent to the substrates than the evaporated films [101, 102]. This improved adhesion may be attributed to the more important penetration of sputtered particles into the substrates and/or the removal of adsorbed gases from the substrate surface and/or the modified film nucleation due to the higher energy of sputtered particles. In IBS films, Schmidt *et al.* [72] and Castellano *et al.* [74] observed lower stress than evaporated films. The stress was found to change from tensile to compressive when the films were exposed to the direct impingement of some primary ions [74]. However, Kane and Alm [77] observed no significant difference in stress between IBS films, rf sputtered films, and evaporated films of 0.1-μm thickness. In these experiments, it should be noted that the IBS films were made under high pressure, 1×10^{-2}–6.5×10^{-2} Pa and so near the deposition conditions of the rf sputtering technique.

Ion-beam sputtered superconducting films of NbTi were found to possess the T_c value of the target material [71]. The value of T_c decreased when the oxygen concentration in the film was increased above 1.5 at. % [83]. The rare

TABLE 5.1

Ion-Beam Sputtering Deposition[a]

Material		Nature and energy of ion beam	Growth rate (nm/min)	Substrate temperature T_s (°C)	Growth pressure (Pa)	Background pressure (Pa)	Remarks and comments	References
Layer	Substrate							
Au, Ta	Pt, Al_2O_3	Ar, Xe, 1.2 keV	30 (Au); 12.5 (Ta)	30	2.7×10^{-2}	6×10^{-5}	Adhesion: IBSD ≡ dc or rf sputtering; IBS film stress < evaporated film stress; Ta: β structure; $\rho = 215$ ohm cm; [RG] = 100 ppm–1 at. %	[72]
Ta		Ar, 2 keV	2.5	100	6×10^{-3}	10^{-6}	bcc: structure preferred (111)	[73]
Ni, Al, Ni_3Al Au	Si (100) + SiO_2 (0.5 μm)	Ar, 0.15–3 keV	4.5–8.5	50–120	10^{-2}	2×10^{-4}	Ni_3Al: composite target; IBS film stress < evaporated film stress; [M.I.] ≃ 2 at. %; [G.I.] ≃ 0–2 at. %	[74]
Ta	Glass	Ar, 5 keV	1–5	80–280	6.5×10^{-2}	10^{-4}	At $T_s = 80$–160°C, $\rho_{ac} =$ 210–195 ohm cm; TCR. very low	[75]
Ag–SiO_2	NaCl, Ge	Ar, 10 keV		20–400	10^{-2}		Cosputtering of Ag and Si targets; At [Ag] = 92–55 wt. %, $\rho = 10^{-6}$–10 ohm m, TCR = 600–3500 ppm/K	[76]
Au, Si, Al, Ta, W, NiFe	Si	Ar, 0.5 keV	2–12		6.5×10^{-3}–5.2×10^{-2}	3×10^{-5}	IBS film stress ≃ evaporated film stress and rf sputtered film stress; NiFe magnetic properties: IBS film ≃ rf sputtering film and sputtering gun film	[77]

(continues)

Table 5.1 *(continued)*

Material		Nature and energy of ion beam	Growth rate (nm/min)	Substrate temperature T_s (°C)	Growth pressure (Pa)	Background pressure (Pa)	Remarks and comments	References
Layer	Substrate							
Ag, Au, Co, Cr_3C_2, Pt, Ni, Cr, Mo, W, Ta_5Si_3, B_4C, CrB_2, SiC	H-13 Steel	Ar, 1.0 keV	1.5–12		4×10^{-2}		Adhesion; IBSD $\geqslant$ rf sputtering deposition and ion plating	[78]
C	C	Ar, 6.0 keV	1		4×10^{-3}		Target: graphite; Film: mixture of graphite and diamond structure; At $\lambda = 350$–1000 nm LT (IBS film) = 20–75 %; LT (Laser evaporated diamond) = 40–75 %	[77]
$CuInSe_2$	SiO_2, NaCl, Cu, Al_2O_3	10 keV		20–260	5–10×10^{-3}		Chalcopyrite structure; $E_g =$ 1.03 eV	[80]
Cr, Mo, Ti, W, Zr, Nb, Ru, Ta, V	Glass, Al_2O_3	Ne, Ar, Kr, Xe, 6 keV	5–30	RT	2.5×10^{-2}	6.5×10^{-6}	Cr film superconducting up to 1.52 K; T_c (IBS film) > T_c (bulk) for Mo, Ti, W, Zr; T_c (IBS Film) < T_c (bulk) for Nb, Ru, Ta, V; Xe → T_c max Ne → T_c min	[81, 82]
NbTi	Si	Ar, Xe, 20 keV		200	1–3×10^{-4}	6.5×10^{-6}	$T_c = 9.7$ K; [O] = 0.6–1.7 at. %; [N] = 0.4 at. %; at P(A) = 10^{-4}–6.10^{-3} Pa; [A] = $\simeq$1 at. %	[71, 83]
Nb	SiO_2	Ar, 6 keV	0.5–1	50–60	1.3×10^{-2}		At P(GI) = 5×10^{-6}–1.3×10^{-4} Pa; $T_c = 7.5$–4. K; T_c max = 8.5 K for laser annealed IBS film	[84]

Nb_3Ge	Al_2O_3	Ar, Kr, 20 keV		500–1000	2×10^{-5}	1.5×10^{-7}	Cosputtering of Nb and Ge targets; T_c max = 8 K at 760°C > T_s > 600°C; T_c max = 16 K for annealed IBS film at 750°C	[67, 85]
Si *p*, *n*	Si *n*, *p*	Ar, 12 keV	40	700–900	10^{-2}		Epitaxial growth above T_s = 730°C; Crystallographic defects	[86]
Si *p*, *n*	Si (100) or (111)	Ar	5–7	450–900	8×10^{-4}	3×10^{-5}	Epitaxial growth above T_s = 650°C; Metallic impurities (Fe, W, Ta); [Ar] = 0.1 at. %	[87–89]
Si *n*, *p*	MgO ∝ Al_2O_3 (100) or (111)	Ar	5–7	450–900	8×10^{-4}	3×10^{-5}	Epitaxial growth above T_s = 850°C; numerous defects; at p = 4×10^{15}–6×10^{17} cm^{-3} (Al doped film with d > 2 μm); μ = 15–80 % μ bulk; at n = 10^{18}–10^{19} cm^{-3} (Sb doped film with d > 2 μm); μ = 15–70 % μ bulk	[87–89]
Si	Si *n* (111)	Ar, 0.6 keV	0.5	800	6×10^{-2}	10^{-7}	Study of growth kinetics	[90]
Si	Si (100)	Ar, Xe, 5–8 keV	0.3–1.9	RT–400	6.5×10^{-4}	2.7×10^{-5}	[Ar] = 0.5–1.2 at. %; [Xe] = 0.09–0.13 at. %; [MI] = 0.2–7 at. %	[91]
Si	Si (100)	Ne, Ar, Xe, 20 keV	5	RT–900	2×10^{-5}	2×10^{-7}	Single crystal growth above T_s = 250°C; Kikuchi lines at $T_s \geqslant$ 700°C; B doped target; at p = 4×10^{16} cm^{-3} and d = 0.4 μm; μ = 80 % μ bulk; [Ar] = 5×10^{-3} at RT	[92–94]
Si (100)	Al_2O_3 (1$\bar{1}$02)	Ar, Kr, 20 keV	2–10	RT–900	5×10^{-5}	5×10^{-7}	Epitaxial growth above T_s = 750°C; B-doped target; at *p* = 5×10^{16} cm^{-3} and *d* = 0.5 μm–μ = 30 % μ bulk	[92]

(*continues*)

Table 5.1 (*continued*)

Material		Nature and energy of ion beam	Growth rate (nm/min)	Substrate temperature T_s (°C)	Growth pressure (Pa)	Background pressure (Pa)	Remarks and comments	References
Layer	Substrate							
GaAs (100)	GaAs (100)	Ar, 20 keV	10	300–500	5×10^{-5}	5×10^{-7}	Target: AsGa film: Ga/As = 1 ± 10^{-2}; epitaxial growth above $T_s = 450$°C; Kikuchi lines at 520°C	[95]
Ge (100)	GaAs (100)	Ar, 20 keV	10	300–500	5×10^{-5}	5×10^{-7}	Epitaxial growth above T_s = 300°C	[95]
GaAs	NaCl, Si, GaAs	Ar, 7 keV				10^{-5}	Epitaxial growth above. T_s = 600°C for GaSa/Si, T_s = 420°C for GaAs/GaAs and T_s = 400°C for GaAs/NaCl	[96]
i-C	Si (111)	1 keV				10^{-4}	Electrical properties of i-C/Si interface	[97]
Si	SiO_2	Ar, 10 keV	4–16		10^{-3}	3×10^{-5}	Grapho-epitaxy; some crystallites observed; after annealing: large single-crystal area	[98]
PTFE	Glass	Ar, Kr, 3–5 keV	80–100		0.07–0.1		10^2 CF_2 groups etched/incident ion; quasi-crystalline structure friction coefficient IBS film ≃ friction coefficient bulk	[99]
PTFE, FEP, CTFE	Glass	Ar, 0.5–2 keV	40–330		$5\text{–}9 \times 10^{-5}$		5–100 CF_2 groups etched/incident ion; quasi-crystalline structure; static friction coefficient 0.3–0.4; (bulk 0.088–0.13); LT (FEP film): 0.81 at $\lambda = 0.33$ μm and 0.95 at $\lambda = 0.60$ μm	[100]

[a] Symbols and abbreviations: [X], concentration of the element X; E_g, energy band gap; TCR, temperature coefficient of resistance; T_c, critical temperature; T_s, substrate temperature; LT, light transmittance; RG, rare gas; GI, gaseous impurity; and MI, metallic impurity.

gas with thermal energy is found to have no influence on the superconducting properties of the thin films. However, at high pressure, the diffusion of the primary beam may result in a non-negligible flux of energetic particles impinging on the growing film. This effect is, of course, strongly dependent on the beam–target–substrate configuration. All of these points may explain the small values of T_c found by Schmidt *et al.* [81] for some materials. Silicon homo- and heteroepitaxy have been the most extensively studied semiconductor films by the IBSD technique. The beginning of monocrystalline growth is observed at low temperature ($\approx$250°C) [92] (Fig. 5.19). The room temperature electrical properties of the bulk material are obtained when the deposition temperature reaches 700°C [87] and for thickness less than 1 μm [93] (Fig. 5.20). In TEM analysis, Weissmantel *et al.* [89] have observed the presence of numerous defects ascribed to the action of impinging energetic neutrals or ions coming from the back-scattering of primary ions and from the energetic tail of the sputtered particle spectrum.

In the IBSD technique, two methods have been used to dope the semiconductor films: cosputtering of a composite target [89] and sputtering of a doped target [92]. By comparison with evaporation techniques, these processes permit the use of all doping materials including those of low vapor pressure and the improvement of the sticking coefficient of the volatile element. In addition, these methods do not require the annealing process, which is needed after epi-implantation of the doping material in molecular beam epitaxy (MBE). In the IBSD technique, when a GaAs target is sputtered a film of equal stoichiometry, to within a few percent, is achieved [95], while in the MBE case an arsenic flux three times larger than the gallium flux is necessary to obtain the same result.

1. Accidental Contamination of the Layer

The majority of the work performed with the IBSD technique shows the presence of metallic or gaseous impurity concentrations up to a few atomic percent in the layers [74, 91].

In the main cases, the metallic impurities are generated in the ion source by sputtering of the internal electrodes by the plasma discharge and, sometimes, by evaporation of the hot cathode. After having been ionized in the plasma discharge, these impurities are extracted from the source like the rare-gas ions and subsequently contaminate the film through the target. The extraction and beam focusing electrodes may also undergo the sputtering by the primary ions [89]. In this case, the sputtered elements will come to deposit on the target or directly on the substrate. However contaminations from the source could be reduced by operating at low discharge voltage to minimize the sputtering process and by using materials that have low

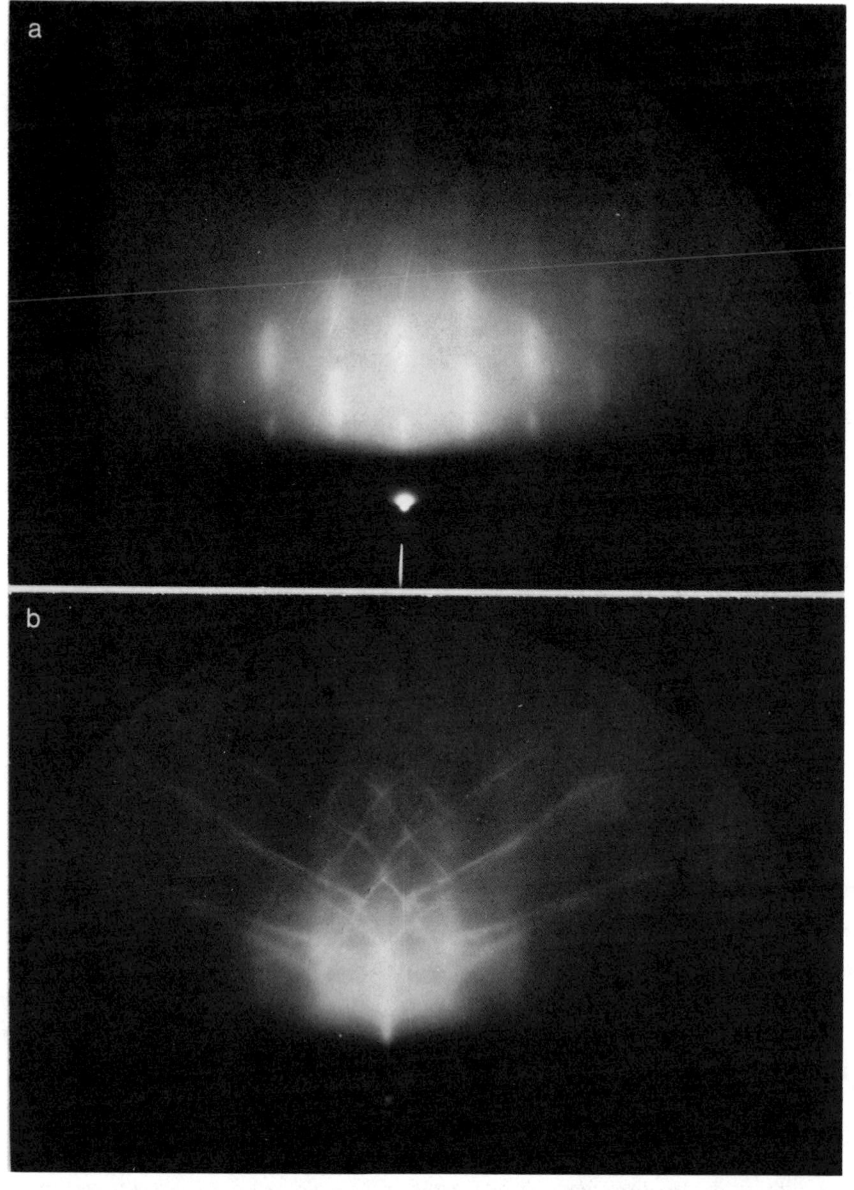

Fig. 5.19. RHEED patterns of 0.5-μm IBSD Si films on (100) Si substrate deposited at substrate temperature (a) $T_s = 250°C$ and (b) $T_s = 850°C$. (From Schwebel *et al.* [94].)

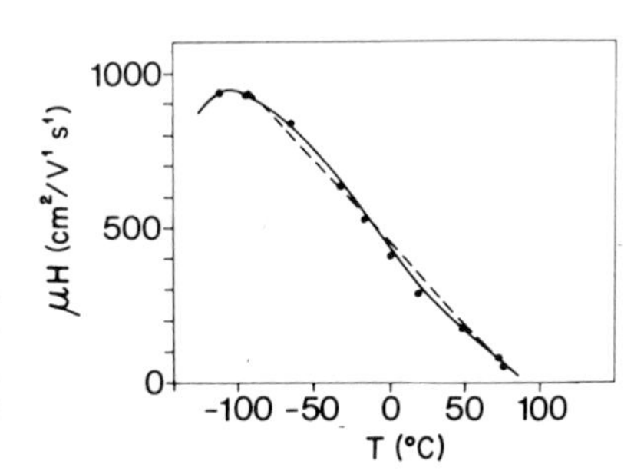

Fig. 5.20. Temperature dependence of the Hall mobility for 0.6-μm IBSD Si epitaxial film deposited at $T_s = 710°C$ ($p = 2 \times 10^{16}$ cm^{-3}). Dashed line represents bulk. (From Schwebel *et al.* [94].)

sputtering yield or are compatible with the desired film. Schwebel *et al.* [94] used an ultra-high-vacuum duoplasmatron source for making silicon epitaxial layers. In this source, the electrode undergoing an erosion is made, in part, with bulk silicon.

Another solution for the elimination of impurities consists of the mass analysis of the ion beam before its interaction with the target. However, due to the space charge phenomena this latter method is difficult to achieve if we want to conserve a large growth rate of the film. In the IBS films the presence of gaseous reactive impurities is often critically correlated to the vacuum performances of the experimental apparatus.

In conclusion, all the results show the important role played by the experimental apparatus. Therefore, with a judicious choice of the ion source and its operating regime, the experimental configuration, and the technology used, the sources of contamination are strongly minimized.

2. Incorporation of Rare Gas in the Layers

The presence of rare gas of a few atomic percent in the layers may be attributed to numerous origins. Since rare-gas atoms with thermal energy possess very low sticking probability, even at room temperature the detected concentration can only be attributed to the trapping of energetic impinging atoms.

Two mechanisms are able to provide these particles: the back-scattering of the primary ions by the target (the importance of this process has been discussed earlier) and the sputtering of implanted primary ions in the target that we now discuss.

The concentrations found in the films are generally smaller than those found in the targets [94]. Therefore, it is clear that only the sputtered rare-gas atoms that belong to the energetic tail of the sputtered particle spectrum with an energy larger than the threshold energy E_s will be incorporated. Comas *et al.* [103] found $E_s \simeq 20$ eV for trapping argon in silicon while Winters *et al.* [104] found $E_s \simeq 100$ eV for argon in nickel.

In conclusion, the inevitable presence of rare gas in IBS layers demonstrated the interaction of energetic particles with the growing film. We may assume that these rare-gas atoms and target atoms that belong to the high-energy part of the sputtered particle spectrum are the cause of certain observed results in the demanding domains of semiconductors and superconductors.

C. Reactive Ion-Beam Sputtering Deposition

Reactive ion-beam sputtering deposition (RIBSD) techniques differ from the basic method by the fact that reactive molecules are introduced during the process of deposition. This injection is accomplished either in the ion beam or in the gaseous phase.

A summary of a large number of works is presented in Table 5.2. In this list, the studied materials include a hydride (a-Si : H), numerous oxides (ITO, Y_2O_3, SiO_2, . . .), and several nitrides (AlN, Si_3N_4, . . .) excluding chloride and fluoride compounds. The domain of application of the work reported in this chapter is principally found in the realm of advanced technologies such as optics, optoelectronics, and microelectronics. Several of the tabulated results are particularly interesting: the conversion efficiency of ITO/semiconductor solar cells (12% for Si and 14% for InP) [118] and the property of Y_2O_3 to act as protective layers due to its resistance to laser-induced damage (up to 2 GW/cm^2) [112] and atmospheric agents. We also note the advantageous characteristics of amorphous hydrogenated silicon, which may be doped by cosputtering [105], and of metal–metal oxide composites, which are favorable for optical recording with an erasure time less than 10 μs [122]. Furthermore, practically ideal MIS structures on silicon may be obtained through a treatment by hydrogen [129]. Finally, it is possible to deposit silicon nitride at room temperature with satisfactory dielectric properties [133].

In the following, we discuss the mechanisms of incorporation of a variety of reactive species. The context of this discussion will follow the semi-quantitative theory, which has been presented in the preceding sections (Sections II and IV.A).

1. Incorporation of Hydrogen

Two cases are to be considered: one with the hydrogen ions in the beam and the other with hydrogen introduced in the gaseous phase. In the first case, the incorporation of hydrogen is clearly explained by the dissociative interaction (reflection and resputtering) of H_2^+ with the target [105, 110].

TABLE 5.2

Reactive Ion-Beam Sputtering Deposition[a]

Material		Ion-beam energy (keV)	Growth rate (nm/min) and/or ion-beam current (mA)	Gases in ion source and reactive gas concentration (%)	Pressure during growth (Pa)	Background pressure (Pa)	Remarks and comments	References
Layer	Substrate							
a-Si : H	Glass	10	10–15 nm/min, 10 mA	Ar + H_2, 50%; Ar, 100%	H_2, 0 or 10^{-3}	10^{-5}	[H] = 4%, $E_g \simeq 1.3$ eV; [H] = 1.4%, doping by cosputtering of Al	[105]
a-Si : H	Quartz, C	0.5	10–50 nm/min	Ar + H_2, 0–75%	1.3×10^{-2}	6.6×10^{-5}	Resistivity up to 10^9 Ω cm, for the higher H concentration	[106]
a-Si : H	Glass	1–5	4–6 mA	Ar + H_2, 0–90%	5×10^{-2}	4×10^{-4}	$1.4 < E_g \leqslant 1.9$ eV (optical) as a function of [H] (0–30%)	[107]
a-Si : H	—	1.0	30–70 mA	Ar + H_2, 0–90%	—	—	Effect of a complementary ion bombardment on the layer morphology	[108]
a-Si : H	Glass	0.8	5 nm/min	Ar + H_2, 50% or 0%	H_2: 0 or 2×10^{-2}	10^{-5}	Measurement of the complex refractive index $n - ik$	[109]
a-Si : H	—	1.0	17 nm/min	Ar + H_2, 0–80%	H_2: 1.3×10^{-2}	—	Discussion on the incorporation mechanisms of H	[110]
Ta_2O_5	Glass	2.0	2.5 nm/min	Ar + O_2	$6–12 \times 10^{-3}$	6×10^{-6}	High optical loss of the deposited guides	[111]
TiO_2–Y_2O_3, Al_2O_3	—	1.0	40 mA	O_2, 0–100%	1.3×10^{-3}	1.3×10^{-4}	High damage threshold from laser irradiation ($\simeq 2$ GW/cm^2)	[112]

(continues)

Table 5.2 *(continued)*

Material		Ion-beam energy (keV)	Growth rate (nm/min) and/or ion-beam current (mA)	Gases in ion source and reactive gas concentration (%)	Pressure during growth (Pa)	Background pressure (Pa)	Remarks and comments	References
Layer	Substrate							
Y_2O_3	Al	—	100 mA	Ar	O_2,—	—	Protective coating on Al mirror; resistance to moisture	[113]
ITO	Si	0.5	7 nm/min	Ar	O_2, 1.3×10^{-2}	—	$\eta = 12\%$ on mono-crystalline Si; $\eta = 6.5\%$ on polycrystalline Si; $[SnO_2] = 10\%$	[114–116]
ITO	Mylar	—	17 nm/min, 50 mA	Ar	O_2, 10^{-3}–10^{-2}	7×10^{-5}	Electrical and optical properties of ITO films as a function of the O_2 pressure	[117]
ITO	InP, GaAs	1–5	8 nm/min	Ar	O_2, 10^{-3}–10^{-2}	—	$\eta = 5\%$ on GaAs substrate; $\eta = 14.4\%$ on InP	[118]
PZT	NiCr, Invar	2	—	Ar	O_2, 1.3×10^{-2}	2×10^{-4}	Ferroelectric films: perovskite structure for $T_s \geqslant 400°C$	[119]
Ta_2O_5, SiO_2, Si_3N_4	GaAs	—	—	—	—	—	$N_{ss} = 2 \times 10^{12}$ eV cm^{-2} in midgap for SiO_2 on GaAs	[120]
Al_2O_3	Steel	1	10 nm/min	Ar	—	10^{-4}	Mechanical application	[121]
Metal–Metal oxide	—	—	—	—	—	—	Application to optical recording threshold energy: 0.25 nJ; contrast ratio: 60	[122, 123]

Pb, Zr, Ti oxides	Si, Au	2	0.5–1 mA/cm^2	Ar	O_2, 10^{-4}–10^{-2}	2×10^{-4}	Deposition rate and stress as a function of the O_2 pressure	[124, 125]
Si_3N_4, AlN	Si, glass, ...	1–10	6.5 mA	N_2	$3–5 \times 10^{-3}$	10^{-4}	Growth rate, composition, and resistivity as a function of the deposition parameters	[126]
Si_3N_4	GaAs	0.5	—	$Ar + N_2$	$2–8 \times 10^{-2}$	7×10^{-5}	Significant O content; encapsulating layers with mechanical stability up to 900°C	[127, 128]
AlO_xN_y	GaAs	0.8	4–5 nm/min	$Ar + N_2$	—	4×10^{-5}	Significant O content; encapsulating layers with mechanical stability up to 900°C	[127, 128]
AlN_xO_y	Si, GaAs	5	0.5 mA, 1.5 nm/min	N_2	10^{-3}–10^{-2}	6×10^{-5} or 6×10^{-6}	$N_{ss} \simeq 10^{11}$ eV cm^2 on Si after an H treatment	[129, 130]
SiO_xN_y	—	10	1 mA/cm^2	$N_2 + O_2$	1.3×10^{-2}	4×10^{-4}	Study of secondary ions from N_2^+ bombarded Si	[131]
Si_3N_4	C	0.5–20	1–5 mA	N_2	10^{-3}	2×10^{-6}	Deposition mechanisms; study of the relation $N/Si = f(E)$	[46]
NbN	NaCl	7.5	0.5 nm/min	$Xe + N_2$	N_2: 13×10^{-2}; Xe: 13×10^{-2}		Epitaxial NbN layers on NaCl; $T_s = 300$ K, $T_c = 11$ K	[132]

[a] Symbols: E_g, energy band gap; T_s, substrate temperature; η, conversion efficiency (solar cells); N_{ss}, density of interface states (MIS devices); T_c, critical temperature (superconducting layers); [X], Concentration of the element X; ITO, Indium–tin oxide: $In_2O_3 + SnO_2$; and PZT, $Pb(Zr, Ti)O_2$.

Since molecular hydrogen does not react with silicon [134], the hydrogen incorporation from the gaseous phase requires an indirect process. This process involves the stimulated adsorption of molecular hydrogen on the target and the subsequent sputtering of the resulting composite. This explanation is confirmed by verifying [107] that the hydrogen content depends strongly on the focusing of the beam as we are able to deduce from the Eqs. (17) and (18).

2. Incorporation of Oxygen

Several experiments tend to indicate that the dominant processes are those involving the transport in the direction of the growing film of the compounds formed at the target surface. For example, the presence of Pb_3O_4 in deposits cannot likely be due to the room-temperature oxidation of deposited metal atoms [123]. In confirmation of this idea, we are clearly able to recognize the transition between a metallic layer to a dielectric layer by observing a decrease in the sputtering yield as a function of the partial pressure of oxygen [124]. The transition partial pressure P_0, beyond which the stoichiometric oxide layers are obtained, is, as expected (see Section IV.A), proportional to the ion-beam current density [124]. Despite the inconsistencies with recent theory on low-pressure oxidation of metals [135], P_0 may still be evaluated by a semi-empirical formula due to Castellano [124].

The effect of oxygen is not only to ensure the oxidation of a bombarded metal target but also to saturate the metal–oxygen bonds that might be broken when we sputter an oxide target [136]. An example of this may be deduced from the work of Castellano and Feinstein [119], in which it is clearly shown that metallic lead is sputtered from a target of $Pb(Zr, Ti)O_3$.

3. Incorporation of Nitrogen

Only the mechanisms of Si_3N_4 deposition will be discussed here, since they have been systematically studied by many authors and seem to be applicable to other nitrides. It can be shown that the nitrogen molecule does not react with silicon (the sticking probability of N_2 is less than 10^{-5} [46]), and consequently for a N_2^+ ionic beam, the processes of sputtering of the formed compound at the surface of the target must predominate. By using Eqs. (15) and (18), we may deduce the sample stoichiometry expressed in the form of the atomic ratio N/Si, namely

$$N/Si = a/S(E/2), \tag{19}$$

where a is the sticking probability of reflected and resputtered nitrogen on the growing film and $S(E/2)$ the sputtering yield of silicon expressed in units of atoms per atomic ion where this value is obtained by taking into account the behavior of the N_2^+ molecular ions (see Section II.D). Note that we may evaluate the value of S as a function of energy from Fig. 5.21a when the coefficient K of Eq. (15) is known. The value obtained at 500 eV is equal to 0.5 atoms/N^+ ion and is significantly larger than the result (0.15) reported by Zalm and Beckers [48].

The predominance of the sputtering processes in the incorporation of nitrogen is clearly proven by the well-established result that the efficiency of nitrogen incorporation stays practically constant over a large range of ionic energy (1–10 keV) (Fig. 5.21b). In this energy range, the sticking coefficient is approximately equal to 0.5. Although an unexplained disagreement exists between our results and those of Erler *et al.* [126], as can be seen in Fig. 5.21c, the results that we were able to obtain for the efficiency of the nitridation reaction are in good agreement with the formal equation proposed by this author for a sputtering rate of 0.5 [126]. Our evaluation of the sticking coefficient is general and not restricted by any assumptions on the state of the sputtered or reflected nitrogen (free atoms or molecules). The assumption that the nitrogen will be re-emitted in atomic form is supported by the results of Winters and Horne [138]. With respect to the secondary ionic emission, Burdovitsin [131] has shown that the proportion of dimers SiN^+ does not exceed 10^{-2}, and this result is also in agreement with the preceding idea. On the other hand, when a thermal desorption is performed from a nitrided silicon surface, the most important desorbed product seems to be the trimer Si_2N [139].

D. Dual-Beam Sputtering

This variation of the basic method involves the use of an ion beam extracted from an additional ion source II that is directed toward the growing film. The effect of this direct ion bombardment is either chemical (with reactive ionized species) or physical (with noble-gas ions). Both these effects are illustrated by the work compiled in Table 5.3. For example, reactive dual beam sputtering with a silicon or aluminum target results in the formation of Si_3N_4 [140] and AlN [141], respectively. The physical effect of the bombardment by noble-gas ions yields essentially the synthesis of metastable phases such as diamondlike carbon [143] and an increase of the deposited atom surface mobility [146], which improves the step coverage [145]. These properties are those we discussed earlier in this chapter, and the ion-assisted nucleation processes are described elsewhere in this book.

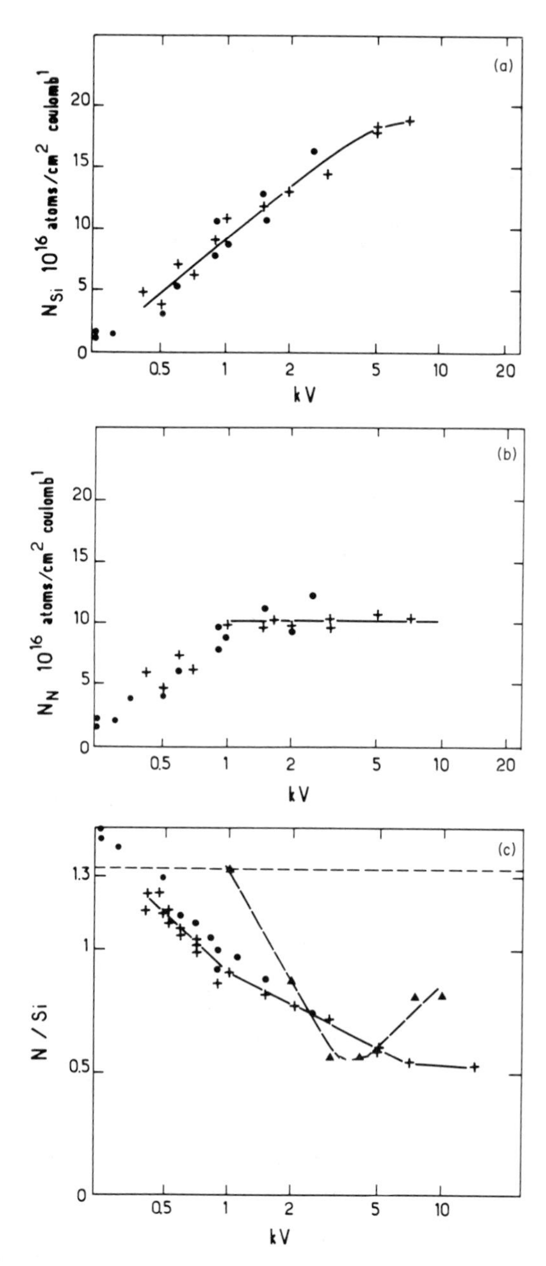

Fig. 5.21. Deposition yields with beam currents 1.5 mA(+) and 0.75 mA (●) of (a) silicon N_{Si}, (b) nitrogen N_N expressed in atoms per square centimeter per coulomb, and (c) the atomic ratio N/Si, as a function of the accelerating voltage V of the N_2^+ ion beam. (Data from [46, 137].) Some results of Erler have been reported. (▲ on dashed line). (Data from Erler *et al.* [126].)

TABLE 5.3

Dual Beam Sputtering Deposition

Material		Ion-beam energy of source I and II (keV)	Beam current (mA)	Gas in source	Background pressure (Pa)	Remarks and comments	References
Layer	Substrate						
Si_3N_4	Si, Glass	10^{-4}	6.5 mA	Ar	I: 8	Nitridation mechanisms; probability for N_2^+ reaction is 0.32 on Si	[140]
			1.35 mA	N_2	II: 0.68		
AlN	C, SiO_2, NaCl		—	Ar	I:	Sticking probability of N_2^+ on Al $\simeq 1$ for N/Al < 1	[141]
			—		II: 0.5		
TiN	Glass	1.3×10^{-5}	—	Ar		Composition of TiN_x layers, $x = 1$ for 2.6 keV	[142]
			—	N_2	II: 2.6		
i-C	—	10^{-4}	—	Ar	I: 8	Hard, transparent C layers; very high microhardness	[143]
			—	Ar + CH_4	II: 1		
i-C	—	4×10^{-3}	55 mA	Ar	I: 1.0	Bombardment of the growing film provided by source I	[144]
SiO_2	Si	7×10^{-5}	40 mA	Ar + O_2,	I: 1.0	Coverage of steps improved by bombarding the growing film; good result for 17% resputtering	[145]
			80 μA/cm^2	75–25%	II: 0.5		
SiO_2, TiO_2	Si		—	Ar + O_2,	I: 1.0	XPS measurement revealing the presence of suboxides in sputtered layers	[136]
		—	0.5 mA/cm^2	75–25%	II: 0.1–0.5		

V. ION-BEAM DEPOSITION—IONIZED CLUSTER BEAM DEPOSITION

A. Ion-Beam Deposition

In the introduction, we have seen that the principle of ion beam deposition involves an ionized form of transport of the desired material from ion source to the substrate.

In such a system, the growth rate will be given by

$$v = j(1 - R_e)(1 - S) \quad \text{atoms/s cm}^2, \tag{20}$$

where j is the current density in ions per square centimeter second, R_e the reflection coefficient, and S the sputtering yield.

The mechanisms that govern the interaction of an ion with a surface are discussed in Section II. For the reflection of the incident ion, in the case of homocondensation and for the incident angles between 0 and 45°, the coefficient R_e would not become larger than a few percent for an energy greater than a few tens of electron volts. Regarding the sputtering, Table 5.4 shows for 4 different elements, the threshold energies and the energy corresponding to the limit at which the coefficient S is equal to 1.

The calculations have been made with the aid of the Sigmund theory, which is good agreement with experiment even outside the strict zone of validity (linear cascade of collisions).

For molecular ions, we use the approach of Zalm and Beckers [48]. This phenomenological theory involves the dissociation of molecular ions in the constituent atoms. If the incident energy is larger than $\simeq$100 eV, these atoms individually conserve, after dissociation, the initial speed of the incident molecule. Table 5.4 clearly shows that for the majority of the elements, IBD implies the use of beams whose kinetic energies will be between 100 and 1000 eV. However, as was mentioned in Section III, the transport of intense beams at weak energies is difficult. Two possible solutions may be used:

(1) Transport of the beam at high energy ($E > 10$ keV) and deceleration immediately before the substrate and

TABLE 5.4

	Material			
E (eV)	B	C	Si	Ge
$E_{\text{threshold}}$	45	60	65	60
$E\ (S = 1)$	$S < 1$	$S < 1$	1200	500

(2) volume neutralization of the space charge by injecting electrons such that the electronic and ionic densities in the beam are equal.

The first solution necessitates a careful calculation of the electrostatic focusing system. Table 5.5 and references therein permit the interested reader to examine the details of the practical applications and the achieved results. The second solution consists of three variations:

(1) Creation of electrons by ionizing collisions between the beam and the residual gas,
(2) electronic emission of a filament or a plasma source situated in or in the neighborhood of the beam, and
(3) simultaneous emission of ions and electrons, a method more commonly known as the "plasma stream transport method."

The creation of electrons by ionization of the residual gas is governed by the cross section curve [171] as shown in Fig. 5.22. For an energy less than 1 keV and for a transport chamber pressure of 10^{-5} mbar, an elementary calculation shows the flux of created electrons is less than 10^{-3} of the ion current for transport distances of a few decimeters. Consequently, effective electronic trapping is required, and therefore excludes all electrostatic focusing. We remark that this solution is inapplicable in the pulsed regime if the duration of the impulses are less than the period of auto-neutralization [173] (the creation time of a sufficient number of electrons—a few milliseconds in the preceding example).

The neutralization by forced electron injection in the ionic beam has been the focus of numerous theoretical and experimental studies, especially in the field of space applications [64, 174–177]. The electrons may be injected from a thermoemissive filament immersed in the beam or from a "plasma bridge." The first approach, attractive by its simplicity, has the inconvenience of a short duration life. This problem is particularly troublesome when we use a gas or vapor in the ion source that chemically reacts with the filament. For the second approach, a hollow cathode [178] is most often utilized as a plasma source.

The plasma stream transport method is derived from plasma physics experiments in the area of controlled fusion. In a very simple picture, it consists of an injection of electrons and ions in a magnetic induction. The Larmor radius of the electrons is very small, and evidently the electric field slows down the radial diffusion of the ions. An illustration of this principle may be found in the work of Tsuchimoto [160]. Figure 5.23 represents a variation of this method. This variation, first used in the work of the authors, permits the separation of the charged particles from the neutrals and the

TABLE 5.5

Ion-Beam Deposition[a]

Ion used (mass analysis)	Energy (eV)	Intensity (μA) and/or density ($\mu A/cm^2$)	Growth rate; (nm/min) and/or area (cm^2)	Notes and comments	References
Cr^+ (no)	—	—	—	Deposition of localized resistance on dielectrics; good thermal stability	[147]
Cr^+ (no)	$230–10^3$	2–10	47 (0.2 mm^2), 0.7 (80 mm^2)	At 230 eV apparent sticking coefficient is 25% (sputtering)	[148]
$Fe^+–Co^+–Ni^+$, $Cu^+–Zn^+–Sn^+$ (yes)	$10–10^5$	10–20 μA; 10–20 $\mu A/cm^2$	1 cm^2	Sticking and self-sputtering coefficients	[149]
$C^+–Si^+$ (no)	4	—	300 nm/min	Important flux of neutrals; near to ion plating; good mechanical, physical, and chemical properties	[150]
In^+ (yes)	100–500	$10^{-2}–1$	1 mm^2	Microbeams	[151]
$Pb^+–Cu^+$ (yes)	45	15 —	0.2 nm/min, 3 cm^2	Damage created by fast neutral (charge exchange)	[152]
$Pb^+–Mg^+$ (yes)	24–500	10–15 —	0.3–1 nm/min, 1 mm^2–1 cm^2	Epitaxy and adhesion studies	[153, 154]
Mn^+ (yes)	$10^2–10^3$	25 (100 eV) —	1–2 cm^2	No study about elaborated materials properties, thickness films $\geqslant 1\ \mu m$	[155]
C^+ (no)	50–100	—	—	Polycrystalline cubic diamond, 5-μm-diameter crystals	[156]

Si^+–Ge^+ (yes)	100–2000	40 μA; 5 $\mu A/cm^2$	0.5–1 nm/min, some cm^2	Ge epitaxy at $T_s = 300°C$; difficulties for Si because of carbonaceous impurities	[157]
Ag^+–Zn^+ (yes)	30–300	8 $\mu A/cm^2$	<1 cm^2	Ion energy effects on defects and diffusion in film	[158]
Ag^+ (yes)	20–200	5–25	<1 cm^2	Sticking and self-sputtering coefficients	[159]
SiH_x^+, $SiH_x^+ + O_2^+$ (no)		102–10^4 μA	5–10 nm/min (Si), 50 nm/min (Si_3N_4)	Study of plasma stream method	[160]
Ge^+–Si^+ (yes)	100–2000	100–200 μA	1 cm^2	Improvement of crystallographic quality for Si films	[161, 162]
Pb^+–Mg^+ (yes)				Improvement of deposition conditions	[163]
Ag^+–Ge^+ (yes)	50–100	8 μA	0.2 cm^2	Sticking coefficient measurement and epitaxy studies; comparison with numerical simulation like MARLOWE	[68, 69]
$B_xN_yH_z^+$ (no)	200–1000	$\simeq 10^4$ μA	6–12 nm/min, 100 cm^2	Cubic BN synthesis	[164]
C^+–As^+ (yes)	10^2–10^3	30–160 (100 eV)	—	Epi plantation for MBE	[165]
Nb^+, N^+, N_2^+, A^+ (no)	qq 10^1	1–500 μA	2–3 nm/min, 3 cm^2	Self-acceleration method (with electrons); superconductivity films (T_c = 16 K) for T_s = 620 K	[166]
Pd^+ (yes)	10–400	1–10 μA	3 cm^2	Ion energy effect on crystallographic and electrical properties; metal–semiconductor contact	[167]
Si^+ (no)	500	1–500 μA	>10 cm^2	Preliminary studies of ion-beam-enhanced CVD	[168]
C^+ (yes)	10^2–10^3	20 (100 eV), 100 (400 eV)	0.5–1 nm/min, 1 cm^2	Physicochemical characteristics near diamond	[169]
C^+ (no)	100		5–50 nm/min	C deposition on optical fibers	[170]

[a] *Symbols*: T_s, substrate temperature and T_c, critical temperature (superconducting films).

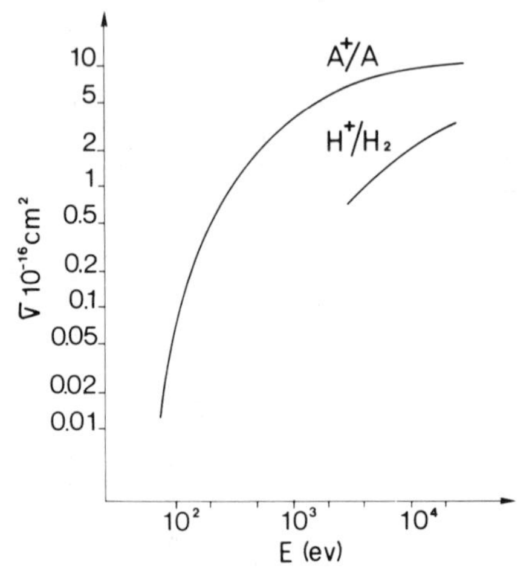

Fig. 5.22. Ionization cross section of various ions in their parent gases; $\bar{X}^+ + Y \rightarrow \bar{X}^+ + Y^+ + e$. (Data from McDaniel [172].)

photons. A beam energy between 50 and 100 eV with a density of 1 mA/cm^2 may be obtained by this method.

1. Results

Table 5.5 summarizes the main work in this field. As we may see, the majority of the studies have been performed at the laboratory stage and have been motivated by the desire to better understand the role of the ions in the deposition of materials in thin films. Of course, the simultaneous action of many mechanisms (sputtering, implantation, nucleation, heating, and migration) results in thin films with unique behavior in a variety of fields, which are summarized in Table 5.5. We will give several examples in the following discussion.

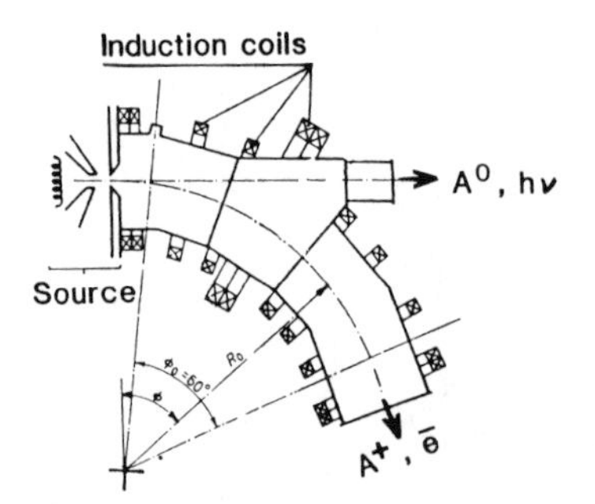

Fig. 5.23. Schematic diagram of the plasma stream transport apparatus [179].

In the field of epitaxy, the substrate temperatures are considerably lower due to the fact that a high mobility of the atoms is obtained by the kinetic energy of the incident ions. The work at Philips Research Laboratories [68] shows that the growth of Ag (111)–Si (111) is always possible at 300 K for ionic energies between 25 and 125 eV. Experiments show that films grown at 25 eV and above 100 eV consistently show a minor contribution of polycrystalline phase superimposed on the single crystalline RHEED pattern. In the same manner, the epitaxial growth of Ge on Si is produced at 230°C instead of 500°C when we use neutrals of thermal energy.

An ion kinetic energy that is too high may engender crystalline defects (dislocation cores of Frenkel point defects), which deteriorate the sample quality as shown by the work of Ohmae *et al.* [158] for the Zn–Al system. An example is given in Table 5.6.

On the other hand, these defects permit a higher diffusion and an enhanced layer adhesion by a larger interfacial zone. Figure 5.24 clearly shows the difference in the depth of diffused layers analyzed by SIMS. Analogous results have been obtained by Amano [154] for samples of Pb and Mg on carbon substrates. For Pb, an ionic energy of 24 eV is sufficient whereas 120 eV are required for Mg.

Some very interesting results have been obtained by Miyazawa *et al.* [169] for carbon films deposited by IBD. Figure 5.25 shows the optical transmission spectrum of an IBD carbon film, 44 nm thick, deposited at 300 eV on a silicon substrate. The spectrum of an amorphous carbon film, 50 nm thick, deposited with an electron-beam gun is also shown for comparison. Some significant differences have also been obtained in the visible spectrum. The optical constants (n, k) at a wavelength of 633 nm, determined by ellipsometry, are (2.4, 0) for the IBD film deposited at 600 eV and are in good agreement with diamond ($n = 2.41$).

Before concluding, we would like to call to the reader's attention, the role of residual pressure on the properties of the materials as evidenced in the

TABLE 5.6

Ion energy (eV)	Dislocation density (cm^{-2})	
	[310]	[211]
30	4.7×10^8	
70	6.1×10^8	1.2×10^9
100	7.8×10^8	1.9×10^9
200		3.7×10^9

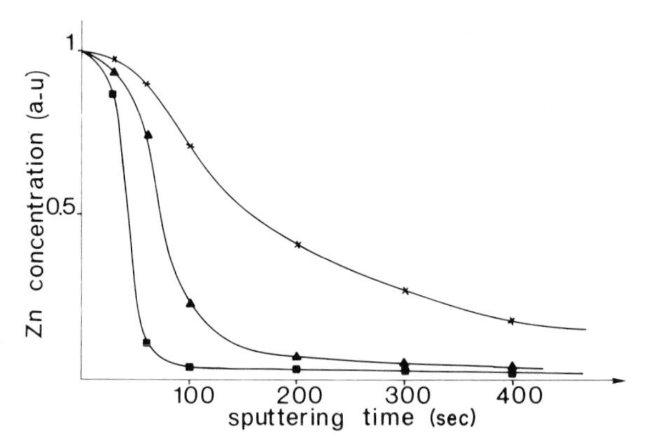

Fig. 5.24. Difference in the depth of diffused layers analyzed by SIMS as a function of the deposited atoms energy; × and ▲ represent ion beam plating at 200 eV and 30 eV on (111), respectively, and ■ represents vacuum evaporation on (111). (Data from Ohmae *et al.* [158].)

work of Amano [163] and Miyake [161]. As in ion-beam sputtering, it is necessary to consider the respective incident fluxes on the substrate.

B. Ionized Cluster Beam Deposition

This technique, which has been described in Section III, has given rise to a large number of applications in semiconductors, metals, dielectrics, optical films, magnetic and thermoelectric films, and solar cells, most notably in the United States and Japan. Table 5.7 provides a summary of the principal materials studied with their results.

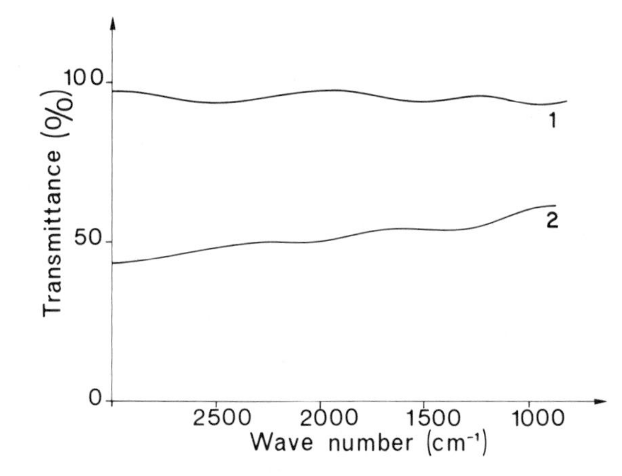

Fig. 5.25. Transmission spectrum of a 44-nm ion-beam deposited carbon film on silicon (1) by using a C^+ beam. This spectrum is compared with that obtained for an evaporated 50-nm carbon layer (2). (Data from Miyazawa *et al.* [169].)

TABLE 5.7

Ionized Cluster Beam Deposition

Material		Accelerating voltage (kV)	Substrate temperature (°C)	Notes and comments	References
Film	Substrate				
Al	SiO_2, Si	0–5	RT–200	Oriented polycrystalline films; epitaxial films on Si for SiO_2, Ohmic contact at low temperature for Si	[180, 181]
Ag	Si *n*	5	RT	Ohmic contact without alloying	[182]
Au, Cu	Glass, kapton	1–10	RT	Strong adhesion; high conductivity in thin layers (15 nm)	[183]
Au	Glass	1–15	RT	Field emission deposition	[184]
PbTe	Glass	0–3	200	Thermoelectric material; high Seebeck coefficient	[185]
Si	Si	6	620	Low-temperature epitaxy	[183]
Si *a*	Glass	2	200	Thermally stable under 400°C; 30 mn anneal	[186]
GaAs	GaAs	?	400	Low-temperature epitaxy	[187]
GaP	Si	4	450	Low-cost LED	[188]
InSb	Sapphire	3	250	Controllable crystal structure (magnetic sensor)	[189]
BeO	Sapphire (0001)	0	400	Single crystal formation; Transparent film; high resistivity and high thermal conductivity	[190]
SiO_2	Si	0–2	200	Low-temperature growth (Si devices)	[191]
SiC	Si, glass	0–8	600	Controlled crystallinity through ion content (energy converter and surface protective coating)	[192]
Anthracene	Glass	0–2	−10	Controllable crystal structure (detectors)	[193]

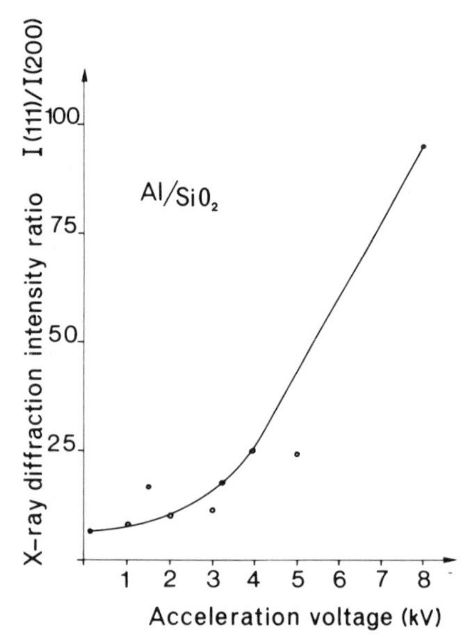

Fig. 5.26. X-ray diffraction intensity ratio I_{111}/I_{200} as a function of the accelerating voltage. Substrate temperature = 60–90°C. (Data from Reader *et al.* [180].)

The growth mechanisms of the layers is very similar to those in IBD, but there exists two important differences. First, the energy range in electron volts per atom is approximately two orders of magnitude weaker. Second, the growth is achieved by the simultaneous flux of thermal neutrals and ionized clusters of the material. This latter point, once again, complicates the analysis of the phenomena.

We give two characteristic examples of this method:

(1) Contacts of weak resistivity (0.1 ohm cm) have been obtained, after annealing at 200°C, from films of Al deposited by ICBD on Si(p)-10 ohm cm. This annealing temperature is significantly weaker than those used in conventional deposition techniques. The grain dimensions may be controlled by adjusting the incident cluster energy. Naturally, larger grain size reduces the penetration of the oxide into the film. Furthermore, the crystallographic orientation may be strongly modified as shown in Fig. 5.26. Several attempts are in progress for determining the way to use ICBD in eliminating electromigration.

(2) When using a reactive gas in the deposition chamber, it is possible to generate oxides, etc. Figure 5.27 compares the x-ray diffraction pattern of sintered BeO to that of the deposited film on glass. Beryllium oxide has been grown epitaxially on both sapphire and silicon; *c*-axis-oriented BeO on (100) silicon substrates has been obtained. After metallization with Al, excellent $C(V)$ characteristics were obtained.

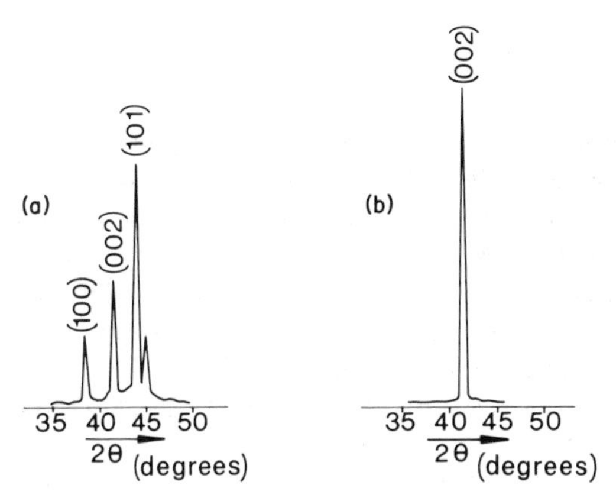

Fig. 5.27. X-ray diffraction patterns for (a) sintered and (b) ICB deposited BeO. (Data from Takagi *et al.* [190].)

VI. CONCLUSION

In this chapter, we have attempted to show the evolution and the recent interest of vacuum deposition methods involving the condensation of superthermal atoms. Several remarkable results may be associated with the thermalization process of the deposited atoms. The achieved results may arise from the high surface mobility of the deposited atoms, from the creation of nucleation sites, or from other effects due to the thermalization processes. We recall, by way of example, the very high hardness of ion-beam deposited carbon layers [169], the resistance to moisture of sputtered Y_2O_3 [113], and the weak oxide penetration in ICB-deposited Al-layers [180]. These properties result from the high density of these samples.

On the other hand, the superthermal energy, which causes the splitting of the molecules, permits the reactive deposition of a large variety of material, notably the nitrides [126, 141]. In addition, we observe at the deposition temperature an increase of the sticking coefficient of the volatile elements, such as the arsenic in GaAs [95] or the dopants (Sb) in silicon [93]. The fact that the nucleation is improved by the high surface mobility of the deposited atoms generates the possibility of epitaxy at low temperature (for example, 400 K for ion-beam deposited silicon and germanium [69] or 520 K for ion-beam sputtered Si [93]). However, the electrical properties, which have been measured only in the case of ion-beam sputtered Si layers, become satisfactory only if the deposition temperature exceeds 700°C [89, 93]. A better knowledge, theoretical as well as experimental, of the basic phenomena may assist in determining the origins of the defects that are responsible for this result. This example illustrates well the two obstacles that still impede the development of the methods described in this chapter. On the one hand, it is regrettable that theoretical knowledge remains insufficient in describing the effects caused by the following:

(1) The impact of a 1–100-eV particle on a surface containing atoms that have been adsorbed or previously deposited: The defects created in the lattice of the growing film [89] or of the substrate [194] may constitute a limitation to the application of superthermal methods. This aspect is particularly crucial for IBSD since the energy spectrum of the sputtered atoms always presents a high energy tail. Let us emphasize that this point has been already investigated by using computer simulation [68].

(2) The arrival of charged particles on the growing film: This point is most important in the case of IBD [65].

(3) The impact of energetic clusters in the context of the ICBD technique: The dissociation processes of the clusters are not known in detail, and more specifically, we ignore the collective movement effect of an important number of atoms arriving on a reduced surface.

On the other hand, a comparison of the results obtained by two techniques (i.e., IBSD and IBD) is often not possible because the deposition conditions are usually poorly controlled and the characterization of the resultant layers is insufficient. This situation is particularly regretable since the methods considered in this chapter are well suited for precise control by UHV techniques. However, it is necessary to remark that rigorous experimental effort and progress is being made by an increasing number of research groups.

In conclusion, certain applications, notably related to the properties of adhesion and density of the achieved layers, already justify the interest paid to these methods, and additional effort addressing the preceding points may permit an enlargement of the area of applications.

ACKNOWLEDGMENTS

The authors would like to thank the members of their research group Microelectronique et Faisceaux d'Ions for useful discussions throughout this work, and M. Meisel for his translation work.

REFERENCES

1. K. Ploog and A. Fisher, *Appl. Phys.* **13**, 111–121 (1977).
2. A. Y. Cho and J. R. Arthur, *in* "Progress in Solid State Chemistry" (G. Somorhai and J. McCaldin, eds.), Vol. 10, Pergamon, New York, 1975; A. Y. Cho and K. Y. Cheng, *Appl. Phys. Lett.* **38**, 360 (1981).

3. R. F. Bunshah, ed., "Deposition Technologies for Films and Coatings," Noyes Publ., Park Ridge, New Jersey, 1982.
4. J. J. Hanak and J. P. Pellicane, *J. Vac. Sci. Technol.* **13**, 406 (1976).
5. G. Carter and J. S. Colligon, "Ion Bombardment of Solids," Butterworth, New York, 1964; G. M. McCracken, *Rep. Prog. Phys.* **38**, 241–327 (1975); P. D. Townsend, J. C. Kelly, and N. E. W. Hartley, "Ion Implantation, Sputtering and Their Applications," Academic Press, New York, 1976.
6. I. R. Berish, ed., "Sputtering by Particle Bombardment," Springer-Verlag, Berlin, 1981.
7. H. F. Winters, *Top. Curr. Chem.* **94**, (1980).
8. R. J. Beuhler and L. Friedman, *J. Appl. Phys.* **48**, 3928 (1977).
9. H. D. Hagstrum, *Phys. Rev.* **96**, 336 (1954); H. D. Hagstrum, *J. Vac. Sci. Technol.* **12**, 7 (1975).
10. D. B. Medved, P. Mahadevan, and J. K. Layton, *Phys. Rev.* **129**, 2086 (1963).
11. J. Lindhard, Thomas Fermi approach and similarity in atomic collisions, *NAS-NRD Publ.* **1**, 1133 (1964).
12. J. Lindhard, M. Scharff, and M. E. Schiott, *Mat. Fys. Medd.* **33**, 14 (1963); J. Lindhard and M. Scharff, *Phys. Rev.* **124**, 128 (1961).
13. J. Bottiger, J. A. Davies, P. Sigmund and K. B. Winterbon, *Rad. Eff.* **11**, 69–78 (1971).
14. K. B. Winterbon, P. Sigmund, and J. B. Sanders, *Mat. Fys. Medd.* **37**, 14 (1970).
15. P. Sigmund, *Phys. Rev.* **184**, 283 (1969).
16. O. B. Firsov, *Sov. Phys. JETP, Engl. Transl.* **6**, 534 (1958).
17. W. D. Wilson, L. G. Haggmark, and J. P. Biersack, *Phys. Rev. B* **15**, 2458 (1977).
18. H. E. Schiott, *Rad. Eff.* **6**, 107–113 (1970).
19. E. V. Kornelsen, *Can. J. Phys.* **43**, 364 (1964).
20. T. M. Buck, Y. S. Chen, G. H. Wheatley, and W. F. van der Weg, *Surf. Sci.* **47**, 244 (1975).
21. E. Taglauer and W. Heiland, *Nucl. Instr. Methods* **132**, 535 (1976).
22. D. Hildebrandt and R. Manns, *Rad. Eff.* **41**, 193–194 (1979).
23. C. Pellet, C. Schwebel, G. Gautherin, and F. Meyer, to be published.
24. M. W. Thomson, *Philos. Mag.* **18**, 377 (1968).
25. R. E. Honig, *RCA Rev.* **23**, 567 (1962); C. Kittel, "Introduction to Solid State Physics," Wiley, New York, 1971; D. P. Stulls and H. Prophet, eds., "JANAF Thermochemical Table," 2nd ed., NSRDS-NBS, No. 37, National Bureau of Standard, Washington D.C. 1971; C. J. Smithells, ed., "Metals Reference Book," Butterworths, London, 1975; I. Barin and O. Knacke, "Thermochemical Properties of Inorganic Substances," Springer-Verlag, Berlin, 1975.
26. D. P. Jackson, *Rad. Eff.* **18**, 185 (1973).
27. Private communication.
28. H. H. Andersen and H. L. Bay, *J. Appl. Phys.* **46**, 2416 (1975).
29. H. L. Bay, H. H. Andersen, W. O. Hofer, and O. Nielsen, *Nucl. Instr. Methods* **132**, 301–305 (1976).
30. J. Bohdanski, J. Roth, and H. L. Bay, *J. Appl. Phys.* **51**, 2861 (1980).
31. J. Bohdanski, *Nucl. Instr. Methods B* **2**, 587 (1984).
32. P. C. Zalm, *J. Vac. Sci. Technol. B* **2**, 151 (1984).
33. H. H. Andersen and H. L. Bay, *in* "Sputtering by Particle Bombardment" (I. R. Berish, ed.), Springer-Verlag, Berlin, 1981.
34. R. V. Stuart and G. K. Wehner, *J. Appl. Phys.* **35**, 1819 (1964).
35. R. V. Stuart, *J. Appl. Phys.* **40**, 803 (1969).
36. I. S. T. Stong and N. A. Yusuf, *Nucl. Instr. Methods* **170**, 357 (1980).
37. G. Blaise and A. Nourtier, *Surf. Sci.* **90**, 495 (1979).
38. V. Gurmin, *Bull. Acad. Sci., URSS Ser.* **33**, 752 (752).

39. H. H. Andersen and P. Sigmund, *Mat. Fys. Medd.* **39**, 3 (1974).
40. P. Sigmund, *J. Vac. Sci. Technol.* **17**, 396 (1980).
41. R. K. Haff and Z. E. Switkowski, *Appl. Phys. Lett.* **29**, 9 (1976).
42. G. W. Reynols, *Nucl. Instr. Methods* **209/210**, 57 (1983).
43. Z. L. Liau, J. W. Mayer, W. L. Brown, and J. M. Poate, *J. Appl. Phys.* **49**, 5295 (1978).
44. R. Collins and G. Carter, *Rad. Eff.* **26**, 181 (1975).
45. J. P. Biersack, *in* "Ion Surface Interaction, Sputtering, and Related Phenomena" (Berish *et al.*, Eds., Gordon and Breach), New York, 1973.
46. D. Bouchier, G. Gautherin, C. Schwebel, A. Bosseboeuf, B. Agius, and S. Rigo, *J. Electrochem. Soc.* **130**, 638 (1983).
47. C. Steinbrüchel, *J. Vac. Sci. Technol. B* **2**, 38 (1984).
48. P. C. Zalm and L. J. Beckers, *J. Appl. Phys.* **56**, 220 (1984).
49. H. H. Andersen and H. L. Bay, *J. Appl. Phys.* **46**, 2416 (1976).
50. J. B. Malherbe, *Rad. Eff.* **70**, 261 (1983).
51. H. H. Andersen and H. L. Bay, *Rad. Eff.* **13**, 67 (1972).
52. P. Blank and K. Wittmack, *Rad. Eff.* **27**, 29 (1975).
53. C. Steinbrüchel and D. M. Gruen, *J. Vac. Sci. Technol.* **18**, 2 (1981).
54. U. Gerlach-Meyer, J. W. Coburn, and E. Kay, *Surf. Sci.* **103**, 177 (1981).
55. M. Cantagrel and M. Marchal, *J. Mater. Sci.* **8**, 171 (1973).
56. W. Wach and K. Wittmack, *J. Appl. Phys.* **52**, 3341 (1981).
57. H. F. Winters and P. Sigmund, *J. Appl. Phys.* **45**, 11 (1974).
58. J. G. Martel, R. Saint-Jacques, B. Terreault, and G. Veilleux, *J. Nucl. Mater.* **53**, 142–146 (1974).
59. H. R. Kaufman, J. J. Cuomo, and J. M. E. Harper, *J. Vac. Sci. Technol.* **21**, 725 (1982).
60. R. G. Wilson and G. R. Brewer (eds.), "Ion Beams," Wiley, New York, 1973.
61. J. Aubert, thesis, University of Paris, 1984.
62. R. Hutter, *in* "Focusing of Charged Particles" (A. Septier, ed.), Academic Press, New York, 1967.
63. P. Grivet, ed., "Electron Optics," Pergamon, Oxford, 1965.
64. G. Spiess, thesis, University of Paris, 1969.
65. T. Takagi, I. Y. M. Kunori, and S. Kobiyama, *Proc. Int. Conf. Ion Sources, 2nd, Vienna, 1972*, 790 (1972).
66. I. Yamada, *Proc. Int. Ion Eng. Congr., ISIAT, Kyoto, 1983*, 1177 (1983).
67. G. Gautherin, C. Schwebel, and C. Weissmantal, *Int. Conf. Solid Surf., 7th, Vienna, 1977*, 1449 (1977).
68. G. E. Thomas, L. J. Beckers, J. J. Vrakking, and B. R. DeKoninge, *J. Cryst. Growth* **56**, 557 (1982).
69. P. C. Zalm and L. J. Beckers, *Appl. Phys. Lett.* **41**, 167 (1982).
70. S. Itoh, K. Morimoto, and H. Watanabe, *Proc. Int. Assisted Technol., ISIAT, Kyoto, 1981*, 459 (1981).
71. D. Bouchier, G. Gautherin, B. Agius, and S. Rigo, *J. Appl. Phys.* **49**, 5896 (1978).
72. P. H. Schmidt, R. N. Castellano, and E. G. Spencer, *Solid State Technol.* **15**, 27 (1972).
73. W. D. Westwood and S. J. Ingrey, *J. Vac. Sci. Technol.* **13**, 104 (1976).
74. R. N. Castellano, M. R. Notis, and G. W. Simmons, *Vacuum* **27**, 109 (1977).
75. S. Yamanaka, M. Naoe, and S. Kawai, *Jpn. J. Appl. Phys.* **16**, 1245 (1977).
76. P. Reinhardt, C. Reinhardt, G. Reisse and C. Weissmantel, *Thin Solid Films* **51**, 99 (1978).
77. S. M. Kane and K. Y. Alm, *J. Vac. Sci. Technol.* **16**, 171 (1979).
78. M. J. Mirtich, *J. Vac. Sci. Technol.* **18**, 186 (1981).
79. S. Fujimori, T. Kusai, and T. Inamura, *Proc. Int. Engineering Congr., ISIAT, Kyoto, 1983*, (1983).

80. F. Bonhoure, E. Dagoury, D. Vignier, and R. Lesueur, *Vide, Couches Minces* **207**, 541 (1981).
81. P. H. Schmidt, R. N. Castellano, H. Barz, A. S. Cooper, and E. G. Spencer, *J. Appl. Phys.* **44**, 1833 (1973).
82. P. H. Schmidt, *J. Vac. Sci. Technol.* **10**, 611 (1973).
83. E. Simonarson, thesis, University of Orsay, Orsay, France, 1982.
84. K. Takei, K. Nagai, and T. Mamura, *Jpn. J. Appl. Phys.* **19**, L392 (1980).
85. D. Bouchier, C. Schwebel, and G. Gautherin, to be published.
86. B. A. Unvala and T. Pearmain, *J. Mat. Sci.* **5**, 1016 (1970).
87. C. Weissmantel, J. Fielder, G. Hecht, and G. Reisse, *Thin Solid Films* **13**, 359 (1972).
88. H. J. Hinneberg, M. Weidner, G. Hecht, and C. Weissmantel, *Thin Solid Films* **33**, 29 (1976).
89. C. Weissmantel, G. Hecht, and H. J. Hinneberg, *J. Vac. Sci. Technol.* **17**, 812 (1980).
90. L. Aleksandrov and R. Lovyagin, *Jpn. J. Appl. Phys.*,Suppl. 2, Pt. 1, 1 (1974).
91. R. S. Bhattacharya, *Jpn. J. Appl. Phys.* **19**, L523 (1980).
92. C. Schwebel, A. Vapaille, D. Bouchier, G. Gautherin, and F. Meyer, *Int. Vac. Congr., 8th, Cannes, 1980*, 125 (1980).
93. C. Schwebel C., F. Meyer, and G. Gautherin, *J. Phys.* **12**, C5–473 (1982).
94. C. Schwebel, F. Meyer, G. Gautherin, and C. Pellet, to be published.
95. C. Schwebel, A. Vapaille, D. Bouchier, G. Gautherin, and F. Meyer, *Coll. Int. de Pulvérisation Cathodique, 3 ème, Nice, 1979*, 361 (1979).
96. Y. Ohsawa, A. Sawabe, and T. Inuzuka, *Proc. Int. Ion Eng. Cong. ISIAT, Kyoto*, 1983 (1983).
97. A. Azim Khan, J. A. Woollam, Y. Chung, and B. Bank, *IEEE EDL* **4** 146 (1983).
98. C. Weissmantel, *Thin Solid Films* **92** 55 (1982).
99. C. Weissmantel, M. Rost, A. Fielder, H. J. Erler, H. Giegengack, and J. Horn, *Jpn. J. Appl. Phys., Suppl. 2*, pt. 1 (1974).
100. J. S. Sovey, *J. Vac. Sci. Technol.* **16**, 813 (1979).
101. D. M. Mattox and J. E. McDonald, *J. Appl. Phys.* **34**, 2493 (1963).
102. K. L. Chopra and M. R. Randlett, *Rev. Sci. Instr.* **38**, 1147 (1967).
103. J. Comas and E. A. Wolicki, *J. Electrochem. Soc.* **117**, 1197 (1970).
104. H. F. Winters and E. Kay, *J. Appl. Phys.* **38**, 3928 (1967); H. F. Winters, E. E. Horne, and E. E. Donaldson, *J. Chem. Phys.* **41**, 2766 (1964).
105. C. Weissmantel, K. Bewilogua, D. Dietrich, H. J. Erler, H. J. Hinneberg, S. Klose, W. Nowick, and G. Reisse, *Thin Solid Films* **72**, 19–31 (1980).
106. G. P. Ceasar, S. F. Grimshaw, and K. Okumura, *Solid State Commun.* **38**, 89–93 (1981).
107. J. Saraie, M. Kobayashi, Y. Fujii, and H. Matsunami, *Thin Solid Films* **80**, 169–176 (1981).
108. A. Kasdan and D. P. Goshorn, *Appl. Phys. Lett.* **42**, 36 (1983).
109. P. J. Martin, R. P. Netterfield, W. G. Sainty, and D. R. McKenzie, *Thin Solid Films* **100**, 141–147 (1983).
110. Y. Suzuki, S. Ogawa, K. Takiguchi, and H. Matsuda, *Proc. Int. Ion Eng. Cong., ISIAT IPAT, 1983, Kyoto, 1983*, (1983).
111. W. D. Westwood and S. J. Ingrey, *J. Vac. Sci. Technol.* **13**, 104 (1976).
112. M. Varasi, C. Misiano, and L. Lasaponara, *Proc. Int. Ion Eng. Cong., ISIAT IPAT, Kyoto, 1983*, (1983).
113. B. E. Cole, J. Moravec, R. G. Ahonen, and L. B. Ehlert, *J. Vac. Sci. Technol. A* **2**. 372 (1984).
114. J. B. Dubow, D. E. Burk, and J. R. Sites, *Appl. Phys. Lett.* **29**, 495 (1976).
115. G. Cheek, N. Inove, S. Goodnick, A. Genis, C. Wilmsen, and J. B. Dubow, *Appl. Phys. Lett.* **33**, 643 (1978).
116. G. Cheek, A. Genis, J. B. Dubow, and V. R. Pai Verneker, *Appl. Phys. Lett.* **35**, 495 (1979).
117. J. C. C. Fan, *Appl. Phys. Lett.* **34**, 515 (1979).

118. K. J. Bachmann, H. Schreiber, Jr., W. R. Sinclair, P. H. Schmidt, F. A. Thiel, E. G. Spencer, G. Pasteur, W. L. Feldmann, and Sree Harcha, *J. Appl. Phys.* **50**, 5 (1979).
119. R. N. Castellano and L. G. Feinstein, *J. Appl. Phys.* **50**, 6 (1979).
120. J. R. Sites, *Thin Solid Films* **45**, 47 (1977).
121. M. E. Königer, G. Reithmeier, and M. Simon, *Thin Solid Films* **109**, 19–25 (1983).
122. A. F. Hebard, G. E. Blonder and S. Y. Suh, *Appl. Phys. Lett.* **44**, 11 (1984).
123. S. Nakahara and A. F. Hebard, *Thin Solid Films* **102**, 345–360 (1983).
124. R. N. Castellano, *Thin Solid Films* **46**, 213–221 (1977).
125. R. N. Castellano, *Proc. Int. Vac. Congr., 7th, Vienna, 1977*, 1449 (1977).
126. H. J. Erler, G. Reisse, and C. Weissmantel, *Thin Solid Films* **65**, 233–245 (1980).
127. L. E. Bradley and J. R. Sites, *J. Vac. Sci. Technol.* **16**, 2 (1979).
128. H. Birey, S. J. Pak, and J. R. Sites, *J. Vac. Sci. Technol.* **16**, 6 (1979).
129. C. Sibran, R. Blanchet, M. Garrigues, and P. Viktorovitch, *Thin Solid Films* **103**, 211–219 (1983).
130. M. Garrigues, R. Blanchet, C. Sibran, and P. Viktorovitch, *J. Appl. Phys.* **54**, 2863 (1983).
131. V. A. Burdovitsin, *Thin Solid Films* **105** 197–202 (1983).
132. K. Takei and K. Nagai, *Jpn. J. Appl. Phys.* **20**, 5 (1981).
133. A. Bosseboeuf, thesis, University of Paris, 1983.
134. D. L. Miller, H. Lutz, H. Wiesmann, E. Rock, A. K. Ghosh, S. Ramamoorthy, and S. Strongin, *J. Appl. Phys.* **49**, 12 (1978).
135. F. P. Fehlner and N. F. Mott, *Oxid, Met.* **2**, 59 (1970).
136. S. M. Rossnagel and J. R. Sites, *J. Vac. Sci. Technol. A* **2**, 376–379 (1984).
137. D. Bouchier and A. Bosseboeuf, *Surf. Sci.* to be published.
138. H. F. Winters and D. E. Horne, *Surf. Sci.* **24**, 587 (1971).
139. A. G. Schrott and S. C. Fain, Jr., *Surf. Sci.* **111**, 39 (1981).
140. C. Weissmantel, *Thin Solid Films* **32**, 11–18 (1976).
141. J. M. E. Harper, H. T. G. Henzell, and J. J. Cuomo, *J. Vac. Sci. Technol. A* **2**, 405 (1984).
142. S. Ogawa, K. Takiguchi, T. Shioyama, T. Yotsuya, and M. Yoshitake, *Proc. Int. Ion Assisted Technol., ISIAT, Kyoto, 1981*, **363**, (1981).
143. C. Weissmantel, G. Reisse, H. J. Erler, F. Henny, K. Bewilogua, U. Ebersbach, and C. Schürer, *Thin Solid Films* **63**, 315–325 (1979).
144. B. A. Banks and S. K. Rutledge, *Meet. Am. Vac. Soc., Yorktown Heights, New York, June 2, 1982, NASA Tech. Mem.* **82**, 873 (1982).
145. J. M. E. Harper, G. R. Proto, and P. D. Hoh, *J. Vac. Sci. Technol.* **18**, 2 (1981).
146. M. Marinov, *Thin Solid Films* **46**, 267–274 (1977).
147. R. A. Volter, *Microelectron. Reliab.* **4**, 101 (1965).
148. B. A. Probyn, *Br. J. Appl. Phys.* (*J. Phys. D*) *ser. 2*, **1**, 457 (1968).
149. A. Fontell and E. Arminen, *Can. J. of Phys.* **47**, 2406 (1969).
150. S. Aisenberg and R. Chabot, *J. Appl. Phys.* **42**, 2953 (1971).
151. R. B. Fair, *J. Appl. Phys.* **42**, 3176 (1971).
152. J. S. Colligon, W. A. Grant, J. S. Williams, and R. P. W. Lawson, *Proc. Int. Conf. Appl. Ions Beams to Met., Warwick, September 1975*, 357 (1975).
153. J. Amano, P. Bryce and R. P. W. Lawson, *J. Vac. Sci. Technol.* **13**, 591 (1976).
154. J. Amano and R. P. W. Lawson, *J. Vac. Sci. Technol.* **14**, 831 (1977); **14**, 836 (1977); **15**, 118 (1978).
155. J. H. Freeman, W. Temple, D. Beanland, and G. A. Gard, *Nucl. Inst. Methods* **135**, 1 (1976).
156. E. G. Spencer, P. H. Schmidt, D. C. Joy, and F. J. Sansalone, *Appl. Phys. Lett.* **29**, 118 (1976).
157. K. Yagi, S. Tamura, and T. Tokuyama, *Jpn. J. Appl. Phys.* **16**, 245 (1977).
158. N. Ohmae, T. Nakai, and T. Tsukizoe, *Proc. Int. Vac. Cong., 7th, Vienna, 1977*, 1607 (1977), and *J. Appl. Phys.* **48**, 4770 (1977).

159. G. E. Thomas and E. E. de Kluizenaar, *Proc. Int. Coll. Surf. Phys. Chem., 3rd,1977*, 136 (1977).
160. T. Tsuchimoto, *J. Vac. Sci. Technol.* **15**, 70 (1978); **15**, 1730 (1978); **17**, 1336 (1980).
161. K. Miyake and T. Tokuyama, *Proc. Symp., Ion Sources and Ion Assisted Technology, 5th, Kyoto, 1981*, 443 (1981); *Thin Solid Films* **92**, 123 (1982).
162. T. Tokuyama, K. Yagi, K. Miyake, M. Tamura, N. Natsuaki, and S. Tachi, *Nucl. Inst. Methods* **241**, 182–183 (1981).
163. J. Amano, *Thin Solid Films* **92**, 115 (1982).
164. S. Shanfield and R. Wolfson, *J. Vac. Sci. Technol. A* **1**, 323 (1983).
165. S. Komiya, *Proc. Int. Ion Eng. Cong., ISIAT-IPAT, Kyoto, 1983*, 715 (1983).
166. N. Terada, Y. Hoshi, M. Naoe, and S. Yamanaka, *Proc. Int. Ion Eng. Cong., ISIAT-IPAT, Kyoto, 1983*, 999 (1983).
167. H. Inokawa, Y. Yamada, and T. Takagi, *Proc. Int. Ion Eng. Cong., ISIAT-IPAT, Kyoto, 1983*, 1765 (1983).
168. O. Tabata, S. Kimura, M. Kishimoto, M. Takahashi, S. Suzuki, and M. Kunori, *Proc. Int. Ion Eng. Cong., ISIAT-IPAT, Kyoto, 1983*, 1119 (1983).
169. T. Miyazawa, S. Misawa, S. Yoshida, and S. Gonda, *J. Appl. Phys.* **55**, 188 (1984).
170. S. Aisenberg, *J. Vac. Sci. Technol. A* **2**, 369 (1984).
171. H. B. Gilbody and J. B. Hasted, *Proc. Roy. Soc. A* **240**, 382 (1957).
172. E. W. McDaniel (ed.), "Collision Phenomena in Ionized Gases," Wiley, New York, 1964.
173. R. Bernas, L. Kaluszyner, and J. Druaux, *J. Phys. Radium* **15**, 273 (1954).
174. J. M. Sellen, *Proc. IRE*, 477 (1960).
175. O. Buneman and G. P. Kooyers, *AIAA Elect. Propul. Conf.* (1963).
176. H. E. Wilhem, *J. Appl. Phys.* **44**, 4562 (1973).
177. H. R. Kaufman and R. S. Robison, *AIAA Pap. 81–0721* (1981).
178. P. D. Reader, D. P. White, and G. C. Isaacson, *J. Vac. Sci. Technol.* **15**, 1093 (1978).
179. G. Gautherin and C. Lejeune, *J. Phys. D. Appl. Phys.* **9**, 1149 (1976).
180. H. Inokawa, K. Fukushima, I. Yamada, and T. Takagi, *Proc. Symp. ISIAT, 6th, 1982*, 355 (1982).
181. H. Inokawa, K. Fukushima, I. Yamada, and T. Takagi, *Proc. Int. Ion Eng. Cong., ISIAT, Kyoto, 1983*, 1127 (1983).
182. T. Takagi, I. Yamada, and A. Sasaki, *Proc. Int. Conf. Ion Plating Allied Techniques, Edinburgh, 1977*, 50 (1977).
183. T. Takagi, I. Yamada, and A. Sasaki, *Thin Solid Films* **45**, 569 (1977).
184. C. Mahony and P. D. Prewett, *Vacuum* **34**, 301 (1984).
185. K. Shigeno, Y. Kuriyama, and T. Takagi, *Proc. Int. Conf. Ion Plating Allied Techniques, Edinburgh, 1977*, 399 (1977).
186. J. Kugo, T. Morishita, K. Iguchi, M. Koba, and K. Awane, *Proc. Int. Ion Symp., ISIAT, 1982*, 355 (1982).
187. Eaton Final Report, Air Force Contract F-33615-83, C113 (1984).
188. K. Morimoto, H. Watanabe, and S. Itoh, *J. Cryst. Growth* **45**, 334 (1978).
189. T. Takagi, K. Inove, S. Mizugaki, A. Sasaki, and I. Yamada, *Vacuum* **22**, 267 (1980).
190. T. Takagi, K. Matsubara, and M. Takaoka, *J. Appl. Phys.* **51** 5419 (1981).
191. Y. Minowa and K. Yamagishi, *Int. Conf. Electron. Ion, Photon Beams, Los Angeles, 1983*, (1983).
192. K. Mameno, K. Matsubara, and T. Takagi, *Proc. Symp., ISIAT, 6th, 1982*, 341 (1982).
193. H. Usui, N. Naemura, I. Yamada, and T. Takagi, *Proc. Symp. ISIAT, 6th, 1982*, 331 (1982); *Proc. Int. Ion Eng. Cong., ISIAT, 1983*, 1247 (1983).
194. B. Sautreuil, B. Bailly, R. Blanchet, M. Garrigues, and P. Viktorovitch, *Rev. Phys. Appl.* **18**, 763 (1983).

6 FORMATION OF THIN SEMICONDUCTING FILMS BY MAGNETRON SPUTTERING

James B. Webb

Semiconductor Research Group
Division of Chemistry
National Research Council Canada
Ottawa, Canada

I. INTRODUCTION

Sputter deposition of thin films for optical and electrical applications has increased markedly in the past few years. This has been mainly a result of the development of high-performance magnetron cathodes. In comparison to conventional diode sputtering, the magnetron cathode provides higher deposition rates at lower operating pressures and the ability to deposit high-quality films at low substrate temperatures. In addition, compared with other vacuum deposition techniques such as evaporation, magnetron sputtered films have greater adherence and greater uniformity over large areas; yet the overall process can still operate as a conventional roll coater with air-to-air sample transfer [1–7].

Although a number of reviews have provided considerable detail about the various target configurations [8–12] and associated discharge characteristics, the magnetron sputtering process is far from being completely understood. A review of the current literature shows that magnetron sputtering is being applied primarily for the deposition of metals and metal alloys. These metal films find applications in integrated-circuit interconnects, heat-reflecting mirror window coatings, and cosmetic coatings such as the chrome films on plated composite automobile bumpers. To a lesser extent the process is finding applications in the formation of thin compound semiconductor and insulator films for active layers in electronic devices. In this role, the demands placed on the process are increased considerably. For instance, small changes in the film stoichiometry, impurity content (intentional or otherwise), and crystallinity/microstructure can result in dramatic changes in the electrical characteristics of the films. Tailoring the layers to the specific application requires a high degree of process control and an understanding of the complex interactions between the plasma discharge, target, and substrate.

It is likely that magnetron sputtering will find increasing use along with techniques such as molecular beam epitaxy (MBE), organometallic vapor-phase epitaxy (OMVPE), vapor-phase epitaxy (VPE), and liquid-phase epitaxy (LPE) for the deposition of high-quality epitaxial layers of compound semiconductors. For instance, multitarget rf sputtering has already been used to deposit high-quality epitaxial layers of GaSb on GaAs substrates [13] and to deposit superlattices of InSb/GaSb [14]. The more immediate application, however, will likely be in the deposition of semiconducting films for less critical or less exotic applications. These applications include:

(1) large-area, low-cost photovoltaic arrays,
(2) large-area electroluminescent panels and displays,
(3) wide-band-gap, heavily doped semiconductors for antistatic window coatings, heat-reflecting mirrors, etc.,
(4) thin film transistor (TFT) arrays for liquid crystal displays, and
(5) surface acoustic wave (SAW) devices.

In these applications, single-crystal epitaxial layers may not be essential, and satisfactory performance may be obtained from polycrystalline films.

At present, the full potential of magnetron sputtering remains largely unexplored. The purpose of this chapter is to outline the recent advances in the magnetron sputtering of elemental and compound semiconductors. The first part of the chapter reviews the basic operating characteristics of the magnetron cathode as well as the current models describing the effects of power, sputter pressure, and gas composition on sputter rate. The latter part of the chapter looks at the properties of materials deposited by reactive and nonreactive magnetron sputtering.

II. THE MAGNETRON CATHODE

Although the increasing number of publications would tend to indicate that magnetron sputtering is a relatively recent development, the concept of magnetically enhanced sputtering is far from new. The magnetron target is based on the work carried out by Penning [15] more than 45 years ago. The concept was subsequently developed by other workers resulting in the sputter gun [16, 17] and cylindrical magnetron cathode [18]. On the other hand, the planar magnetron—which is perhaps the most widely used target assembly today—was not introduced until the early 1970s by Chapin [19], although the basic configuration was demonstrated by Kesaer and Pashkova [20] more than 15 years earlier. The planar magnetron cathode has many obvious advantages over other target assemblies, particularly its simplicity and its suitability for in-line production systems and coating of large-area planar substrates. A number of cathode geometries are shown in Fig. 6.1. The discussions that follow will be confined to the planar structure; however, in general, the behavior of the various cathodes is similar.

A. Cathode Geometry and Electron/Ion Motion

The planar target in its simplest form is shown schematically in Fig. 6.2. It consists of the target material backed by permanent magnets that provide a toroidal confinement field with the field lines forming a closed tunnel on the target surface. The field strength is chosen to provide effective confinement for electrons while allowing heavier ions considerable freedom. The transverse field component is typically >0.01 T.

Secondary electrons emitted from the target during the sputtering process are accelerated across the cathode dark space towards the highly charged plasma sheath. Their path is modified by the $\mathbf{E} \times \mathbf{B}$ Lorentz force. One component of their motion is a helical path about the magnetic field lines. The electrons traveling along these helical lines toward the center of the target are reflected due to the higher density of field lines in this region and the repulsive electric field encountered. After reflection the electrons eventually reach the perimeter of the target where the field lines again intersect the surface. An anode placed in this region effectively collects these electrons and prevents them from reaching the substrate. A second component of their

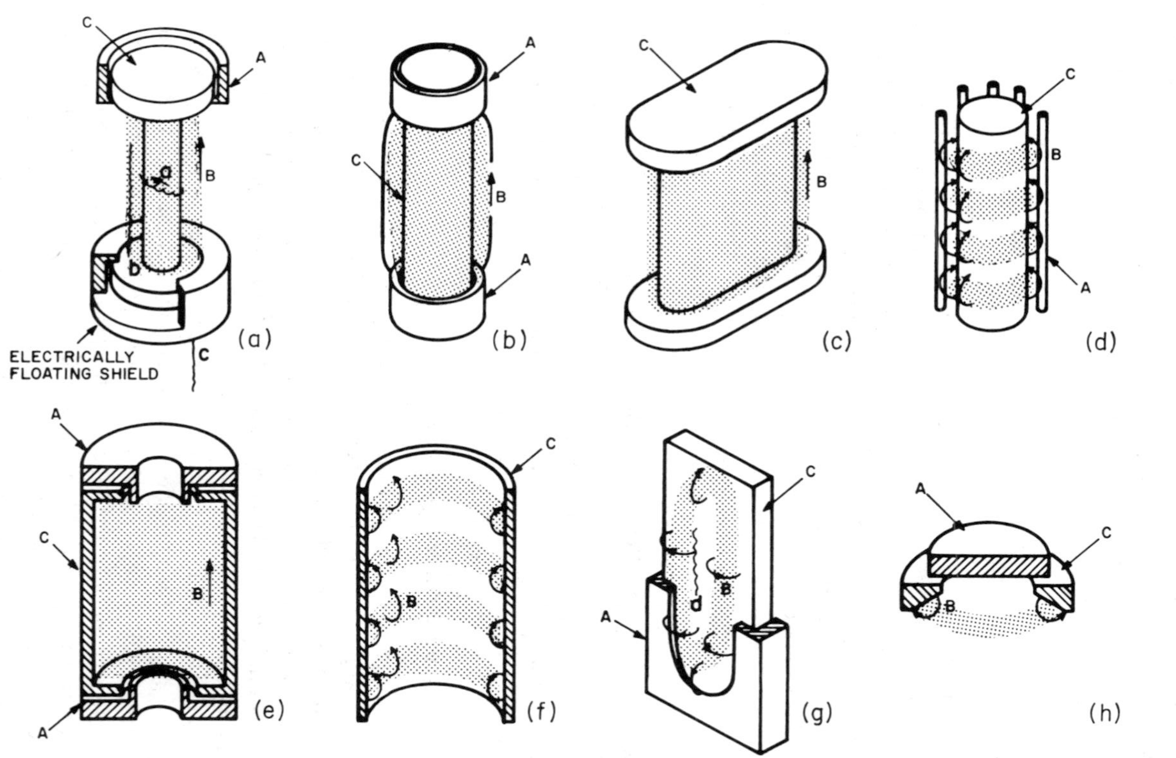

Fig. 6.1. Schematic representations of various magnetron assemblies. **A**, Anode; **B**, direction of magnetic field; **C**, cathode. (a) Cylindrical postmagnetron with electrostatic end confinement (EEC); (b) cylindrical postmagnetron with magnetic end confinement (MEC); (c) rectangular postmagnetron with EEC; (d) ring discharge postmagnetron; (e) cylindrical-hollow magnetron with EEC; (f) ring discharge hollow cathode magnetron; (g) planar magnetron; (h) "gun" type magnetron. (From Thornton [11].)

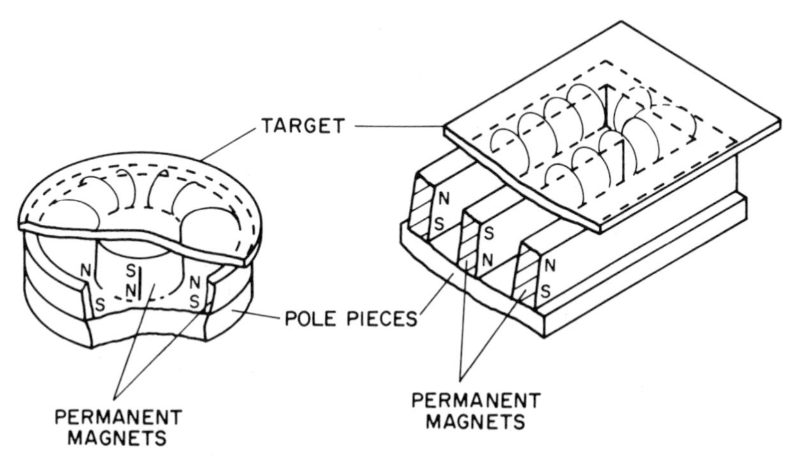

Fig. 6.2. Two variations of the planar magnetron cathode. The confinement field is indicated by the curved lines. (From Waits [8].)

motion is a drift from one field line to another resulting in a race track effect about the toroidal tunnel on the target surface. The combined motion gives an extended path length resulting in a large number of collisions of the electrons with gas atoms. Ideally, their energy is dissipated by these ionizing collisions before reaching the anode ring. Since not all field lines terminate on the cathode surface or the anode ring, some electrons may escape from the trap. These electrons can reach the substrate; however, in general, their density is low. The net effect of this trap is an increase in plasma density close to the target, which in turn leads to higher deposition rates. The low density of electrons reaching the substrate also results in lower substrate temperatures.

The ions, of course, experience the same Lorentz force as the electrons; however, due to the much higher mass, their motion is not as restricted. The Larmor radius of an electron in a magnetic field typical of a magnetron cathode is 1–3 mm, whereas that of the ions is at least an order of magnitude higher. Where the field lines are parallel to the cathode surface the motion of the ions is perpendicular to the target surface. Ions extracted from the plasma sheath, are accelerated across the cathode dark space (typically 3 mm) and strike the target resulting in removal or sputtering of the target material. Typical operating characteristics of the magnetron cathode are given in Table 6.1.

B. DC and RF Magnetron Sputtering

Most magnetron targets operate in the pressure range from 1 to 10 mtorr with cathode potentials of 300–500 V. The cathode voltage follows only a

TABLE 6.1

Characteristics of the Magnetron Cathode

Discharge parameters	
Operating pressure	0.1–10 mtorr
Direct current	
Discharge current	0.25–100 A
Discharge voltage	300–800 V
Current densities	up to 2000 A/m^2
Radio frequency	
Self-bias voltage	50–500 V
Target power	50–10 kW
Deposition rates	
High-rate metals	1–1.5 μ/min
Oxides, etc.	~0.1 μ/min
Magnetic field	>0.01 T
Plasma parameters	
Electron temperature	2–20 eV
Ion temperature	0.25–2 eV
Ion/electron gyromagnetic radius	~10
Cathode dark space	~3 mm

weak dependence on current of the form $I = kV^n$, where n is typically 5–9 (see Fig. 6.3). The larger n values correspond to a greater electron trapping efficiency of the magnetic field. The characteristics are essentially independent of magnetron type, but higher voltages and smaller n values are observed for rf operation. Measurements [21] at lower operating currents, however, have shown that the preceding expression overestimates the current at low voltages and that the I–V characteristics follow more closely an expression of the form $I = \beta(V - V_0)^2$, where V_0 is the minimum voltage necessary to maintain a discharge. This dependence was observed for both dc and rf operation and for a variety of cathode types. The V^2 dependence was shown to be a result of a space-charge-limited electron current in the magnetron geometry, which reduces the electron mobility by several orders of magnitude from the zero magnetic field value.

The basic operation of the magnetron cathode is altered somewhat in the rf case. In rf sputtering, the rf field is capacitively coupled to the cathode. The difference in mobility of the ions and electrons results in a negative self biasing of the cathode, which in turn gives the necessary potential for sputtering. In general the rf field makes both electrodes a cathode during half of the rf cycle. However, since the area of the one electrode (anode/substrate

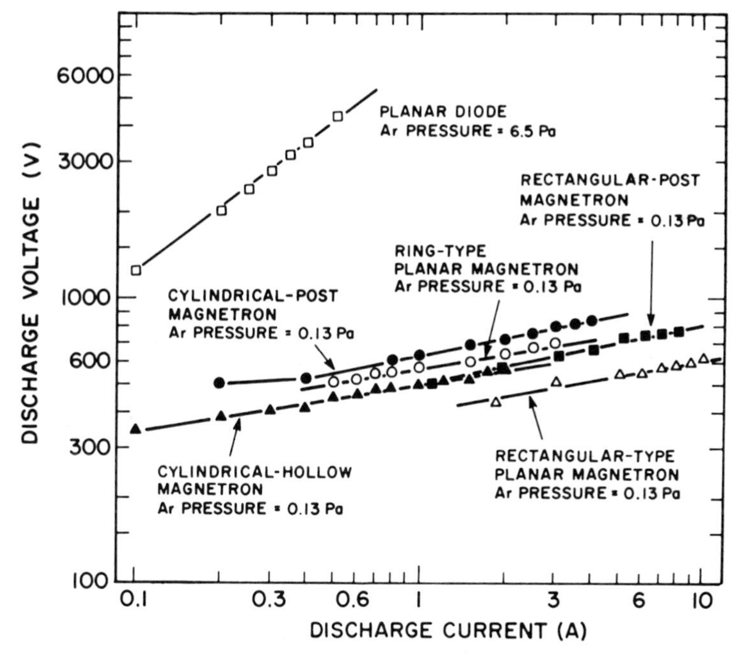

Fig. 6.3. Discharge I–V characteristics of several types of magnetron cathodes. Also shown are the characteristics for a conventional diode system. All targets are aluminum. The operating pressures are indicated. (From Thornton [11].)

holder/chamber walls) is very much larger than the other (cathode assembly), nearly the entire applied potential appears between the plasma sheath and the cathode assembly. This effectively eliminates sputtering from the chamber walls. The main advantage of rf sputtering is the ability to sputter nonconducting materials.

C. Target Erosion and Film Uniformity

All magnetron targets are inherently nonuniform sources. Oscillating magnetic fields have been used to enhance sputter uniformity and target utilization; however, in general, slotted masks and/or substrate motion are employed [8]. The planar magnetron behaves essentially as a ring source [22] owing to the shape of the confined plasma. The highest rate of material removal occurs where the magnetic field lines are parallel to the cathode surface. At the low pressures employed, the mean free path of the sputtered atoms is long enough that it reaches the substrate unimpeded. In this case the deposition profile follows essentially a cosine distribution [22].

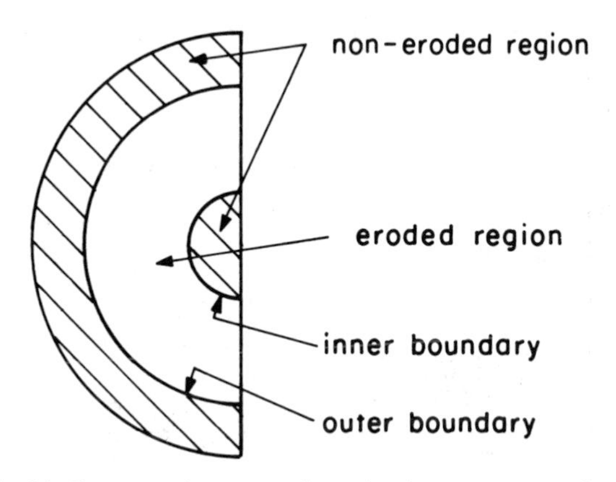

Fig. 6.4. Target erosion pattern for a circular magnetron cathode.

Fukami and Sakuma [23] have developed a simple method for observing the erosion region in a planar magnetron and have determined the behavior of the erosion pattern as a function of the magnetic field strength, applied voltage, and sputter pressure. The method used was to either precoat an aluminum target with a few nanometers of copper oxide or to reactively sputter a copper target in a partial pressure of oxygen. Due to the dark color of the copper oxide, a high contrast ratio was obtained between the oxide and the underlying material. The erosion pattern is shown schematically in Fig. 6.4. It was found that increasing the magnetic field caused a shift of both boundaries of the erosion ring towards the center of the target. The boundaries of the ring were determined by the vertical component of the magnetic field. A very different effect was observed as the pressure was changed. At low operating pressures, the erosion ring was broad and extended well into the center of the target. At sufficiently low pressures (<1.0 mtorr), the outer boundary exceeded the target radius and the discharge was extinguished. Increasing the pressure caused the boundaries to move together, thus decreasing the erosion ring. The effect was attributed to a shorter mean free path of the electrons at higher pressures. Changing the cathode potential had a similar effect to changing the pressure. For increasing voltages both boundaries spread out. This was correlated with the greater energy of the secondary electrons that results in a greater effective mean free path at higher cathode potentials. These observations show that to minimize contamination of the growing film, i.e., from sputtering the guard or anode ring components at the perimeter of the target, the sputtering parameters must be chosen carefully to maintain the erosion ring well within the target boundaries.

III. EFFECTS GOVERNING DEPOSITION RATE OF ELEMENTAL AND COMPOUND SEMICONDUCTORS

A. DC versus RF Sputtering

The film growth rate of any material is determined by the flux of sputtered material arriving at the substrate, as well as by complex surface interactions between the bombarding ions and target material, substrate effects from secondary ions or neutral particles, electromagnetic radiation (x-rays and UV photons), substrate heating, sticking coefficient of sputtered particles, chemical dissociation and/or bonding, adatom mobility, crystallinity, and microstructure and reemission or backscattering of particles from the substrate. Magnetron sputtering allows considerable control over many of these variables, particularly those affecting the growing film at the substrate. For example, substrate temperature and ion bombardment are inherently low compared to conventional diode sputtering. In general, the sputtering yield for elemental semiconductors is less than that of most metals and that of compound semiconductors is less than that of the individual elemental components [24]. Typical sputtering rates for a number of semiconductors and insulators are shown in Table 6.2. In addition to the nature of the target material, the deposition rate is affected by factors such as pressure, applied power, and gas composition.

Although it was earlier believed that there was no real difference in deposition rate between dc and rf sputtering [24], results have shown that rf sputtering has a rate that is approximately $\frac{1}{2}$ that observed for the dc case for the same effective applied power. Nyaeish and Holland [40] carried out a detailed measurement of power and rate for both dc and rf magnetron sputtering of an aluminum cathode. Their results are shown in Fig. 6.5. Similar effects were observed for a copper target [41]. The lower efficiency for rf sputtering was attributed to power dissipation in the ion sheath regions.

In general, however, in either the dc or rf case the deposition rate will be given by some function dependent on substrate temperature, source-to-substrate distance, applied power, and effects due to pressure and gas composition. Webb [25] has carried out measurements of the deposition rate of amorphous hydrogenated silicon where these effects have been modeled.

B. Pressure Effects

Even though magnetron sputtering has a lower working pressure limit than conventional sputtering, in some cases it is desirable to deposit at higher pressures [42, 43]. For instance, higher pressures can enhance gas-atom

TABLE 6.2

Deposition Rates for Various Compounds Prepared by Magnetron Sputtering

Material	Target	Sputter gases	Power density (W/cm^2)	Deposition rate (Å/s)	Pressure (mtorr)	System	Reference
a-Si : H	Si(poly)	$Ar + H_2$	5.0	4.0	1–10	RF planar	[25]
a-Si : H	Si(III)	Hr/Ar (40/60)	3.8	3.0	5	RF planar	[26]
CdSe	CdSe hot pressed	Ar	0.6	25	3–6	RF and dc planar	[27]
CdS : In	Indium doped CdS	Ar/H_2S (1/2)	>2.8	10	3	cylindrical dc	[28]
Cu_xS	Copper	Ar/H_2S (~1/2)	>3.0	5	5.3	cylindrical dc	[29]
$Pb(Zr, Ti)O_3$	$Pb(Zr, Ti)O_3$ ceramic	100% O_2	3.8	1.1–2.7	2.0–100	RF planar	[30]
SiO_2	Fused silica	$Ar + O_2$ ($>1 \times 10^{-5}$ torr)	5.6	6.6	—	RF planar	[31]
$K_3Li_2Nb_5O_{15}$	Sintered	Ar/O_2 (50/50)	1.3	0.22	90	RF planar	[32]
AlN	Aluminum	Ar/N_2 (1/1)	1.3–2.6	0.56–2.2	3–10	RF planar	[33]
AlN	Aluminum	10^{-2} torr N_2	2.0	1.7	10	DC planar	[34]
$BaTiO_3$	$BaTiO_3$ ceramic	Ar/O_2 (80/20)	2.1	1.5	1–20	RF planar	[35]
TiO_x	Titanium	Ar + 30% O_2	19	1.5	5.0	RF and dc planar	[36]
Al_2O_3	Al_2O_3	O_2/Ar (1/1)	7.0	5.8	1.3	RF planar	[37]
Al_2O_3	Aluminum	10% O_2 + Ar	1.13	0.47	20	DC planar	[38]
		10% O_2 + Ar	4.5	6.7		RF planar	
GaN	Gallium	$O_2/N_2 = 0.08$			1.0	RF planar	[39]

interactions in the plasma. At higher pressures mean free path effects become important in determining the deposition rate. The dependence of sputter rate on the deposition pressure has been modeled for nonmagnetron diode sputtering by Keller and Simmons [44]. The percentage of sputtered atoms reaching the substrate in terms of the scattering of particles undergoing supersonic diffusion (i.e., their velocity is greater than the average thermal velocity) and those undergoing normal diffusion is given by

$$\beta = (M\lambda/D)[1 - \exp(-D/M\lambda)], \tag{1}$$

where M is the Mach number of the sputtered atom, D the source-to-substrate distance, and λ the mean free path given by [45]

$$\lambda^{-1} = n'\sqrt{2}\pi N_s\sigma_s^2 + \tfrac{1}{4}\pi N_g(\sigma_s + \sigma_g)^2(1 + M_s/M_g)^{1/2}, \tag{2}$$

where n' is the average number of collisions required to reduce the velocity of the sputtered atom to the thermal velocity, N_s and N_g are the number per unit

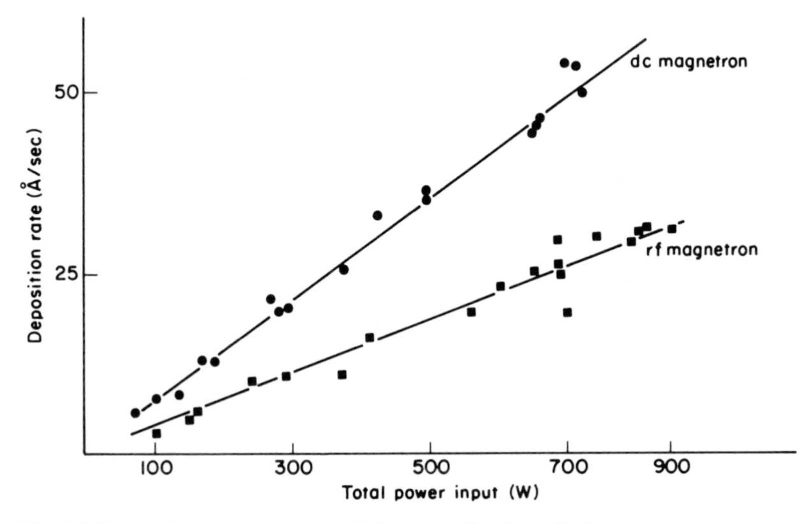

Fig. 6.5. Deposition rate versus applied power for dc and rf magnetron sputtering of aluminum in argon at 5 mtorr total pressure. The deposition rate is monitored by using a quartz crystal microbalance at the center of the target. (From Nyaiesh, and Holland [40]. *Vacuum* **31**. Reprinted with permission. Copyright 1981 Pergamon Press, Ltd.)

volume of sputtered and gas atoms, respectively, σ_s and σ_g the corresponding atom diameters, and M_s and M_g the atom masses. This equation does not take into account effects due to re-emission at the substrate and target. By fitting Eq. (1) to the results observed for amorphous silicon, Webb [25] has shown that the re-emission of material appears to be small in the magnetron case. Furthermore, the parallel-plate geometry assumed in the derivation of this equation fits, at least to a first approximation, with the magnetron configuration. The deposition rate for a-Si as a function of sputtering pressure for various power levels, is shown in Fig. 6.6. The solid lines in the figure are calculated from Eq. (1).

C. Effects Due to Gas Composition (Nonreactive)

The sputter yield/rate, to a large extent, depends on gas composition. For instance, sputtering with helium is far less efficient than with argon due to the much lower atomic mass of helium. The case is complicated further when the gas is reactive, since changes in target surface composition can occur, which, in turn, may cause rapid changes in the sputter rate. Furthermore, reactions in the plasma or at the substrate surface may also result in a change in deposition rate. Thus, the deposition rate may be considerably different than that predicted by the target erosion rate. One of the first observations of the

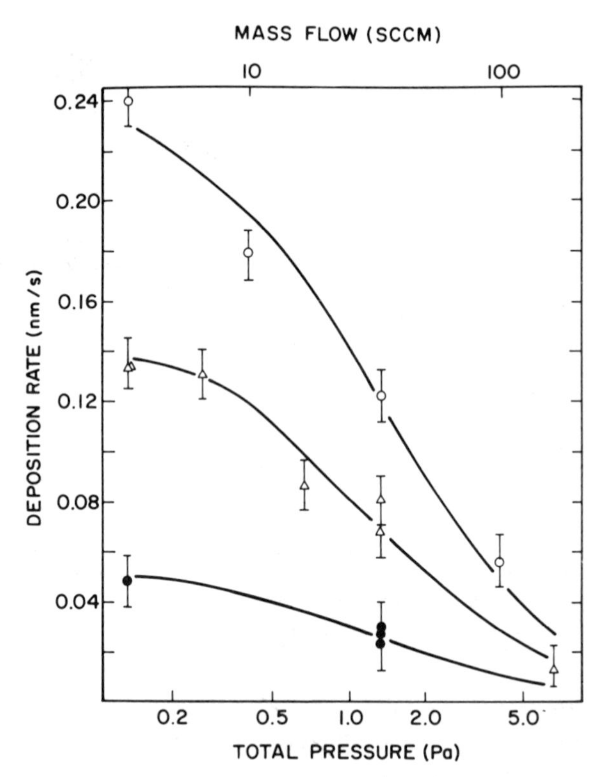

Fig. 6.6. Deposition rate versus total sputtering pressure. Sputter gas is argon: ●, 100 W; △, 200 W; and ○, 300 W. Solid curves are theoretical fits to the data using Eq. (1).

influence of relatively large additions of gases of low atomic mass to the magnetron discharge indicated that there was only a relatively small change in the sputter rate. This is in direct contrast to conventional diode sputtering, where even small additions of gases, such as hydrogen, to the discharge caused large decreases in deposition rate. Since water vapor is almost always present in a vacuum system, small amounts of ionized hydrogen could potentially limit the sputter rate. The effect of hydrogen on the deposition rate of aluminium has been studied by Maniv and Westwood [46] for both dc and rf planar magnetron sputtering and by Webb [25] for the deposition of silicon. In both cases compositional changes of the target surface were small, although in the latter case reactions in the plasma and/or substrate occurred.

Maniv and Westwood [46] showed that the insensitivity of the deposition rate to additions of hydrogen resulted from the low concentration of H^+ ions in the discharge. Their approach was to model the reactions in the plasma, calculate the relative concentrations of the various species, and fit these results to experimental values of the deposition rate as a function of hydrogen concentration. The reactions considered were:

$$Ar + e \longrightarrow Ar^+ + 2e, \quad (3)$$

$$Ar^+ + H \longrightarrow ArH^+, \quad (4)$$

$$ArH^+ + e \longrightarrow Ar + H^+ + e, \quad (5)$$

$$H^+ + e \longrightarrow H. \quad (6)$$

To obtain a manageable solution, a number of reactions were omitted such as the formation of H_2^+ by impact ionization and H^+ by dissociative ionization of H_2. Experimentally the effect of either H_2^+ or H^+ is small, and thus the omission of these reactions is not serious. From the rate equations governing the concentrations of these species, the authors derived an expression for the ion concentrations

$$N^2[x^2 + Ax) - N[(a + B)x + u + Ab] = -ab, \quad (7)$$

where $N = n(\mathrm{Ar})$, $x = u(1 + W)/z$, $A = 1 - uW/z$, $u = n_e r_1/r_2$, $W = r_3/r_4$, $z = n_e r_3/r_2$, and n_e is the number of electron collisions required for thermalization and r_1–r_4 are the various rate constants for the reactions given in Eqs. (3)–(6). The hydrogen to argon ratio is b/a. Since the sputter rate is directly proportional to the concentration of argon ions, fitting Eq. (7) to the data for both the diode and magnetron discharge yields values of a, b, W and u/z in terms of N. The concentration of each species is then readily calculated. Maniv and Westwood fitted the results of Stern and Caswell [47] for the conventional diode system and found that the calculated ionic concentrations as a function of the hydrogen to argon ratio compared well with the corresponding deposition rates up to a hydrogen concentration of approximately 10%. A less satisfactory fit was obtained at higher concentrations where the model indicated 85% H^+ at 30% added hydrogen. This concentration predicts a much lower sputter rate than experimentally observed.

A similar calculation for the magnetron cathode indicated that less than 0.01% of all ion species were other than Ar^+ for a b/a ratio of 0.01. Increasing the b/a ratio resulted in little change. This is in agreement with the experimental results, where large additions of hydrogen to the plasma had little effect on the rate.

Webb [25] has studied the deposition of silicon in an argon–hydrogen discharge and found a similar result. Webb modeled the rate directly by extending the model of Stern and Caswell [47]. The deposition rate is given by

$$R/R_0 = [1 - (K_H p_H / K_A p_A)]^n, \quad (8)$$

$$R/R_0 \approx 1 - n(K_H p_H / K_A p_A), \quad (9)$$

where R_0 is the deposition rate in the absence of hydrogen, K_A and K_H represent the product of ion mobility and ionization probability for each of the gases, p_A and p_H are the corresponding partial pressures for argon and hydrogen, respectively. Here n is related to the power dependence of the deposition rate for silicon by the expression

$$R = AP^n, \tag{10}$$

where P is the applied power and A a constant related to system geometry, sputter pressure, and sputter gas.

A plot of R/R_0 versus the hydrogen to argon ratio for the deposition of silicon was found to give a straight line with slope $(3\mu_H I_H/2\mu_A I_A) = 0.16$, where μ_H, μ_A, and I_H, I_A are the corresponding mobilities and ionization probabilities of the two ions, respectively. If μ_H/μ_A is taken as approximately 20 [47], then the relative ionization probabilities of hydrogen and argon are $I_H/I_A < 0.5\%$. Since the only differentiation made among the various ionic contributions in this model lies in the relative sputtering abilities of the ions, the ionization probabilities of hydrogen and argon represent all ions of hydrogen, such as H^+, H_2^+, and argon (Ar^+, ArH^+). The 0.5% upper limit for the ratio of hydrogen to argon ions compares closely with that determined by Maniv and Westwood (at 30% added hydrogen, 99.5% of the ions were Ar^+).

By comparing the equations for power, pressure, and hydrogen/argon ratio, the deposition rate for various power levels and hydrogen concentrations can be compared to experimentally observed rates. Figure 6.7 is a plot of deposition rate as a function of total sputter pressure at various power levels and hydrogen partial pressures. The dashed lines are calculated from Eqs. (1), (8), and (10). The solid lines are for no added hydrogen. At low sputtering pressures where the hydrogen/argon ratio is large, the deposition rate is low. As the total pressure is increased (for fixed power and hydrogen partial pressure), the rate increases as more of the power is carried in the argon ion current. At higher pressures the effect of a reduced mean free path dominates, and the rate falls. Perhaps the most interesting observation is the maxima in deposition rates for various power levels. The maxima depend on both total and partial hydrogen pressures. Figure 6.8 shows the deposition rate as a function of hydrogen partial pressure for various power levels at constant total pressure. The calculated curves are indicated by the solid lines. As shown in the figure, when the partial pressure of hydrogen becomes a significant fraction of the total sputter pressure, the sputter rate for all power levels decreases rapidly.

D. Effect of Gas Composition (Reactive)

Gases that are highly reactive in terms of the target material can give rather abrupt changes in sputter rate as a function of time. The effect is

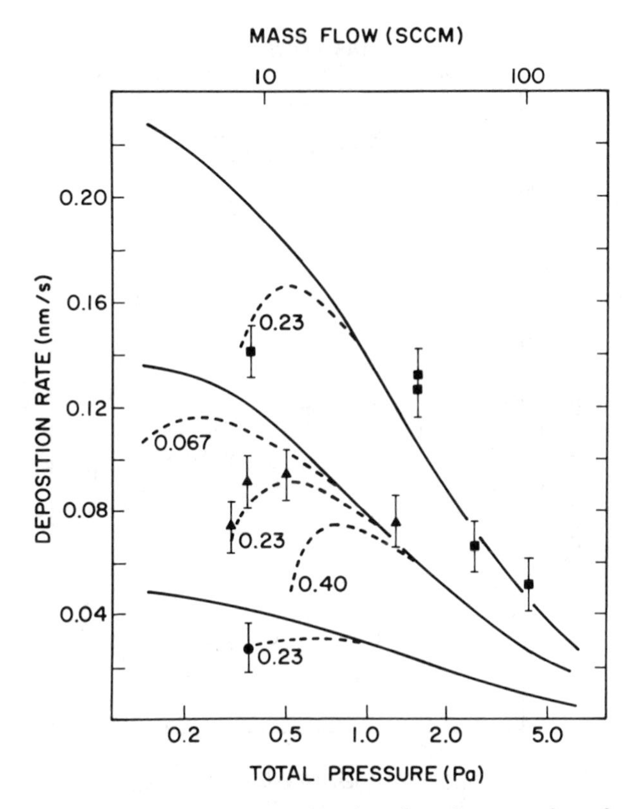

Fig. 6.7. Changes in deposition rate as a function of total pressure for a fixed hydrogen partial pressure of 0.23 Pa at power levels of 100 W (●), 200 W (▲), 300 W (■). The dashed lines are derived from theory for the hydrogen partial pressures indicated. The solid curves are for no added hydrogen.

essentially due to differences in the sputter rate of different materials (e.g., the composition of the target can be time dependent). The initial effect of adding a reactive gas to the discharge is to change the rate by changing the atomic mass of the sputter gas (see Section III.C). However, if the added gas is gettered by the target, then a change in target composition can occur. For instance, reactively sputtering a metal target with oxygen will result in target oxidation (depending on the relative oxidation and sputter rates of the oxidized and nonoxidized targets). Since the sputter rate for most metals is greater than that of the metal oxides [24], a dynamic feedback between oxidation and material removal can arise. As oxidation proceeds, the sputter rate decreases, which in turn causes oxidation to proceed even further. During the target transition between pure metal and metal oxide, the deposited film will show a compositional change with thickness. The dynamics of this process are discussed in Section IV.

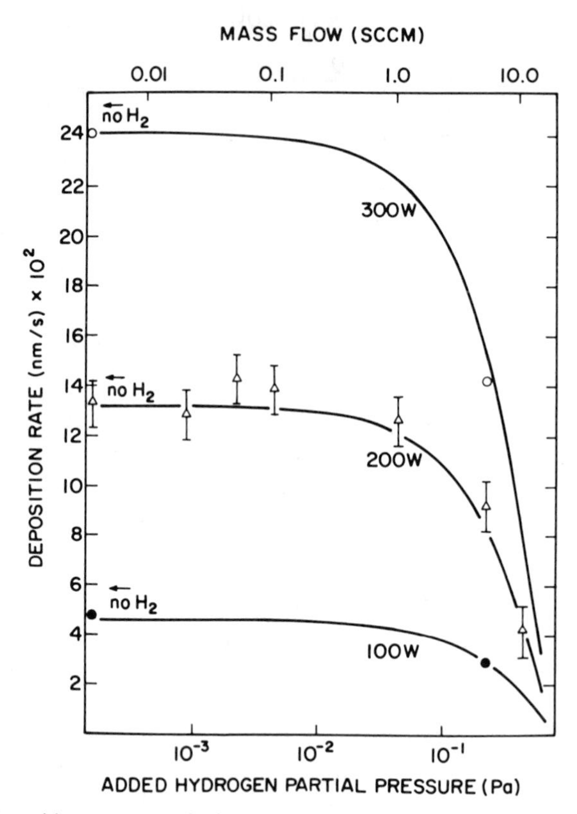

Fig. 6.8. Deposition rate versus hydrogen partial pressure for a fixed argon pressure of 0.13 Pa. The solid lines are theoretical fits to the data.

IV. DEPOSITION OF COMPOUND SEMICONDUCTORS

A. Time Dependence of Target Oxidation (Reactive Sputtering)

The discharge characteristics and target oxidation kinetics have been studied in some detail for both metal and metal-oxide targets [48–50]. In general, for a metal target that is sputtered in an argon discharge, there is no observable change in system pressure as the oxygen flow rate is increased until some critical value is reached. At this point the oxygen partial pressure (and, thus, system pressure) rises to some new stable value. Increasing or decreasing the flow rate at this point causes corresponding changes in system pressure. If, however, the flow rate is decreased to some minimum critical value, then the system pressure decreases abruptly to the original value prior

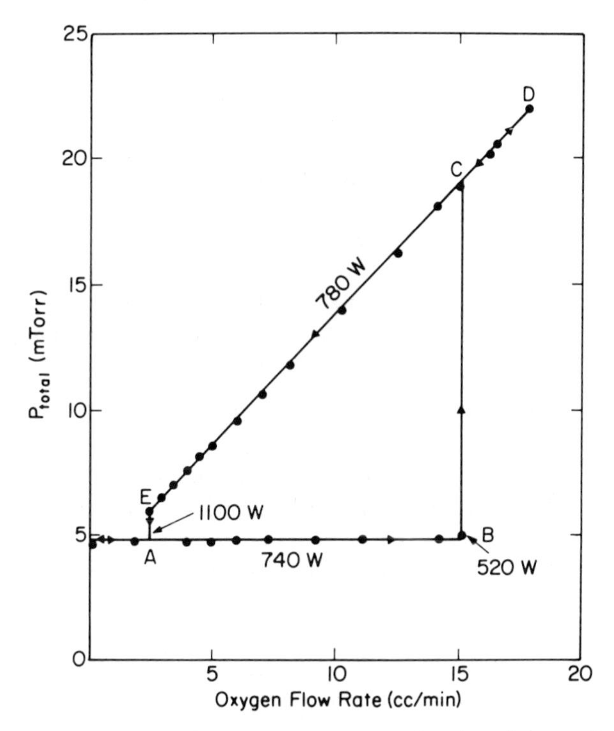

Fig. 6.9. Steady-state values of total gas pressure as a function of oxygen flow rate for dc sputtering of an aluminum target with a fixed argon flow rate of 3.1 sccm. The regions denoted by the letters A–E are discussed in the text. (From Maniv and Westwood [48]. Copyright © 1980 Bell-Northern Research Ltd.)

to oxygen admission. The characteristics are shown in Fig. 6.9 for an aluminum target sputtered in an argon–oxygen discharge. Region A–B in the figure corresponds to the deposition of oxygen doped metal films and D–E to Al_2O_3 films. For region A–B, the rate of removal of metal and metal oxide from the target exceeds the oxidation rate and the target remains essentially metallic. At a critical oxygen flow rate (point B), the oxidation rate exceeds the sputter rate and the target surface becomes increasingly oxidized (B–C). The oxygen partial pressure rises since a smaller flow rate is required to maintain the oxidation state of the target. At this point increasing or decreasing the oxygen flow rate has little effect on the film composition unless a minimum flow rate is reached (point E). At this point, the rate of material removal exceeds the oxidation rate and the target returns to its original oxide-free state. A similar behavior is observed for other sputter parameters such as deposition rate and target self-bias. Maniv and Westwood [48] have used these parameters to probe the oxidation kinetics of the target. A model was developed starting from the rate of change in oxide thickness given by

$$dx/dt = (dx/dt)_{\mathrm{ox}} - (dx/dt)_{\mathrm{etch}}. \tag{11}$$

This can be related to the system pressure by using

$$p_s = A' - B'(dx/dt), \tag{12}$$

where A' and B' are system parameters. If the power remains constant during the oxidation, as in the rf case, then the equations can be combined to give an empirical expression of the form

$$p_s(t) = u - v/[\exp(wt) - z], \tag{13}$$

where u, v, w, and z are constants used to fit the data. This expression has been used to fit the experimentally observed time dependence of the system pressure at the critical flow rate as shown in Fig. 6.10.

To calculate the target surface oxide thickness the form of the oxidation

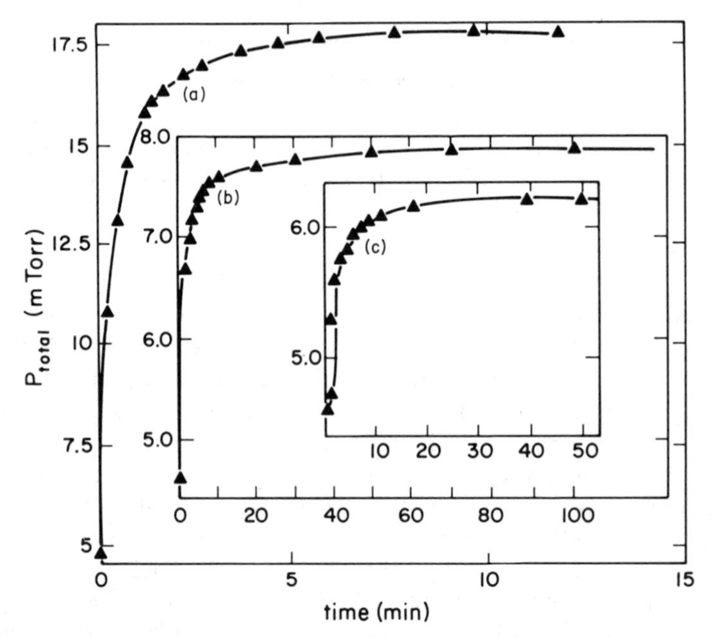

Fig. 6.10. Variation of total system pressure with sputter time at an oxygen flow rate corresponding to point B of Fig. 6.9. The full lines are calculated from Eq. (13). (a) dc power of 560 W; (b) rf power of 500 W, shutter open; and (c) rf power 500 W, shutter closed. (From Maniv and Westwood [48]. Copyright © 1980 Bell-Northern Research Ltd.)

rate needs to be considered. Two forms were used, namely, an exponential rate first proposed by Greiner [51] and a parabolic oxidation rate

$$(dx/dt)_{ox} = K \exp(-x/x_0), \tag{14}$$

$$(dx/dt)_{ox} = K_1/2x. \tag{15}$$

By using these expressions, oxide thicknesses for various powers and pressures were calculated along with the time necessary for oxide removal. These values were compared to the experimentally determined values. Reasonable agreement was found for the parabolic oxidation rate, while more than an order of magnitude difference was observed for the exponential rate dependence. Typical equilibrium values of oxide thicknesses were 50–100 nm depending on the sputter conditions. For instance, the time taken for oxidation of the target varied with applied power. For a dc power of 560 W, (5 × 8-in. target), the time taken for complete oxidation was approximately 10 min, whereas for an rf power of 500 W the time taken was approximately 25 min. During this time, the composition of the film changed from metallic to oxygen-doped to oxide.

Target oxidation effects have also been observed by using oxide targets. Buchanan *et al.* [50] looked at the deposition of rf sputtered films of indium tin oxide (ITO) from a sintered disk of In_2O_3 : Sn (3 mole %). No secondary discharges were used, and the substrates were unheated. For the case with no added oxygen, the films were amorphous with poor transparency but increased in conductivity and transparency when small amounts of oxygen were added to the discharge (5×10^{-5} torr in 4 mtorr argon). This was primarily a result of the films becoming polycrystalline. For higher partial pressures of oxygen and after prolonged sputtering, the films became highly insulating. The kinetics of the process was studied from the properties of the deposited layers as a function of sputter time and film thickness. Although changes in target self-bias, sputter rate, etc. were not significant in this regime (i.e., the target is essentially stoichiometric within 1 part in 1000), the sensitivity of the electrical properties of the films clearly reflected small changes in target oxidation. A simple kinetic model was proposed in which the rate of change of fully oxidized metal atoms on the target surface was given by

$$dx/dt = A''(z_0 - x) - Ex, \tag{16}$$

where A'' is the rate constant for oxidation of the free metal atoms with density z_0 at $t = 0$ and E the etch rate constant for oxidized and unoxidized atoms (assumed to be equal). The density of unoxidized species is thus

$$m(t) = z_0 - dx/dt, \tag{17}$$

$$m(t) = z_0\{1 - [A''/(A'' + E)](1 - \exp(-A'' + E)t)\}. \tag{18}$$

Since the sputtered material will reflect the target composition and since the concentration of oxygen vacancies is proportional to the density of unoxidized metal atoms, the carrier concentration of the film is

$$n_{\text{meas}} = \int_0^t n_0\left[1 - \frac{A''}{A'' + E}(1 - \exp(-A'' - E)t)\right], \tag{19}$$

$$n_{\text{meas}} = \frac{n_0}{A'' + E}\left[E + \frac{A''}{(A'' + E)t}(1 - \exp(-A'' - E)t)\right]. \tag{20}$$

Using this equation, the authors were able to fit the observed dependence of carrier concentration on sputter time (Fig. 6.11a), which gave values of $n_0 = 1 \times 10^{21}/\text{cm}^3$ and $A'' = 3.28 \times 10^{-3}/\text{s}$ with $E \ll A''$, thus indicating that the density of free carriers was governed by the oxidation rate. Assuming the oxidation rate to be proportional to oxygen partial pressure in Eq. (20), (i.e., $A'' = ap_s$, where a is the proportionality constant), the carrier concentration as a function of added oxygen (Fig. 6.11b) could be fitted satisfactorily. This simple model can be extended to the case of a metal target [48] since the original state of target oxidation is arbitrary. If the oxidation rate exceeds the etch rate and if sufficient oxygen is present (i.e., point B in Fig. 6.9) then the partial pressure of oxygen (hence, the system pressure) will rise as the target oxidizes. By using Eq. (19) the time dependence of the system pressure will be of the form

$$p_s(t) = A_1 - B_i \exp(-\gamma t). \tag{21}$$

This equation has the same form as Eq. (13) for $z \ll \exp(wt)$. However, unlike the previous model [48], the oxidation rate is assumed to depend only on the available density of unoxidized metal sites, independent of their spatial distribution, rather than a spatial oxidation rate as assumed by Maniv and Westwood [48]. In spite of these approximations, both models provide a satisfactory description—at least to first order—of the oxidation kinetics. A more fundamental understanding of the oxidation process is likely to lie in the basic microscopic effects associated with the sputter process (see, for instance, Sigmund [52]).

B. Film Composition

Control of semiconductor purity and/or stoichiometry is critical in defining the electrical and optical properties of the deposited films. For elemental semiconductors this means that target purity and purity of sputter gases are essential. To date, the most widely studied elemental semiconductor prepared by magnetron sputtering has been amorphous hydrogenated silicon. Hydrogen is added to the sputter gas to control the defect

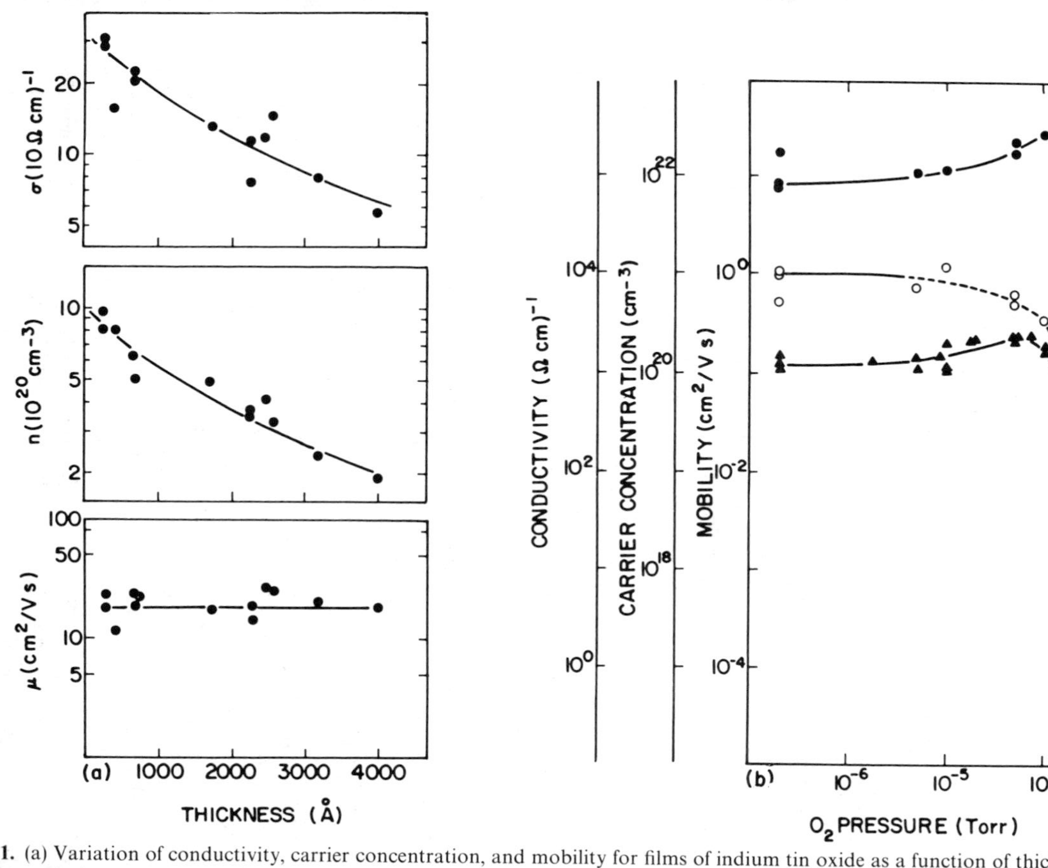

Fig. 6.11. (a) Variation of conductivity, carrier concentration, and mobility for films of indium tin oxide as a function of thickness (sputter time). The data shown is for an added oxygen partial pressure of 5×10^{-5} torr. The solid lines are calculated from Eq. (20). (b) Variation in carrier density (○), mobility (●), and conductivity (▲) of indium tin oxide films as a function of added oxygen for a constant thickness of 800 Å. The dashed line is a theoretical fit to the data. (From M. Buchanan *et al.* [50].)

density of the material; however, it would appear that the addition has only a minimal effect on target surface composition. The sputter characteristics are essentially those of pure silicon.

For compound semiconductors the target is a metal that is sputtered in a reactive gas or a sintered disk of the compound. The latter may be of low density with a large effective surface area for gas adsorption. Degasing of the target before deposition is likely to be critical, especially if the system has been vented to the atmosphere. Furthermore, the purity of these sintered targets is, in general, lower than that of the pure elements. Hot pressing even high-purity compound semiconductor powders can result in nonstoichiometric targets, with additional impurities from the die used in manufacturing. Single-crystal or polycrystalline targets are either expensive or simply not available.

If, however, we start with a target of known and satisfactory composition, the sputtering process assures stoichiometric removal of material. When one component of the target is preferentially sputtered, then a surface depletion of this component occurs. This depletion continues until the concentration of the component with the lower yield compensates for the lower rate of removal. A dynamic state of equilibrium is set up and the rate of removal of both components at any instant is proportional to the target composition. In contrast to this are processes such as thermal evaporation, in which a highly nonstoichiometric vapor can occur, particularly if there is a large difference in vapor pressure of the components of the compound.

Of course, life is never this simple and although the sputtered material is of the same composition as the target, what actually sticks to the substrate may be quite different. Substrate temperature and substrate material can alter the composition of the growing film in the same fashion as that observed for vacuum deposited films [53]. Other factors that are poorly understood, such as chemical reactions in the plasma discharge, bombardment of the growing film by energetic neutral species, and the effect of secondary discharges at the substrate, can also alter the film stoichiometry and structure.

In spite of these factors, growing a film of stoichiometric composition is much easier to achieve than films in which we want to define the electrical conductivity through the controlled addition of vacancies or interstitials (such as the transparent conducting oxides). As an example, sputtering a metal target in a partial pressure of oxygen can produce an oxide target that is sputtered stoichiometrically. An overpressure of oxygen assures that the metal oxide being deposited is close to stoichiometry even though the sticking coefficients of the metal and oxygen atoms may be quite different. The same is true for the III–V and II–VI semiconductors. Sputtering an InP target with added PH_3 in the discharge can make up for losses of phosphorous in the growing film.

The magnetron system, of course, can be highly controlled and thus, in spite of the complex interactions during deposition, high-quality semiconductor compound films can be produced (for example, highly transparent and conductive films of metals oxides).

V. COMPOUND SEMICONDUCTORS DEPOSITED BY MAGNETRON SPUTTERING

A. Transparent Conducting Oxide Films

The term transparent conducting oxide refers to heavily doped oxide semiconductors that have a band gap sufficiently large to make them "transparent" over the visible spectral range and a conductivity high enough such that they exhibit metallike behavior. Typical conductivities are in the range of 10^3 to 10^5/ohm cm with optical transmittance of 80–95% averaged over the visible spectrum. Doping is generally achieved by adding either donor impurities (e.g., Sn in indium oxide) or through controlled changes in stoichiometry, which create oxygen vacancies that act as donors. Due to their high conductivity, the films also show high reflectivity in the near infrared. Figure 6.12 gives the observed and calculated transmission and reflection of films of cadmium stannate (CTO) and indium tin oxide (ITO). The films are highly transparent in the wavelength region of 0.3–1/μm while showing high reflectivity at wavelengths greater than 4 μm for the CTO and 2 μm for the ITO. Heat-reflecting mirror coatings for architectural glass and for solar heat traps take advantage of these properties.

A number of reviews [55–58] and a book [59] have been published detailing fundamental studies of these layers and various methods of preparation including sputtering (both diode and magnetron), spray pyrolysis, chemical vapor deposition (CVD), and thermal evaporation. Although all these techniques have produced high-quality layers, most films require heated substrates and/or post annealing treatments, whereas magnetron sputtering has achieved comparable or superior films on unheated substrates. The most commonly prepared conducting metal oxides are given in Table 6.3 for various preparation techniques. These have found applications as

(1) transparent electrodes for liquid crystal displays,
(2) heat reflecting mirrors,
(3) antistatic window coatings,
(4) defoggers for auto and aircraft windows,
(5) antireflection coatings and collector and/or junction electrodes for SIS photovoltaic cells.

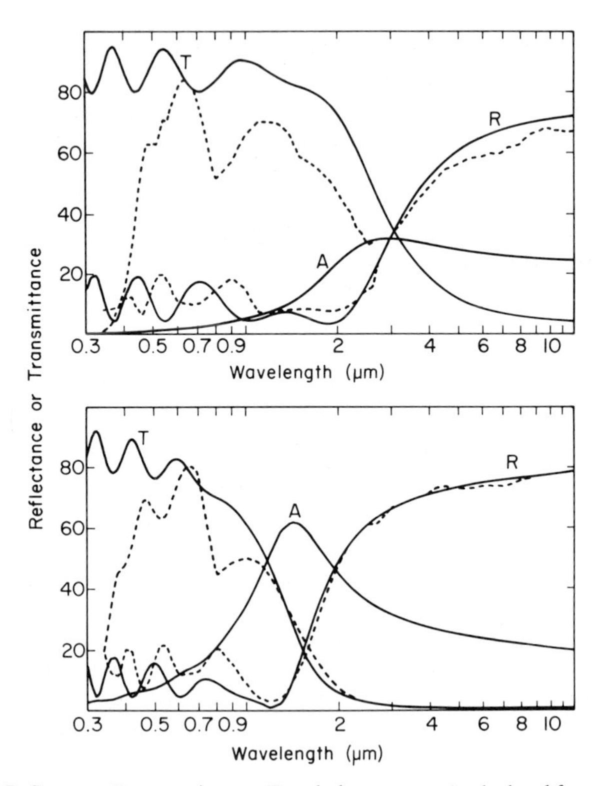

Fig. 6.12. Reflectance R, transmittance T, and absorptance A calculated from the measured film parameters of an indium tin oxide film on glass (solid lines, upper graph). Measured R and T are given by the dotted lines. Lower graph shows the corresponding measurements for a film of cadmium stannate on glass. (From Howson and Ridge [54].)

These applications require high volume and large surface area coverage at low cost, features that magnetron sputtering can provide.

Controlling the state of target oxidation is, of course, critical to achieving uniform films of high conductivity and transparency. This has been discussed in Section II.A. Changes in target self-bias, oxygen partial pressure, or sputter rate during deposition have all been used as a feedback parameter to control the quality of the film as a function of sputter time.

Enjouji *et al.* [81] used a spectroscopic analysis technique to monitor the emission from the plasma discharge during the deposition of indium tin oxide. The films were deposited from an indium–tin metal target in an oxygen–argon discharge. It was found that the emission from one particular indium line was directly proportional to the deposition rate. By using the intensity of this line in a feedback circuit, stable operation of the sputtering system could be maintained for all oxygen flow rates. However, to obtain

TABLE 6.3

Properties of Transparent Conducting Oxides Prepared by Various Deposition Techniques[a]

Material	Technique	t (μm)	n (cm^{-3})	μ (cm^2/V s)	ρ (ohm cm)	T (%)	Reference
In_2O_3	Evaporation	0.27	4.7×10^{20}	74	1.8×10^{-4}	~90	[60]
In_2O_3	Spray pyrolysis and Ar anneal at 400°C	0.14	7.8×10^{19}	30	2.7×10^{-3}	~85	[61]
In_2O_3	Planar magnetron	0.10	4.0×10^{19}	19	8.2×10^{-3}	>80	[57]
In_2O_3 : Sn	Evaporation	0.25	3.8×10^{20}	43	3.7×10^{-4}	~85	[62]
In_2O_3 : Sn	Active reactive evaporation		$\sim 10^{21}$	~30	$\sim 7 \times 10^{-5}$	>90	[63]
In_2O_3 : Sn	RF sputtering	~0.7	$\sim 6 \times 10^{20}$	35	$\sim 3 \times 10^{-4}$	~90	[64]
In_2O_3 : Sn	RF planar magnetron	0.08	$\sim 10^{21}$	10	4×10^{-4}	>85	[50]
In_2O_3 : Sn	DC planar magnetron and rf on substrate	0.33	$>10^{20}$	20	$<1 \times 10^{-3}$	>80	[65]
In_2O_3 : Sn	Spray pyrolysis	0.46	4.9×10^{20}	51	$\sim 2.5 \times 10^{-4}$	85	[66]
In_2O_3 : Sn	CVD (vacuum annealed at 400°C)	0.95	8.8×10^{20}	43	$\sim 1.8 \times 10^{-4}$	90	[67]
SnO_2	Spray pyrolysis (Ar annealed at 400°C)	0.59	9.5×10^{19}	13	5×10^{-3}	85	[61]
SnO_2	CVD	0.60	9.0×10^{18}	10	7×10^{-2}	90	[68]
SnO_2	RF planar magnetron (RT substrates)	0.10	3.0×10^{20}	13	1.6×10^{-3}	85	[69]
SnO_2	RF planar magnetron (400°C substrates)	—	3×10^{20}	10	2×10^{-3}	80	[70]
SnO_2 : Sb	Active reactive evaporation	0.5	—	—	$\sim 8 \times 10^{-5}$	85	[71]
SnO_2 : Sb	Spray pyrolysis	0.35	$\sim 5 \times 10^{20}$	~10	$\sim 10^{-3}$	85	[72]
SnO_2 : Sb	CVD	0.36	1.2×10^{20}	23	$\sim 2 \times 10^{-3}$	85	[73]
SnO_2 : Sb	OMCVD	0.50	—	—	$<1 \times 10^{-3}$	>80	[74]
SnO_2 : F	Spray pyrolysis	0.50	$\sim 5 \times 10^{20}$	23	5×10^{-4}	85	[75]
ZnO	Active reactive evaporation	0.25	$\sim 10^{20}$	~30	$\sim 8 \times 10^{-4}$	88	[76]
ZnO	RF planar magnetron	0.10	$\sim 2 \times 10^{20}$	~16	$\sim 2 \times 10^{-3}$	>90	[77]
ZnO	Spray pyrolysis (H_2 annealed at 350°C)	—	—	—	$\sim 10^{-3}$	~90	[78]
Cd_2SnO_4	RF sputtering	1.0	2×10^{21}	30	$\sim 1.7 \times 10^{-4}$	85	[79]
Cd_2SnO_4	DC planar magnetron and rf on substrate	—	—	—	4.5×10^{-4}	85	[80]
Cd_2SnO_4	DC planar magnetron and rf on substrate	0.28	$>10^{20}$	35	$<1 \times 10^{-3}$	>80	[65]

[a] Here t is the thickness, T the percent transmittance at ~400–800 nm, n the carrier concentration, μ the mobility, and ρ the resistivity.

high-quality films with high optical transparency and low resistivity, a post heat treatment of the layers was required.

Chandler *et al.* [82] used a computer-controlled magnetron system that incorporated both a quartz crystal microbalance and optical thickness monitor as feedback sensors. High-quality optical films were deposited although no examples of conducting oxides were reported.

McMahon *et al.* [83] developed a computer-controlled planar magnetron system for the deposition of aluminum nitride. As with any reactive gas, the same effects observed for target oxidation were also observed. In this case an aluminum target was sputtered in a mixture of argon and nitrogen. Three modes of deposition were studied, namely,

(1) controlling gas flow at constant power,
(2) controlling the power at constant gas flow, and
(3) controlling the cathode voltage at constant gas flow.

The first two methods were found to give runaway behavior (i.e., uncontrolled nitriding of the target), while the third method gave stable operation for all degrees of target nitriding.

A different approach to the problem of changes in target composition was the development of techniques to desensitize the magnetron target from oxidation or "poisoning." Maniv *et al.* [84] modified the magnetron target assembly by enclosing the target in a stainless steel box with a slotted exit mask (Fig. 6.13). Argon was admitted to the box to provide the necessary sputter discharge. At the same time oxygen was introduced near the substrate surface. By carefully controlling the system pressure and box overpressure, oxygen contamination of the target could be reduced while still maintaining sufficient oxygen at the substrate for deposition of the metal oxide. The metal target was dc sputtered, but it was found that a second ionizing rf discharge was necessary at the substrate to enhance oxidation of the arriving metal atoms. The low power in this discharge and the low observed self-bias assured that no resputtering of the film occurred. Unfortunately the target enclosure reduced the deposition rate considerably, thus degrading one of the principal advantages of the magnetron source.

Ridge *et al.* [85] also deposited a number of conducting oxides using dc sputtering of a metal cathode in an Ar/O_2 mixture. Unlike the results of Maniv *et al.* [84], the oxides were deposited onto a continuously moving substrate where the quality of the film was monitored as deposition progressed. By changing the sputter power in response to film quality, films of low resistivity and high optical transparency could be deposited over large areas on a continuous basis. Although a secondary rf discharge at the substrate was not necessary [86], use of a second discharge broadened the resistivity minimum, thus making target power and oxygen flow rate control

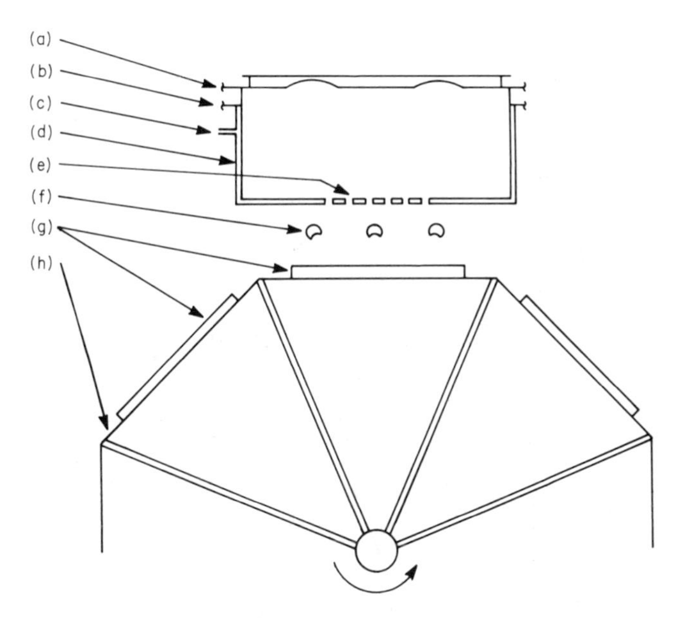

Fig. 6.13. Target and substrate configuration for the "box" arrangement. (a) and (b) Target and cathode assembly; (c) argon gas inlet; (d) target enclosure; (e) exit mask; (f) oxygen gas inlet; (g) substrates; (h) substrate carousel. (From Maniv *et al.* [84]. Copyright © 1983 Bell-Northern Research Ltd.)

less critical, while at the same time increasing the deposition rate. Even with this, the power to the target and oxygen admission rates had to be controlled to within 0.05%. The measured mobilities and carrier concentrations as functions of deposition rate (applied power) for various oxides prepared by this technique are shown in Fig. 6.14.

Oxide targets have also been used to deposit high-quality layers of ITO, ZnO, and SnO using an rf discharge with small additions of oxygen or hydrogen to the argon plasma ($<5 \times 10^{-5}$ torr). Buchanan *et al.* [50] found that for indium tin oxide the optimum value of resistivity was obtained for additions of oxygen in excess of that necessary to maintain the target in a reduced state. As a result, thin films (<1000 Å) with high optical transparency and low resistivity (4×10^{-4} ohm cm) could be deposited, whereas for films of increasing thickness—and, hence, longer sputter time—the resistivity also increased. Less than optimal values of resistivity, however, could be obtained with lesser amounts of added oxygen on a continuous basis.

The reported results for transparent conducting films of zinc oxide, however, show rather unusual behavior. Webb *et al.* [87–89] using an oxide

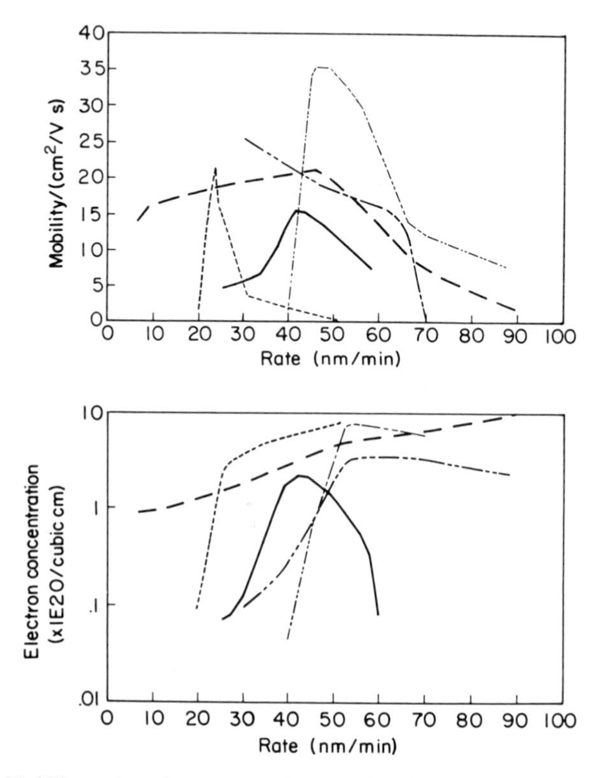

Fig. 6.14. Mobility and carrier concentration as a function of rate of deposition for In–Sn and Cd–Sn films prepared at 100 W and a system pressure of 5 mtorr (1.5 mtorr oxygen): ——, In–Sn bias of 0 V; – – –, In–Sn bias of −140 V; ------, Cd–Sn bias of 0 V; — — — —, Cd–Sn bias of −100 V; —··—··—, Cd–Sn bias of −125 V. (Bias voltage developed on substrate as a result of rf field applied to substrate.) (From Howson and Ridge [65].)

target, were successful in depositing highly conducting films by rf sputtering in an argon plasma by adding small amounts of hydrogen (1×10^{-5} torr). For substrate temperatures less than 50°C, all films were highly insulating ($>10^8$ ohm cm). For substrates heated to 130°C the films were highly transparent with a resistivity of 10^{-2} ohm cm independent of thickness. To increase the reactivity between the hydrogen and sputtered zinc oxide (and, thus, increase the density of oxygen vacancies which, in turn, results in an increase in carrier concentration), a second low-frequency discharge was added at the substrate (Fig. 6.15a). The power in this discharge was less than 1 W making resputtering and substrate heating negligible. As shown in Fig. 6.15b, with the discharge, the resistivity of the films decreased by approximately one order of magnitude to 2×10^{-3} ohm cm for the same hydrogen partial pressure and target power. Furthermore, films with the

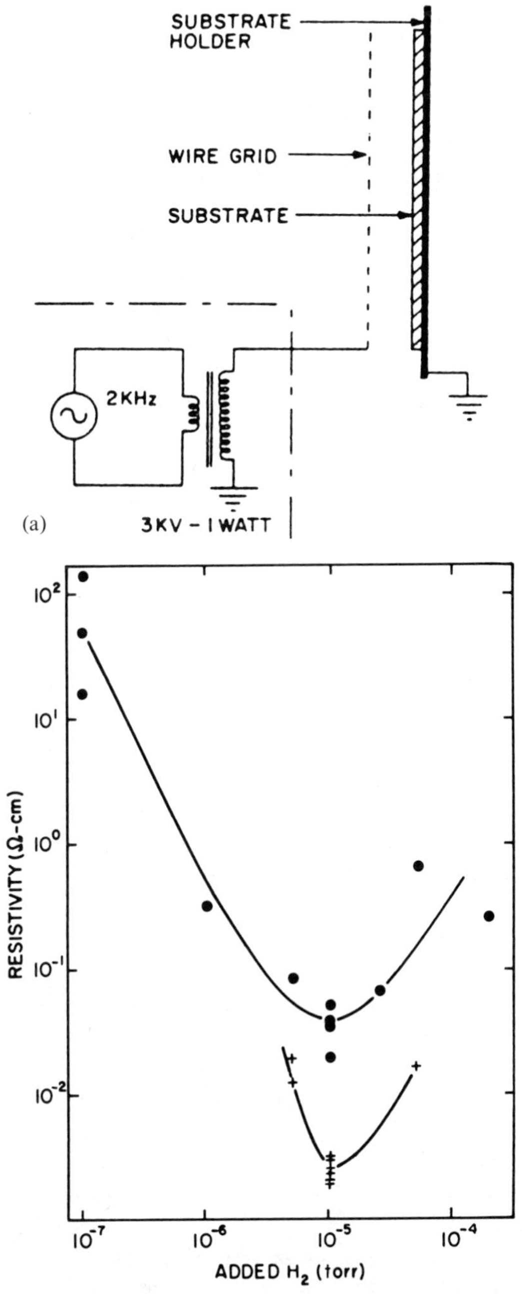

Fig. 6.15. (a) Schematic representation of experimental setup to achieve secondary discharge on the substrate for the deposition of ZnO and (b) the measured dependence of film resistivity versus added hydrogen partial pressure; ●, power = 200 W, T_{SUB} = 400 K; +, power = 100 W, T_{SUB} = 315 K, second discharge used to initiate growth (see text).

lowest resistivity were obtained on unheated substrates. The most startling observation, however, was the fact that this second discharge could be switched off after initiating growth of the low-resistivity film, and the film would continue to grow in the low-resistivity state. It appeared that this effect was related to a nucleation phenomenon although the process is far from being completely understood. Other authors have reported equally unusual results with zinc oxide. For instance, Minami *et al.* [90] deposited high-quality zinc oxide films from an oxide target (5×10^{-4} ohm cm) by sputtering in pure argon with a second external magnetic field perpendicular to the target surface. The substrates were unheated (estimated surface temperature during deposition was 90°C) and also placed perpendicular to the target surface. Substrates placed parallel to the target produced films having several orders of magnitude higher resistivity.

Recent results on the film properties as a function of substrate position relative to the magnetron cathode have suggested possible reasons for such diverse results as those indicated earlier. Webb [91] has shown that the resistivity of zinc oxide films deposited without the second discharge to be highly dependent on the target–substrate geometry. Film resistivity was measured as a function of position in a plane parallel to the target surface at a fixed deposition distance. Films showing the lowest resistivity were produced directly above the target center (where the component of the magnetic field is perpendicular to the target surface) while higher resistivity values were observed for depositions directly above regions of highest target erosion (field parallel to target surface). In fact, the resistivity imaged the target erosion pattern. Up to an order of magnitude change in resistivity was recorded while thickness variations were less than 7%.

A similar type of effect was observed for sputtered films of zinc oxide produced for SAW devices. Krupanidhi and Sayer [92] deposited highly insulating and oriented films from a zinc metal target in an rf discharge using 100% pure oxygen. Below a sputtering pressure of 10 mtorr, the film structure, orientation, and physical properties were dependent on substrate position with respect to the target. Films with optimal properties were produced for substrate positions directly above the target center. For deposition pressures greater than 10 mtorr, the position of the substrate became unimportant.

Tominaga *et al.* [93] used a time-of-flight technique to analyze the energy distribution of neutral particles from a dc planar magnetron cathode. The sputter targets were either hot pressed disks of ZnO or $BaTiO_3$. Sputtering was in a pure argon discharge or an argon–oxygen discharge. For an applied target voltage of 350 V, the primary neutral sputtered particles were oxygen atoms with an energy of 350 eV. No other high-energy neutral particles were observed. It was concluded that oxygen ions formed at the target were

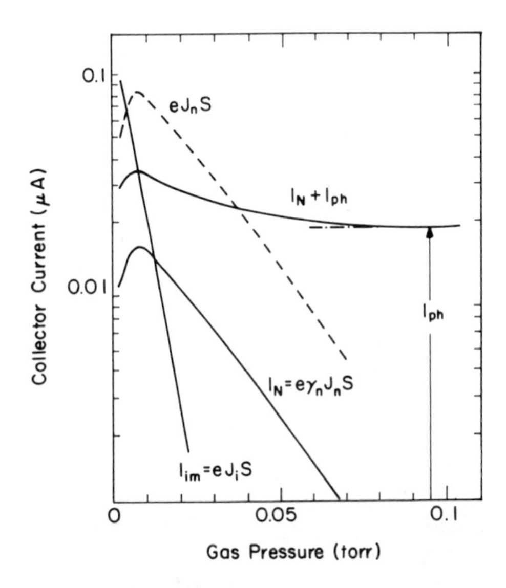

Fig. 6.16. Collector currents due to negative oxygen ions I_{im}, neutral oxygen atoms I_n and photons I_{ph} as a function of gas pressure for dc planar magnetron sputtering of ZnO in argon. (From Tominaga *et al.* [93].)

neutralized in traversing the sputter chamber. Measurements were also performed to determine the density of negative ions bombarding the substrate. Presumably these are ions that do not undergo electron stripping and travel directly to the substrate. Only oxygen ions were observed. The measured neutral and ion particle density as a function of sputter pressure is shown in Fig. 6.16. At low sputter pressures (<1.0 mtorr) the density of negative oxygen ions is comparable to that of the neutral species, whereas at higher pressures the neutral species dominate. In an earlier work Tominaga *et al.* [94] looked at the detailed spatial distribution of these particles for a ZnO target. It was found that the density of neutral oxygen atoms arriving at the substrate was highest directly above the target erosion ring (Fig. 6.17). Since the high conductivity of the zinc oxide films depends on the introduction of oxygen vacancies to increase the density of free carriers, enhanced bombardment of the growing film by oxygen neutrals is likely to reduce the conductivity in these regions considerably. This is consistent with the results of Webb [91]. Similarly, bombardment of the growing film is also likely to cause considerable changes in structure (increased defect density, etc.); an effect observed by Krupanidhi and Sayer [92]. These results demonstrate clearly the complex effects of ion or neutral particle bombardment on a growing film. They also demonstrate the sometimes selective

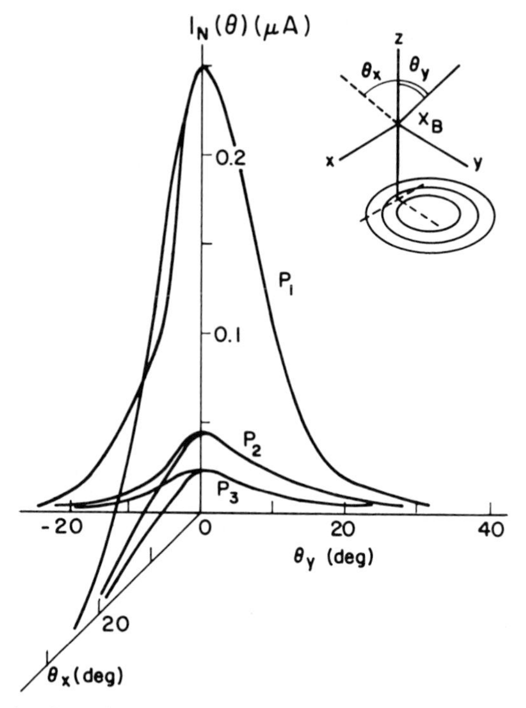

Fig. 6.17. Angular dependence of high-energy neutral oxygen atom distribution in planar magnetron sputtering of ZnO for various sputtering pressures: $P_1 = 0.01$ torr, $P_2 = 0.05$ torr, and $P_3 = 0.1$ torr. (From Tominaga *et al.* [94].)

nature of the process as a result of the nonuniform sputter characteristics of the magnetron cathode. Of course, not all ion or neutral particle bombardment is detrimental to the deposition of high quality films. For instance rf bias sputtering has been used to enhance substrate ion bombardment during deposition in order to improve the photoresponse in layers of amorphous hydrogenated silicon (a-Si : H) [95]. Although bias sputtering is a convenient way to control substrate ion bombardment, no effective method is available for controlling the high-energy neutral particle flux. If the results of Tominaga *et al.* [94] hold for other materials, then substrate placement and cathode size may be critical factors in determining overall film quality.

Table 6.4 gives a listing of the various transparent conducting oxides deposited using magnetron sputtering. Comparing these results with those shown in Table 6.3 illustrates the advantages of magnetron sputtering over other techniques. Material of equal or higher quality is obtained with no high-temperature processing or postdeposition treatments.

TABLE 6.4

Properties and Deposition Parameters for the Transparent Conducting Oxides Prepared by Magnetron Sputtering

Material	System	Target	Gas	T_s (°C)	Power density (W/cm^2)	Rate (Å/s)	ρ (ohm cm)	n (cm^{-3})	μ (cm^2/V s)	t (Å)	T (%)	Reference
ITO	Planar dc, rf on substrate and box	$In_{0\cdot9}Sn_{0\cdot1}$	Ar + O_2	RT	3.2	1.0	1×10^{-4}				>79	[84]
ITO	Planar dc, rf on substrate	$In_{0\cdot9}Sn_{0\cdot1}$	Ar + O_2	RT	0.31	8.3	$<1 \times 10^{-3}$	$>10^{20}$	20	3300	>80	[65]
ITO	DC planar	$In_{0\cdot9}Sn_{0\cdot1}$	Ar + O_2	RT		1–16	400 ohm/□				>85	[96][a]
ITO	DC planar	$In_{0\cdot9}Sn_{0\cdot1}$	Ar + O_2	RT		22.3	25 ohm/□[b]					[58]
ITO	RF planar	In_2O_3, 3% SnO_2	Ar + O_2	RT	2.5	2.7	4×10^{-4}	10^{21}	10	800	>85	[50]
ITO	RF planar	In_2O_3 90%		50			4.9×10^{-4}	4.4×10^{20}	29			
		SnO_2 10%	Ar	370	0.625		6.8×10^{-5}	2.7×10^{21}	36	1000	>82	[97]
In_2O_3	Planar dc, rf on substrate	In	Ar + O_2	RT	0.31	1.7	8.2×10^{-3}	4×10^{19}	19	1000	>80	[57]
SnO_2	RF planar	Sb/Sn, 2.3% oxide	Ar + O_2	400	4.7	18.3	2×10^{-3}	3×10^{20}	10		80	[70]
SnO_2	RF planar	SnO_2	Ar + O_2	RT	0.64	2.0	1.6×10^{-3}	3×10^{20}	13	1200	>85	[69]
Cd_2SnO_4	Planar dc, rf on substrate and box	Cd_2Sn	Ar + O_2	RT	2.9	2.5	4.5×10^{-4}				85	[80]
Cd_2SnO_4	Planar dc, rf on substrate	Cd_2Sn	Ar + O_2	RT	0.32	8.3	$<1 \times 10^{-3}$	$>10^{20}$	35	2800	>80	[65]
Cd_2SnO_4	DC plasmatron	Cd_2Sn	Ar + O_2	RT	5.0		4.5×10^{-4}	5×10^{20}	40	2500	90	[98]
ZnO	RF planar, external magnetic field	ZnO	Ar	RT	0.6–1.6	0.42–4.2	4.5×10^{-4}	1.1×10^{20}	120	2000	>85	[99][c]
ZnO	RF planar, secondary discharge	ZnO	Ar + H_2	RT	1.27	1.0	1.4×10^{-3}	1.1×10^{20}	8	1000	>90	[51]
ZnO	DC planar, rf on substrate and box	Zn	Ar + O_2	RT		8.0	2×10^{-2}				~80	[100]

[a] Annealed. Here T_s is the substrate temperature, T the percent transmittance at ~400–800 nm, and t the thickness.

[b] Annealed at 300–500°C in 5% H_2 and 95% N_2.

[c] Substrate perpendicular to target.

B. Piezoelectric Materials

A variety of oxides and nitrides have been prepared by magnetron sputtering for various acousto-optic and acousto-electric devices, such as surface acoustic wave delay lines and filters. The most widely studied are zinc oxide and aluminum nitride and to a lesser extent, oxides such as $K_3Li_2Nb_5O_{15}$ [32], PZT [30], and $BaTiO_3$ [35].

1. Zinc Oxide

The large electromechanical coupling coefficient of ZnO has made it one of the more important materials for the manufacture of surface acoustic wave devices [101–107]. In addition, applications to optical waveguides has also progressed owing to the excellent optical transparency and optical flatness of the deposited films. Crystallographic orientation and surface flatness are of obvious importance for these applications, and a number of studies based on the preparation of high-quality films using both rf and dc magnetron sputtering have been performed [108–117]. Unlike the conducting oxides, oxides and nitrides for SAW devices must be highly resistive (stoichiometric) and exhibit characteristics close to that of the bulk crystalline material.

Takeda *et al.* [118] studied the effect of hydrogen gas on the *c*-axis orientation of the films prepared by rf magnetron sputtering. In a pure oxygen ambient the films showed *c*-axis orientation normal to the substrate surface, while for a mixture of oxygen and hydrogen a *c*-axis orientation parallel to the film plane was observed.

Other authors have studied angle of incidence effects of the sputtered material on film properties [119] and the effect on *c*-axis orientation [120] as a result of film bombardment by energetic atoms during growth. Controlling these factors has resulted in transducers with electromechanical coupling coefficients close to that of the bulk material [121]. These same factors controlling the orientation and crystallinity of the films are, of course, equally important in determining the mobility of carriers in the highly conducting oxides.

2. Aluminum Nitride

Besides having a high ultrasonic velocity and fairly large electromechanical coupling coefficient, aluminum nitride is also much more chemically stable than zinc oxide; it has a very high decomposition temperature (2490°C) and it has a large (6.3-eV) band gap. These factors have made AlN an important material for the passivation of semiconductor surfaces and for surface acoustic wave devices. Until recently, the methods used for preparation, such as CVD, reactive sputtering, and reactive evaporation, required relatively high temperatures (1000–1300°C). This had severely restricted the type of substrate material employed. More recently, magnetron sputtering has been shown to produce high-quality layers of AlN for substrate temperatures less than 300°C [122, 123]. Effects similar to those observed for zinc oxide, relating to gas composition on *c*-axis orientation, have also been reported [124]. Excellent films have been deposited on a variety of substrates including glass [124], silicon [125], and sapphire [126]. More recently Kline and Lakin [127] reported the fabrication of an aluminum nitride 1.0-GHz thin film bulk acoustic wave resonator on GaAs.

C. Amorphous Hydrogenated Silicon

There has been a virtual explosion in the number of reported research studies on amorphous hydrogenated silicon [127] in the last few years. This interest stems from the potential technological importance of this material for the fabrication of large-area low-cost photovoltaic arrays. The best material to date (for photovoltaic applications) has been prepared by the glow discharge decomposition of silane. Currently, solar panels with an area of 900 cm^2 and efficiency of about 6% are in production using this method. However, it should be recognized that a-Si has potential applications in such diverse areas as field-effect transistors for LCDs [128], high-current diodes [129], photosensors [130], vidicon targets [131], electrophotography [132], and optical recording and photolithography [133]. Each of these applications requires a specific but widely different set of structural, electrical, and optical properties of the a-Si : H films. For these applications, magnetron sputtering provides a high level of control over all preparation parameters, thus giving one of the largest ranges in film properties. For instance, films with widely differing microstructures can be reproducibly deposited [134]. The band gap of these films can also be varied from 1.1 to 2.2 eV and the electrical conductivity by more than 11 orders of magnitude from 10^{+1} to 10^{-10}/ohm cm [135] (see Fig. 6.18). Because of these features, magnetron sputtering is an excellent research tool in the study of the basic properties of the amorphous state, in general, and in particular the factors related to the deposition of device-quality films. One of the earliest reports (1981) on magnetron sputtered films of a-Si was concerned with the development of baseline parameters for the deposition of layers for solar cell fabrication [136]. More recent studies have investigated the nature of the hydrogen bonding [137–140] as well as the effect of sputter gas on internal stress [141], the thermal stability of films prepared by rf bias sputtering [142], and the effect of hydrogen on the performance of thin film field-effect transistors [143]. In general, the material most suited for devices has been deposited at relatively high substrate temperatures (>200°C). This material has a low gap

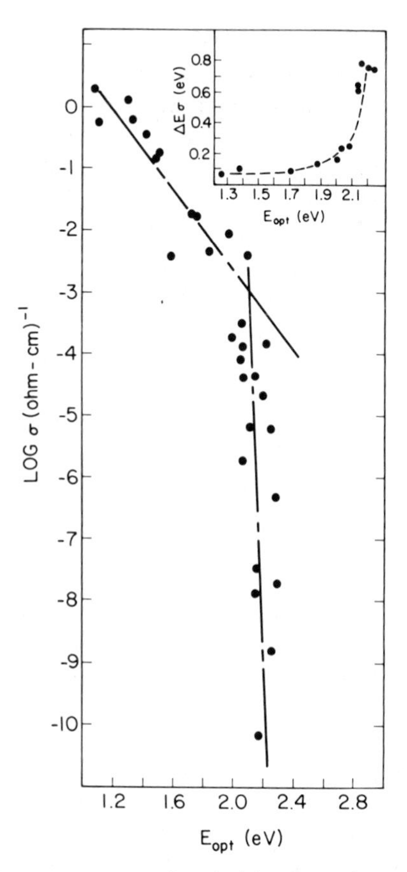

Fig. 6.18. Conductivity versus measured optical band gap for films of a-Si : H prepared under various deposition conditions. Inset shows the variation in conductivity activation energy versus optical gap. All measurements at 293 K.

state density with a band gap of 1.7 eV. Recently material with a band gap of ~1.9 eV with over four orders of magnitude change in photoconductivity (light intensity = 100 mW/cm^2) has been reported [144]. The substrates were unheated during deposition.

The relatively recent application of magnetron sputtering to the preparation of a-Si : H has meant that it is unlikely the full potential of this technique has been realized.

D. Compound Semiconductors for Device Fabrication

The success of today's high speed transistors, LEDs, and laser diodes depends on the growth of high-quality layers of III–V compound semiconductors and associated ternary and quaternary alloys (e.g., $In_xGa_{1-x}As_yP_{1-y}$ on InP) [145]. Deposition techniques such as MBE have generally been used: however, this technique, while providing high-quality epitaxial layers, has an inherently low throughput. In addition, the growth of phosphorus compounds is difficult to achieve due to the low sticking coefficient of the phosphorus atoms. Many of these difficulties have been overcome by using techniques such as OMVPE and VPE. Much less attention has been given to sputter deposition, even though it has been demonstrated that high-quality epitaxial layers of the III–V compounds can be deposited by using this technique [13]. Sputter deposition has considerable flexibility such as its use in a reactive mode with solid, gaseous, as well as organometallic sources. The enhanced features of magnetron sputtering could prove valuable in such cases (owing to the reduced damage caused by energetic particle bombardment of the growing film); however, its use in epitaxial or, indeed, polycrystalline semiconductor deposition remains essentially unexplored.

The few results reported have been for polycrystalline films of Cu_xS, CdS, and CdSe($CdSe_{1-x}O_x$). These studies were initiated as a result of the potential technological importance of these materials for the fabrication of low-cost, large-area photovoltaic arrays and thin film transistor arrays for large-area liquid crystal displays. Coupled with the capability of depositing high-quality transparent conducting oxides, dielectrics, and metallic layers, the magnetron system offers the potential of fabricating an all-thin-film solar cell at low cost [146].

Films of CdS prepared by thermal evaporation have been studied extensively, but it usually has been difficult to control the stoichiometry of the films by using a single evaporation source. The deposited films are usually cadmium rich with correspondingly low resistivities ($<10^4$ ohm cm). The excess cadmium acts as an electron donor, thus making the material strongly *n*-type. Controlling the carrier concentration by changes in film stoichiometry, however, is less than satisfactory. In addition, compensation in the II–VI compounds is of concern, in that effective doping by addition of impurities is sometimes difficult to achieve. For instance most II–VI semiconductors show only unipolar conduction.

Thornton *et al.* [147] investigated the control of film resistivity via indium doping of cylindrical magnetron sputtered films of CdS. An indium–cadmium metal target was used in a sputter discharge of Ar and H_2S. Samples were prepared with indium–cadmium ratios up to a maximum of 1% (target composition). Substrate temperature and H_2S injection rates were the primary parameters studied. Figure 6.19 gives a plot of film resistivity and deposition rate as a function of the H_2S injection rate for various target compositions. For low H_2S injection rates, the deposition rate was limited by

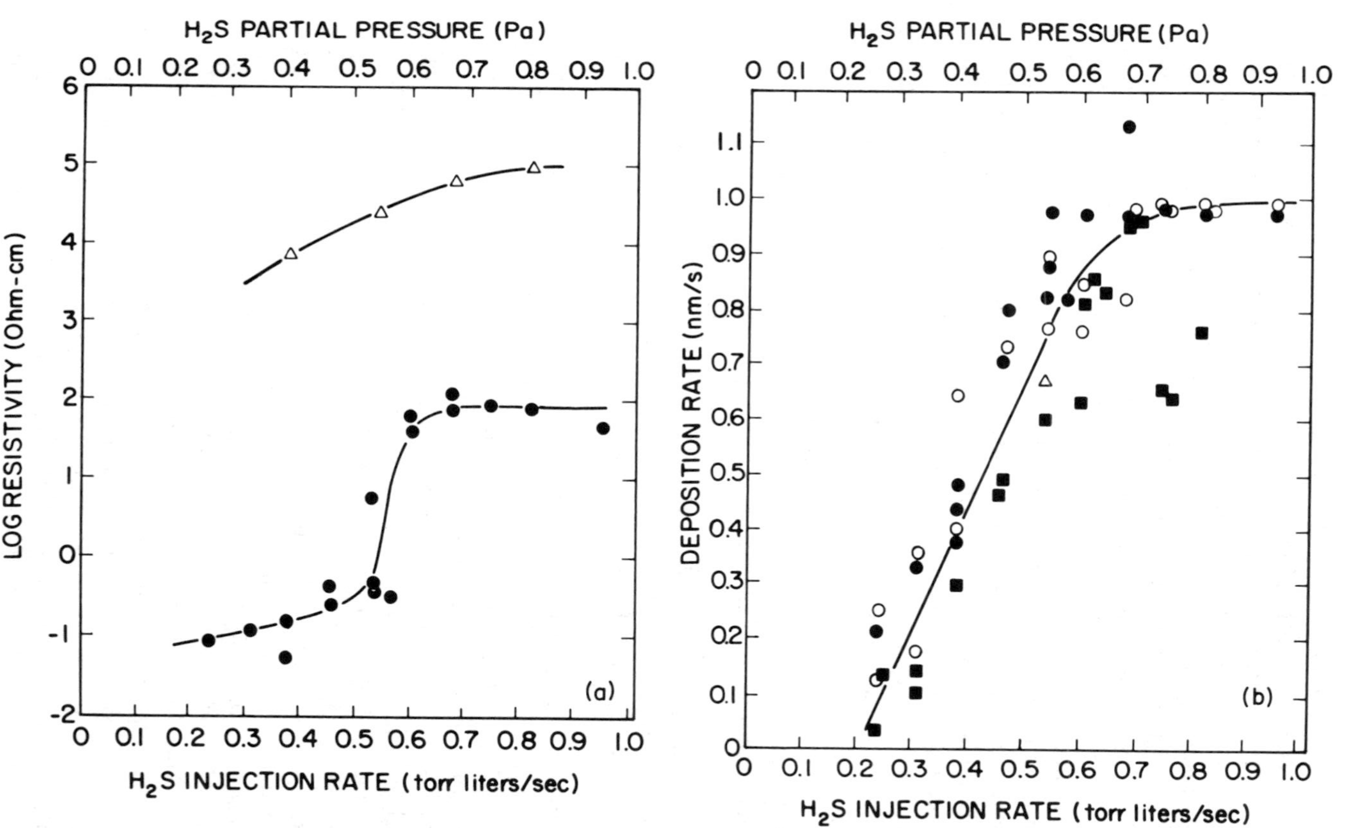

Fig. 6.19. Dependence of resistivity and deposition rate on H_2S injection rate for CdS films sputtered from Cd/In targets with the concentrations of indium indicated in the figures. Substrate temperature 250°C; ●, 1 at. % In; ○, 0.3 at. % In; ■, 0.1 at. % In; and △, undoped. (From Thornton *et al.* [147].)

the available sulfur flux, while the availability of cadmium vapor limited the rate at high H_2S injection rates. Near stoichiometric films were obtained with resistivities of 10^4–10^5 ohm cm ($\mu = 2$ cm^2/V s and $n = 3 \times 10^{13}$/cm^3) compared to single-crystal values of 5 ohm cm ($\mu = 300$ cm^2/V s and $n = 4 \times 10^{15}$ cm^3). Doping with indium decreased the resistivity to values as low as 10^{-1} ohm cm ($\mu = 14$ cm^2/V s and $n = 4 \times 10^{18}$/cm^3). The effectiveness of indium doping was observed to have a strong dependence on H_2S injection rates—increasing with decreasing H_2S. This was believed to be the result of the compensating effect from Cd vacancy formation. Maximum doping efficiencies of 1–3% were obtained. This was essentially limited by the inability to maintain a sufficient Cd vapor pressure over the film during deposition.

The same researchers have studied the deposition of Cu_xS for possible use in CdS–Cu_xS heterojunction solar cells [148, 149]. The control of stoichiometry is extremely critical for Cu_xS in that a number of distinct phases of copper sulfide exist for various ratios of copper/sulfur. It has been shown that the desired phase for solar cell applications is the chalcocite phase with $x = 1.999$ [150]. This combination yields a *p*-type semiconductor (copper vacancies acting as acceptors) of low resistivity and large minority carrier diffusion length. Films were deposited from a copper target in a sputter discharge of Ar and H_2S on substrates heated to either 130 or 35°C. For substrate temperatures of 130°C, resistivities of 100 ohm cm were observed, while for the lower substrate temperatures the resistivity decreased to 0.01 ohm cm. Hall effect mobilities were relatively low (1–6 cm^2/V s) for both of the copper sulfide phases studied. The low-resistivity material (a mixture of chalcocite and djurleite) with $x = 1.98$–1.99 was dominated by the chalcocite phase making it particularly attractive for solar-cell fabrication. Based on these studies, Thornton and Anderson [151] have recently reported results of an all-sputter deposited Cu_xS/CdS solar cell with a conversion efficiency of ~4% (no antireflection coating). The results show that cells with performance characteristics equivalent to those fabricated by the topotaxial ion-exchange method can be prepared without introducing significant interfacial or bulk defects from the magnetron process.

Maniv *et al.* [152] investigated the use of planar rf magnetron sputtering for the deposition of CdSe and $CdSe_{1-x}O_x$ films for various applications. Like CdS, the resistivity of the CdSe films depended strongly on stoichiometry. Only unipolar conduction was observed (the material was *n*-type) with the undoped free carrier concentration determined by the density of selenium vacancies, which act as shallow donor levels. The films were prepared by reactively sputtering a hot pressed CdSe target in a mixture of argon and oxygen. This allowed the incorporation of atomic oxygen into the films to form acceptor levels or to minimize selenium vacancy formation by substitutional doping on selenium sites. At low oxygen flow rates the deposited films were lightly doped, increasing at high flow rates, to a maximum stable composition of $CdSe_{0.8}O_{4.5}$. Corresponding changes in resistivity from 1 ohm cm to greater than 10^6 ohm cm were observed as well as a decrease of 1% in the hexagonal parameter for the wurtzite phase. The total oxygen content of the films however was 25%, which suggested the presence of a second oxygen-rich amorphous phase. All films prepared at high oxygen flow rates were amorphous. Films deposited in pure argon, however, showed a strong (0001) orientation of the hexagonal wurtzite phase. Thin-film transistors were fabricated from the various layers, with saturated characteristics observed for the oxygen free films after a 370°C anneal in flowing nitrogen. Devices prepared from the amorphous films did not show transistor behavior before or after annealing.

VI. CONCLUSION

Magnetron sputtering has become a commercially accepted technique for depositing metal films for use as cosmetic coatings, coatings for architectural glass, and for the metallization of integrated circuits. The potential also exists for the preparation of a wide range of semiconducting compounds with characteristics rivaling those prepared by more currently accepted techniques. The characteristics of low substrate heating, high deposition rates, large area coverage, and low deposition-induced damage in the growing film have made it particularly attractive.

In 1978 it was pointed out [8] that there were more questions than answers when it came to understanding planar magnetron sputtering. These questions, concerning sputtered particle energy distributions, optimization of the magnetic field pattern, effect of substrate bias, and the feasibility of depositing high-quality dielectric layers still require more complete answers.

Perhaps it can be said, that at least our present understanding has increased to the point where we have become more aware of the full potential of magnetron sputtering.

ACKNOWLEDGMENTS

The author wishes to thank Dr. D. F. Williams and Dr. S. R. Das for reading the manuscript and offering their critical comments.

REFERENCES

1. A. Aronson and S. Weinig, *Vacuum* **27**, 151–153 (1977).
2. B. Singh and P. R. Denton, *Opt. Spec.* 76–79 (1981).
3. J. A. Thornton, *Met. Finish.* 83–87 (1979).

4. N. Hosokawa, *J. Electron. Eng.* 86–88 (1981).
5. J. J. Bessot, *Met. Finish.*, 63–69 (1980).
6. P. van Vorous, *Photonics Spectra*, 66–67 (1982).
7. R. J. Hill, R. Cormia, and P. McLeod, *Proc. Int. Vac. Congr., 7th, Vienna, 1977*, 1583–1586 (1977).
8. R. K. Waits, *in* "Thin Film Processes" (J. L. Vossen and W. Kern, eds.), Chapter II-4, Academic Press, New York, 1978.
9. K. Wasa and S. Hayakawa, *Thin Solid Films* **52**, 31–43 (1978).
10. A. R. Nyaiesh, *Thin Solid Films* **86**, 267–277 (1981).
11. J. A. Thornton, *Thin Solid Films* **80**, 1–11 (1981).
12. R. K. Waits *in* "Thin Film Processes" (J. L. Vossen and W. Kern, eds.), p. 131, Academic Press, New York, 1978.
13. J. E. Greene and A. H. Eltoukhy, *Surf. Interface Anal.* **3**, 34–54 (1981).
14. A. H. Eltoukhy and J. E. Greene, *J. Appl. Phys.* **50**, 505–517 (1979).
15. F. M. Penning, *Physica* **3**, 873 (1936).
16. P. J. Clarke, U.S. patent 3, 616, 450 (1971).
17. R. P. Riegert and P. Clarke, *Electron Packag. Prod.* **13(2)**, 85 (1973).
18. A. S. Penfold and J. A. Thornton, U.S. patent 3, 884, 793 (1975).
19. J. S. Chapin, *Res. Develop.* **25**(1), 37 (1974).
20. I. G. Kesaer and V. V. Pashkova, *Sov. Phys.-Tech. Phys.* **4**, 254 (1959).
21. W. D. Westwood, S. Maniv, and P. J. Scanlon, *J. Appl. Phys.* **54**, 6841–6846 (1983).
22. L. I. Maissel and R. Glang, eds., "Handbook of Thin Film Technology," p. 158, McGraw-Hill, New York, 1970.
23. T. Fukami and T. Sakuma, *Jpn. J. Appl. Phys.* **21**, 1680–1683 (1982).
24. S. Maniv and W. D. Westwood, *J. Vac. Sci. Technol.* **17**, 743–751 (1980).
25. J. B. Webb, *J. Appl. Phys.* **53**, 9043–9048 (1982).
26. M. Usui, T. Kujirai, T. Nagatomo, and O. Omoto, *Vac. Soc. Jpn.* **25**, 1–484 (1982).
27. S. Maniv, W. D. Westwood, F. R. Shepherd, and P. J. Scanlon, *J. Vac. Sci. Technol.* **20**, 1–6 (1982).
28. J. A. Thornton and D. G. Cornog, *J. Vac. Sci. Technol.* **18**, 199–202 (1981).
29. A. D. Jonath, W. W. Anderson, J. A. Thornton, and D. G. Cornog, *J. Vac. Sci. Technol.* **16**, 200–203 (1979).
30. S. B. Krupanidhi, N. Maffei, M. Sayer, and K. El-Assal, *J. Appl. Phys.* **54**, 6601–6609 (1983).
31. K. Urbanek, *Solid State Technol.*, April, 87–90 (1977).
32. T. Shiosaki, M. Adachi, and A. Kawabata, *Thin Solid Films* **96**, 129–140 (1982).
33. T. Shiosaki, T. Yamamoto, T. Oda, K. Harada, and A. Kawabata, *Ultrason. Symp.* 451–454 (1980).
34. E. V. Gerova, N. A. Ivanov, and K. I. Kirov, *Thin Solid Films* **81**, 201–206 (1981).
35. T. Nagatomo, T. Kosaka, S. Omori, and O. Omoto, *Ferroelectrics* **37**, 681–684 (1981).
36. A. R. Nyaiesh and L. Holland, *J. Vac. Sci. Technol.* **20** 1389–1392 (1982).
37. R. S. Nowicki, *J. Vac. Sci. Technol.* **14**, 127–133 (1977).
38. C. Deshpandey and L. Holland, *Thin Solid Films* **96**, 265–270 (1982).
39. K. Matsushita, Y. Matsuno, T. Hariu, and Y. Shibata, *Thin Solid Films* **80**, 243–247 (1981).
40. A. R. Nyaiesh and L. Holland, *Vacuum* **31**, 315–317 (1981).
41. G. Samuel and L. Holland, *Vide, Couches Minces*, Suppl. Rev. **N201**, 34–37 (1980).
42. J. B. Webb and S. R. Das, *J. Appl. Phys.* **54**, 3282–3285 (1983).
43. S. R. Das, D. F. Williams, and J. B. Webb, *J. Appl. Phys.* **54**, 3101–3105 (1983).
44. J. H. Keller and R. G. Simmons, *IBM J. Res. Develop.* **1**, 24 (1979).
45. W. D. Westwood, *J. Vac. Sci. Technol.* **15**, 1 (1978).
46. S. Maniv and W. D. Westwood, *J. Vac. Sci. Technol.* **17**, 403–406 (1980).

47. E. Stern and H. L. Caswell, *J. Vac. Sci. Technol.* **4**, 128–132 (1967).
48. S. Maniv and W. D. Westwood, *J. Appl. Phys.* **51**, 718–725 (1980).
49. S. Maniv and W. D. Westwood, *J. Vac. Sci. Technol.* **17**, 743–751 (1980).
50. M. Buchanan, J. B. Webb, and D. F. Williams, *Thin Solid Films* **80**, 373–382 (1981).
51. J. H. Greiner, *J. Appl. Phys.* **42**, 5151 (1971).
52. P. Sigmund, *Phys. Rev.* **184**, 383–416 (1969).
53. P. S. Vincett, W. A. Barlow, and G. G. Roberts, *J. Appl. Phys.* **48**, 3800–3806 (1977).
54. R. P. Howson and M. I. Ridge, *Proc. SPIE, Int. Soc. Opt. Eng.* **324**, 16–22 (1982).
55. J. C. Manifacier, *Thin Solid Films* **90**, 297–307 (1982).
56. R. P. Howson and M. I. Ridge, *Proc. SPIE, Int. Soc. Opt. Eng.* **325**, 46–51 (1982).
57. R. P. Howson, J. N. Avaritsiotis, M. I. Ridge, and C. A. Bishop, *Thin Solid Films* **63**, 163–167 (1979).
58. J. F. Smith, A. J. Aronson, D. Chen, and W. H. Class, *Thin Solid Films* **72**, 469–474 (1980).
59. J. L. Vossen, *in* "Physics of Thin Films" (G. Hass, M. H. Francombe, and R. W. Hoffman, eds.), p. 9, Academic Press, New York, 1977.
60. C. A. Pan and T. P. Ma, *Appl. Phys. Lett.* **37**, 163 (1980).
61. J. C. Manifacier, L. Szepessy, J. F. Bresse, M. Perotin, and R. Stuck, *Mater. Res. Bull.* **14**, 163 (1979).
62. M. Mizuhashi, *Thin Solid Films* **70**, 91 (1980).
63. P. Nath, R. F. Bunshah, B. M. Basal, and O. M. Staffsud, *Thin Solid Films* **72**, 463 (1980).
64. J. C. C. Fan, F. J. Bachner, and G. H. Foley, *Appl. Phys. Lett.* **31**, 773 (1977).
65. R. P. Howson and M. I. Ridge, *Thin Solid Films* **77**, 119–125 (1981).
66. R. Groth, *Phys. Status Solidi* **14**, 69 (1966).
67. L. A. Ryabova, V. S. Salun, and I. A. Serbinov, *Thin Solid Films* **92**, 327 (1982).
68. J. Kane, H. P. Schweizer, and W. Kern, *J. Electrochem. Soc.* **122**, 1144 (1975).
69. R. G. Goodchild, J. B. Webb, and D. F. Williams, *J. Appl. Phys.* **57**, 2308–2310 (1985).
70. K. Suzuki and M. Mizuhashi, *Thin Solid Films* **97**, 119–127 (1982).
71. H. S. Randhawa, M. D. Mathews, and R. F. Bunshah, *Thin Solid Films* **83** 267 (1981).
72. E. Shanthi, V. Dutta, A. Banerjee, and K. L. Chopra, *J. Appl. Phys.* **51**, 6243 (1980).
73. J. Kane, H. P. Schweizer, and W. Kern, *J. Electrochem. Soc.* **123**, 270 (1976).
74. T. S. Chow, M. Ghezzo, and B. J. Baliga, *J. Electrochem. Soc.* **129**, 1040–1045 (1982).
75. E. Shanthi, A. Banerjee, V. Dutta and K. L. Chopra, *J. Appl. Phys.* **53**, 1615 (1982).
76. D. E. Brodie, R. Singh, J. H. Morgan, J. D. Leslie, L. J. Moore, and A. E. Dixon, *Proc. IEEE Photovoltaic Spec. Conf., 14th, San Diego, 1980*, 468 (1980).
77. J. B. Webb, D. F. Williams, and M. Buchanan, *Appl. Phys. Lett.* **39**, 640–642 (1981).
78. J. Aranovich, A. Oritz, and R. H. Bube, *J. Vac. Sci. Technol.* **16**, 994 (1979).
79. G. Haacke, *Appl. Phys. Lett.* **28**, 622 (1976).
80. S. Maniv, C. Miner, and W. D. Westwood, *J. Vac. Sci. Technol.* **18**, 195–198 (1981).
81. K. Enjouji, K. Murata, and S. Nishikawa, *Thin Solid Films* **108**, 1–7 (1983).
82. P. J. Chandler, P. R. Meek, and L. Holland, *Thin Solid Films* **86**, 183–191 (1981).
83. R. McMahon, J. Affinito, and R. R. Parsons, *J. Vac. Sci. Technol.* **20**, 376–378 (1982).
84. S. Maniv, C. J. Miner, and W. D. Westwood, *J. Vac. Sci. Technol.* **1**, 1370–1375 (1983).
85. M. I. Ridge, R. P. Howson, and C. A. Bishop, *Proc. SPIE Int. Soc. Opt. Eng.*, **325**, 46–51 (1982).
86. R. P. Howson, private communication, 1984.
87. J. B. Webb, D. F. Williams, and M. Buchanan, *Appl. Phys. Lett.* **39**, 640–642 (1981).
88. J. B. Webb and D. F. Williams, *J. Vac. Sci. Technol.* **20**, 467–468 (1982).
89. J. B. Webb, M. Buchanan, and D. F. Williams, U.S. patent 442, 8810 (1984).
90. T. Minami, H. Nanto, and S. Takata, *Appl. Phys. Lett.* **41**, 958–960 (1982).
91. J. B. Webb, to be published.

92. S. B. Krupanidhi and M. Sayer, *J. Appl. Phys.* **56**, 3308–3318 (1984).
93. K. Tominaga, S. Iwamura, Y. Shintani, and O. Tada, *Jpn. J. Appl. Phys.* **21**, 688–695 (1982).
94. K. Tominaga, N. Ueshiba, Y. Shintani, and O. Tada, *Jpn. J. Appl. Phys.* **20**, 519–526 (1981).
95. M. Suzuki, T. Maekawa, S. Okano, and T. Bandow, *Jpn. J. Appl. Phys.* **20**, 485–487 (1981).
96. P. J. Clarke, *J. Vac. Sci. Technol.* **14**, 141–142 (1977).
97. S. Ray, R. Banerjee, N. Basu, A. K. Batabyal, and A. K. Barua, *J. Appl. Phys.* **54**, 3497–3501 (1983).
98. S. Schiller, G. Beister, E. Buedke, H. J. Becker, and H. Schicht, *Thin Solid Films* **96**, 113–120 (1982).
99. T. Minami, H. Nanto, and S. Takata, *Appl. Phys. Lett.* **41**, 958–960 (1982).
100. S. Maniv, C. J. Miner, and W. D. Westwood, *J. Vac. Sci. Technol.* **1**, 1370–1375 (1983).
101. M. D. Ambersley and C. W. Pott, *Thin Solid Films* **80**, 183–195 (1981).
102. S. Onishi, M. Eschwei, and W. C. Wang, *Appl. Phys. Lett.* **38**, 419–421 (1981).
103. J. S. Wang and K. M. Lakin, *Appl. Phys. Lett.* **42**, 352–354 (1983).
104. G. S. Kino, *Ultrasonics Symp., IEEE, 1979*, 900–910 (1979).
105. K. Setsune, T. Tanaka, O. Yamazaki, and K. Wasa, *Proc. Symp. Ultrasonic Electronics, 1st, Tokyo, 1980; Jpn. J. Appl. Appl. Phys.* **20**, Suppl. 20–23, 137–139 (1981).
106. S. Fujishima, S. Arai, and H. Ieki, *Proc. of ISCAS*, 625–628 (1979).
107. A. C. Anderson and D. E. Oates, *Utrasonic Symp., IEEE, 1982*, 329–333 (1982).
108. T. Yamamoto, T. Shiosaki, and A. Kawabata, *J. Appl. Phys.* **51**, 3113–3120 (1980).
109. T. Hata, T. Minamikawa, E. Noda, O. Morimoto, and T. Hada, *Jpn. J. Appl. Phys.* **18**, 219–224 (1979).
110. T. Shiosakai, S. Ohnishi, Y. Murakami, and A. Kawabata, *J. Cryst. Growth* **45**, 346–349 (1978).
111. T. Shiosaki, A. Adachi, and A. Kawabata, *Thin Solid Films* **96**, 129–140 (1982).
112. B. T. Kkhuri-Yakub, J. G. Smits, and T. Barbee, *J. Appl. Phys.* **52**, 4772–4774 (1981).
113. T. Hata, E. Noda, O. Morimoto, and T. Hada, *Appl. Phys. Lett.* **37**, 633–635 (1980).
114. T. Hata, F. Takeda, H. Yamamoto, and O. Morimoto, *Proc. Electr. Electron. Insul. Conf. Jpn., 1981*, 314–317 (1981).
115. T. Shiosaki, T. Yamamoto, T. Oda, K. Haradaand, and A. Kawabata, *Jpn. J. Appl. Phys.* **20**, 149–152 (1981).
116. T. Hata, T. Minamikawa, O. Morimoto, and T. Hada, *J. Cryst. Growth* **47**, 171–176 (1979).
117. T. Hata, E. Noda, O. Morimoto, and T. Hada, *Ultrasonics Symp., IEEE, 1979*, 936–939 (1979).
118. F. Takeda, T. Mori, and T. Takahashi, *Trans. Inst. Electron. Comm. Eng. Jpn. E* **65**, 601 (1982).
119. S. Maniv, W. D. Westwood, and E. Colombini, *J. Vac. Sci. Technol.* **20**, 162–170 (1982).
120. K. Tominaga, S. Iwamura, I. Fujita, Y. Shintani, and O. Tada, *Jpn. J. Appl. Phys.* **21**, 999–1002 (1982).
121. B. T. Khuri-Yakub, J. G. Smits, and T. Barbee, *Proc. Ultrasonics Symp., Boston, 1980*, 801–804 (1980).
122. L. G. Pearce, R. L. Gunshor, and R. F. Pierret, *Appl. Phys. Lett.* **39**, 878–879 (1981).
123. F. Takeda and T. Hata, *Jpn. J. Appl. Phys.* **19**, 1001–1002 (1980).
124. F. Takeda, T. Mori, and T. Takahashi, *Jpn. J. Appl. Phys.* **20**, 169–172 (1981).
125. E. Gerova, N. A. Ivanov, and K. I. Kirov, *Thin Solid Films* **81**, 201–206 (1981).
126. T. Shiosaki, T. Yamamoto, T. Oda, K. Harada, and A. Kawabata, *Ultrasonics Symp., IEEE, 1980*, 451–454 (1980).
127. *Solar Cells* **4**(3) and (4) (1981).
128. P. G. LeComber, W. E. Spear, and A. Ghaith, *Electron. Lett.* **15**, 179 (1979).
129. R. A. Gibson, P. G. LeComber, and W. E. Spear, *Appl. Phys. Lett.* **21**, 307 (1980).

130. A. J. Snell, W. E. Spear, P. G. LeComber, and K. MacKenzie, *Appl. Phys. A* **26**, 83 (1981).
131. I. Shimizu, T. Komatsu, K. Saito, and E. Inoue, *J. Non-Cryst. Sol.* **35**, 773 (1980).
132. I. Shimizu, S. Oda, K. Saito, H. Tomita, and E. Inoue, *Proc. Int. Conf. Amorphous Liqu. Semicon., Grenoble, 1981* (D. Kaplan, ed.), *J. Phys. Paris Colloq.* **42**, Chapter 4, 1123 (1981).
133. M. Janai and F. Moser, *J. Appl. Phys.* **53**, 1385 (1982).
134. S. R. Das, D. F. Williams, and J. B. Webb, *J. Appl. Phys.* **54**, 3101–3105 (1983).
135. J. B. Webb and S. R. Das, *J. Appl. Phys.* **54**, 3282–3285 (1983).
136. A. D. Jonath, W. W. Anderson, J. L. Crowley, H. F. MacMillan, Jr., and J. A. Thornton, "Amorphous Thin films for Solar Cell Applications" (1981), SERI-XS-0-9237-1.
137. D. R. McKenzie, N. Savvides, R. C. Mcphedran, L. C. Botten, and R. P. Netterfield, *J. Phys. C* **16**, 4933–4944 (1983).
138. M. Usui, T. Kujirai, T. Nagatomo, and O. Omoto, *Vacuum Soc. Jpn.* **25**, 56–62 (1982).
139. A. R. Mizra, A. J. Rhodes, J. Allison, and M. J. Thompson, *J. Phys., Colloq. (Orsay, Fr)*, 659–662 (1981).
140. S. R. Das, J. B. Webb and D. F. Williams, *Proc. Can. Semi. Technol. Conf., 1st, Ottawa, Canada, 1982*, 125 (1982).
141. J. A. Thornton and D. W. Hoffman, *J. Vac. Sci. Technol.* **18**, 203–207 (1981).
142. M. Suzuki, M. Suzuki, M. Kanada, and Y. Kakimoto, *Jpn. J. Appl. Phys.* **21**, 89–91 (1982).
143. M. C. Abdulrida and J. Allison, *Appl. Phys. Lett.* **43**, 768–770 (1983).
144. Y. Z. Sun, S. R. Das, D. F. Williams, and J. B. Webb, submitted to *J. Appl. Phys.*
145. H. Kressel, M. Ettenberg, J. P. Wittke, and I. Ladany, *in* "Semiconductor Devices for Optical Communication" (H. Kressel, ed.), Vol. 39, Springer-Verlag, Berlin and New York, 1982.
146. K. L. Chopra and S. R. Das, "Thin Film Solar Cells," Plenum Press, New York, 1983.
147. J. A. Thornton, D. G. Cornog, and W. W. Anderson, *J. Vac. Sci. Technol.* **18**, 199–202 (1981).
148. A. D. Jonath, W. W. Anderson, J. A. Thornton, and D. G. Cornog, *J. Vac. Sci. Technol.* **16**, 200–203 (1979).
149. J. A. Thornton, D. G. Cornog, R. B. Hall, and L. C. DiNetta, *J. Vac. Sci. Technol.* **20** 296–299 (1982).
150. H. Rau, *J. Phys. Chem. Solids, Suppl.* **28**, 903 (1967).
151. J. A. Thornton and W. W. Anderson, *Appl. Phys. Lett.* **40**, 622–624 (1982).
152. S. Maniv, W. D. Westwood, F. R. Shepherd, P. J. Scanlon, and H. Plattner, *J. Vac. Sci. Technol.* **20**, 1–6 (1982).

7 SILICON CARBIDE FILMS

Hiroyuki Matsunami

Department of Electrical Engineering, Faculty of Engineering
Kyoto University
Kyoto, Japan

I. INTRODUCTION

Silicon carbide (SiC) has been used as an abrasive and as firebricks utilizing its mechanically hard, refractory, and chemically inactive properties. In electronic applications, SiC has been used for arrestors and varistors. The materials for these applications were produced by the Acheson method in which a mixture of SiO_2 and coke is heated electrically using graphite electrodes. As a semiconducting material, SiC has advantages since it can be used at high temperatures and under harsh environments. Moreover, the large energy gap of SiC allows it to emit visible light. Actually, electroluminescence of SiC was observed for the first time in 1923, thus making it the first luminescent material [1]. Much attention has been given to this material for developing solid-state electronic devices. Progress in this field was surveyed in the proceedings of the past three international conferences on SiC [2–4].

Growth of SiC single crystals of semiconducting grade is rather difficult since there is no molten phase of SiC at 1 atm pressure in the Si–C binary system. Hence, single crystals for semiconductor applications have been prepared by sublimation, growth from solution, and thermal decomposition of compounds containing silicon and carbon. The most common method of growing SiC bulk single crystals with high purity is by sublimation above 2500°C as shown by Lely [5]. Devices such as rectifiers, particle detectors, tunnel diodes, and light-emitting diodes based on *p–n* junctions were fabricated using single crystals made by the Lely method. Processing techniques for SiC such as polishing, cutting, etching, and oxidation have been developed together with junction formation by sublimation, diffusion, and alloying methods.

However, diffusion processes require very high temperatures (above 1900°C) and a very complicated furnace in order to avoid sublimation of SiC during processing. Alloying can be carried out in a short time, but it also needs very high temperature. Such difficulties in technology have prevented the development of SiC devices in electronics where high production rates are required.

Thin film technology has been developed for obtaining good single crystals of SiC at rather low temperatures and for easy control of impurity doping. Epitaxial techniques are capable of producing materials of superior quality reproducibly, which is necessary for electronic applications. For nonelectronic applications, thin films of SiC are often used as surface coatings for passivation or hardening.

In this chapter, thin film deposition methods of SiC are reviewed. Physical properties of SiC are briefly summarized, and film depositions from the vapor phase and liquid phase including epitaxial growth are described. Amorphous SiC thin films are surveyed and current progress is discussed. Applications of SiC thin films in new fields are presented.

II. PHYSICAL PROPERTIES

Silicon carbide is a stable compound of silicon and carbon. In a crystalline form, the smallest distance between Si and C atoms is 1.89 Å. Each Si (or C) atom is surrounded by four C (or Si) atoms tetrahedrally with sp^3 hybrid bonds (covalency is about 88%). Crystallographically, there are a number of polytypes in SiC. In the close packing arrangement of atoms shown in Fig. 7.1, the second layer B is arranged always in the same way on the first layer A. The third layer takes one of two possible arrangements. The first is the same arrangement of atoms in the layer A and the second is a new arrangement (layer C). In accordance with the order of the three layers, there

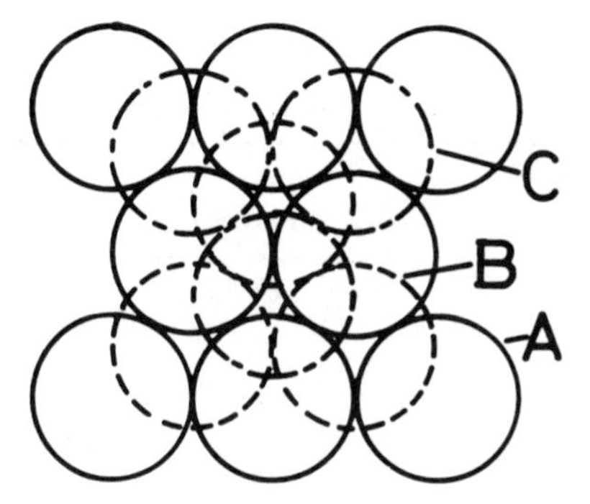

Fig. 7.1. Close packing arrangement of atoms.

are a great number of versions of sequence from the simplest types of AB AB (wurzite) and ABC ABC (zinc blende) to structures consisting of 100 layers.

In Table 7.1, the lattice constants and the energy gaps of common SiC polytypes are shown [4]. The first number in the expression of the polytype shows the repetition of periodicity and the letter indicates the crystal symmetry system such as cubic (C), hexagonal (H), and rhombohedral (R). The cubic structure of the polytypes exists only as 3C-SiC named β-SiC, whereas there are a great number of polytypes, altogether named α-SiC, show hexagonal or rhombohedral structures. Physicochemical and mechanical properties of the polytypes are very similar [4]. Some of the properties are

TABLE 7.1

Lattice Constants and Energy Gaps of Common SiC Polytypes[a]

Polytype	Lattice constant (Å)		Energy gap[b] (eV at 4.2 K)
	a	c	
2H (wurzite)	3.076	5.0480	—
4H	3.076	10.046	3.765
6H	3.08075	15.11738	3.023
8H	3.079	20.147	2.80
15R	3.073	37.30	2.986
21R	3.073	52.78	2.853
33R	—	82.5	3.002
3C (zinc blende)	4.3596	—	2.390

[a] Data from Marshall *et al.* [4].

[b] The energy gap shown here is an exciton energy gap, and the true energy gap is given by the sum of this value and the exciton binding energy.

TABLE 7.2

Physicochemical and Mechanical Properties of SiC[a]

Properties	Polytype[b]	Value	Temperature (°C)
Density (g/cm^3)	β	3.210	20
	α	3.211	
Thermal expansion coefficient (cm/°C)	β	5.8×10^{-6}	1000
	α	5.12×10^{-6}	25–1000
Hardness (Mohs)		9.2–9.3	20
Young's modulus (kg/m^2)	α	37.0×10^3	20
Debye temperature (K)	β	1430	—
	α	1200	—
Specific heat (cal/g °C)	β	0.17	20
	α	0.165	27
Thermal conductivity (W/cm °C)	β	0.255	200
	α	0.410	20

[a] Data from Marshall *et al.* [4].
[b] Polytypes are shown if necessary.

shown in Table 7.2. Although electronic and optical properties of the polytypes are considerably different, the details are not within the scope of this chapter, and the reader is referred to the literature [2–4].

III. FILM FORMATION BY CHEMICAL VAPOR DEPOSITION

Films of SiC can be synthesized by chemical reactions of volatile gases on the substrate surface, which is kept at high temperatures. The reactant gases are introduced with a carrier gas, usually H_2, of high purity. Various combinations of reactant gases have been reported for chemical vapor deposition (CVD). These include $SiCl_4$–CCl_4 [6–8], $HSiCl_3$–C_6H_{14} [9], SiH_4–C_3H_8 [10–13], $SiCl_4$–C_6H_{14} [14, 15], $SiCl_4$–C_3H_8 [16–18], and decomposition of CH_3SiCl_3 [19]. When the reactant gas is thermally decomposed, argon gas can be used as a carrier gas, which has the advantage of not etching the substrate and/or susceptor.

An example of a CVD apparatus is shown schematically in Fig. 7.2. Epitaxial growth is usually carried out in the temperature range of 1200 to 1750°C using radio frequency (rf) induction heating of the susceptor. Since the susceptor is in the reaction chamber and can be either a major source of the constituent, etchant vapor, or contaminant, the material of the susceptor should be carefully selected. Although graphite susceptors are often used, they contribute significantly to the partial pressure of C in the reaction

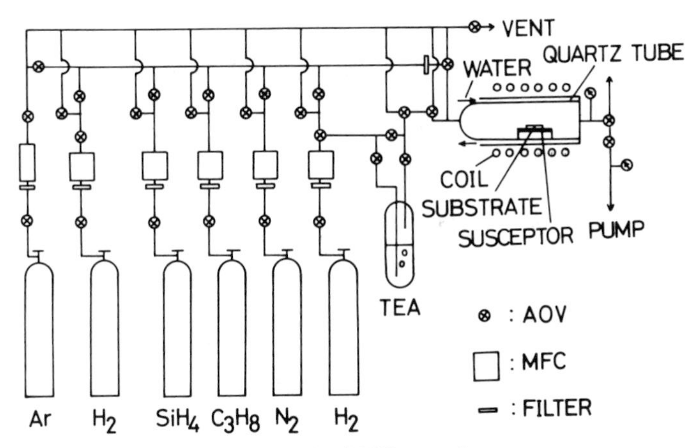

Fig. 7.2. Schematic of CVD growth system.

chamber, particularly in the case of H_2 carrier gas. Alternative materials such as W, WC [9], Ta [11–13], and TaSi-coated Ta [8], have been used as susceptors. A disadvantage of these susceptor materials is possible contamination in the growing layer. One of the most useful susceptors is SiC-coated graphite [16–18].

The substrate temperature is limited by its melting point. Polytypes of the grown layer are changed with substrate temperature. The polytype 6H-SiC grows at high substrate temperatures and 3C-SiC grows at low substrate temperatures. Hereafter, the growth on SiC substrates and on foreign substrates is described separately.

A. Growth on SiC Substrates

Epitaxial growth of 6H-SiC on 6H-SiC substrates prepared by the Lely method has been studied as a possible way to make solid-state devices. The substrates should be polished and etched before crystal growth in order to obtain good epitaxial layers. Chemical etching with concentrated alkaline solution at 600–700°C, electrolytic etching [20], and H_2 gas etching at high temperatures [21] can be used. Gas etching using a mixture of O_2 and Cl_2 is suitable for processing during device fabrication [22]. Immediately after etching, growth is started by introducing the reactant gases into the reaction chamber.

Chemical reactions around substrates are different depending on the combination of reactant gases and carrier gases. Theoretical analyses of equilibrium partial pressure in the hydrogen–silicon–carbon system were carried out [11, 23]. The model is based on the idea that the substrate surface is in equilibrium with the reactant species coming through the boundary layer and the crystals grow only on the substrate surface [11]. According to the calculation, the efficiency of SiC formation decreases with increasing substrate temperature and the Si/C ratio, which agrees with their experimental results. However, these results cannot be readily applied to the other systems because the model is very limited. In the case of H_2 as a carrier gas, single crystals can grow within a wide range of Si/C ratios and substrate temperatures, but single crystals grow only for extremely selected conditions in the case of Ar as carrier gas [23]. This is due to the fact that H_2 gas may carry away the excess carbon in the form of hydrocarbon, whereas Ar gas cannot.

The mole ratio of the reactant gases is determined experimentally, and the optimum value of the Si/C ratio ranges between 1/3 and 5/1 depending on the reaction system. When the mole ratio of the reaction gas to H_2 gas is less than 10^{-3}, single crystals are obtained for a wide range of Si/C ratios. When the mole ratio becomes larger than 10^{-3}, the Si/C ratio should be chosen precisely to obtain single crystals of good quality. The growth rate ranges between 0.2 and 1.0 μm/min within the substrate temperature range of 1500–1800°C. Smooth layers are generally obtained by using lower growth rates [19].

Single crystals of 6H-SiC were epitaxially grown on 6H-SiC substrates using $SiCl_4$–C_3H_8–H_2 by the author's research group [17, 18]. High temperatures of 1500–1700°C were obtained by rf induction heating of a graphite susceptor in a water-cooled quartz reaction chamber. A growth rate of 0.1–0.15 μm/min was obtained at 1600°C by keeping a standard gas composition ($H_2 \simeq 1$ liter/min, $SiCl_4 \simeq 1$ ml/min, $C_3H_8 \simeq 0.05$ ml/min, and Si/C $\simeq 6.7$) [18]. The grown layer showed a transparent greenish blue color and the surface was mirrorlike and smooth. Epitaxial growth was confirmed by x-ray and reflection electron diffraction (RED) analyses. The growth rate on the $(000\bar{1})$ C face was larger than that on the (0001) Si face. At lower temperatures ($\sim$1550°C), the surface morphology was different depending on the surface orientation, as shown in Fig. 7.3. The growth mechanism was investigated qualitatively by noting the differences in the surface morphology.

The crystal quality of SiC was examined by x-ray topography, which indicated the presence of defects, twinning, and the change in polytype [24]. Layers of good quality exhibited a lower surface defect density, as observed by etch pits using an optical microscope. If the substrate was a single polytype, then the epitaxial layer usually was of the same polytype. Random nucleation occurred at the edge of the crystal and spread inward, which resulted in polycrystal growth. Change in polytype could occur, for example, when 3C-SiC nucleated and spread over the entire substrate.

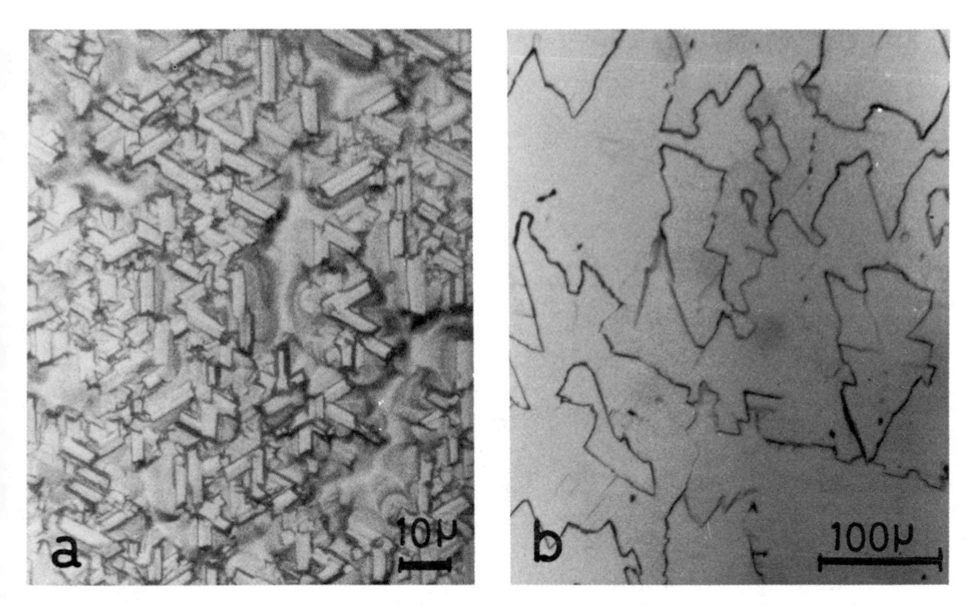

Fig. 7.3. Surface morphology of 6H-SiC on 6H-SiC: (a) (0001) Si face and (b) (0001) C face. (From Nishino *et al.* [18]. Copyright 1978 North-Holland Publ. Co.; Amsterdam.)

Impurity doping has been investigated together with vapor-phase epitaxial growth of 6H-SiC on 6H-SiC substrates [8, 12, 14, 15]. The conductivity type and carrier concentration of the grown layer were controlled by using B_2H_6 as a *p*-type dopant and N_2, AsH_3, or PH_3 as *n*-type dopants. Epitaxial *p–n* junction structures have been examined [8, 12]. Doping of Al using $AlCl_3$ was attempted and the doping effect was examined by inspection of the photoluminescence spectrum [17]. Doping efficiencies of B_2H_6, NH_3, and N_2 were studied [15]. The most effective doping was obtained with B_2H_6, and the lowest with N_2. Aluminum doping using organometallic aluminum compounds such as $(CH_3)_3Al$ was suggested [15]. The Al doping effect using $(C_2H_5)_3Al$ was checked by cathode luminescence [25].

B. Growth on Foreign Substrates

Growth of 3C-SiC on foreign substrates such as Si or sapphire has been investigated extensively for the following reasons: (1) applicability of 3C-SiC for new electronic devices, (2) availability of large substrates of good crystal quality, and (3) lower temperature growth than for 6H-SiC. Carbonization and chemical vapor deposition methods have been widely used in these studies.

Initially 3C-SiC films were prepared by flowing CH_4 gas over Si single crystals heated at 1300°C [26]. Studies on film formation of 3C-SiC by heating Si above 900°C in a hydrocarbon atmosphere (i.e., carbonization) were actively carried out during 1965–1973 [27–34]. The formation mechanism of SiC was studied extensively. Silicon carbide was thought to be formed by the inward diffusion of C derived from thermal decomposition of hydrocarbon [27, 28]. On the contrary, outward diffusion of Si has been claimed as the dominant growth mechanism [30]. Based on experiments dealing with SiC formation using radioactive ^{14}C, the predominate mechanism was found to be Si diffusion through lattice defects in SiC [32]. The fact that grain growth was influenced by the partial pressure of hydrocarbon is well explained by this model. By the observation of the surface morphology and growth rate, Si diffusion through growing SiC was pointed out as the dominant mechanism [33].

Although SiC films several micrometers thick can be obtained, the crystal quality becomes worse with increasing thickness due to the dominant diffusion mechanism in crystal growth. A good epitaxial layer is obtained for thicknesses less than 1 μm.

Thick films of 3C-SiC with good crystal quality can be obtained by chemical vapor deposition [35–40]. Epitaxial films of 3C-SiC several micrometers thick were obtained with a speed of 0.3 μm/min on Si (111) and (100) substrates using the $SiCl_4$–C_3H_8–H_2 system. Electrical conduction types were controlled by doping impurities during film formation using AsH_3, PH_3, N_2 for *n*-type and B_2H_6 for *p*-type doping [35].

The influence of the orientation of Si substrates for the growth of 3C-SiC was studied using the decomposition of $(CH_3)_2SiCl_2$. Crystals of good quality were obtained on Si (100) substrates, but on (111) or (110) substrates twinning ocurred [38]. Fast growth rates of 4 μm/min (faster by one order of magnitude than ordinary growth rates) could be obtained by this method. Crystal quality becomes better with increasing film thickness, and it becomes best at around 12 μm.

The influence of substrate temperature and orientation on crystal quality was studied extensively using the $SiCl_4$–C_3H_8–H_2 system in the author's laboratory [40]. As shown in Fig. 7.4, RED patterns of 3C-SiC on the Si (111) substrate prepared at lower substrate temperature show Debye rings indicating the polycrystalline nature of the films. Strong spots and streak lines in the films prepared at 1275°C show a single-crystalline nature including twins or stacking faults. To obtain a good epitaxial layer, the substrate temperature should be kept at 1390°C. Better crystal quality was obtained for the 1-μm-thick films on the (100) or (110) substrates rather than on the (111) substrates. Stacking sequences of the [111] direction consist of the Si layer and the C layer, alternatively. Stacking faults are frequently

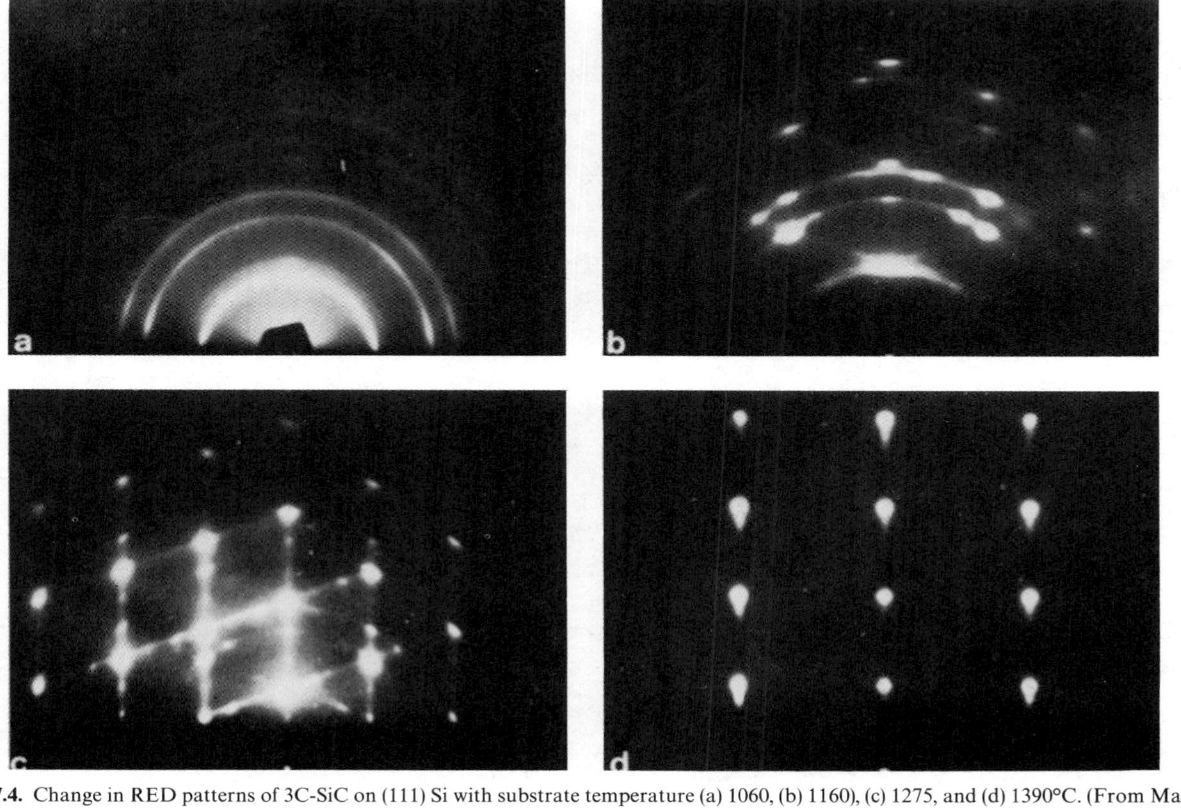

Fig. 7.4. Change in RED patterns of 3C-SiC on (111) Si with substrate temperature (a) 1060, (b) 1160), (c) 1275, and (d) 1390°C. (From Matsunami *et al*. [40]. Copyright 1978 North-Holland Publ. Co., Amsterdam.)

observed on the (111) surface of 3C-SiC naturally grown by sublimation. Thus, the layer of 3C-SiC grown on the Si (111) substrate might have inferior crystal quality compared with those grown on Si (100) and (110) substrates. Prior etching of Si substrates by HCl gas at 1160°C before SiC deposition is desirable, and the optimum gas flow rates are $H_2 \simeq 1$ liter/min, $SiCl_4 \simeq 3$ ml/min, $C_3H_8 \simeq 1$ ml/min, and Si/C $\simeq 1$, which allows a growth rate of 0.2 μm/min [40]. Growth of 3C-SiC on sapphire [38] or Si on sapphire [39] was also reported.

Epitaxial layers of 3C-SiC could be obtained on Si substrates as described earlier, but the thickness was limited to several micrometers and the problem of reproducibility was not solved. Difficulty in heteroepitaxial growth comes from the large lattice mismatch between the growing layer and the Si substrate ($a_{SiC} = 4.358$ Å, $a_{Si} = 5.430$ Å). One of the approaches to overcome this difficulty is to put a buffer layer between the substrate and the deposit. Based on previous findings, sputtered 3C-SiC layers were suggested as good buffer layers [41]. Single crystals of 3C-SiC were grown reproducibly by CVD at 1360°C on Si substrates incorporating sputtered 3C-SiC intermediate layers using the SiH_4–C_3H_8–H_2 system. Crystal quality of the grown layer was examined by RED and x-ray Laue methods. Overall it was found that sputtering should be carried out at a substrate temperature between 800 and 1000°C with the thickness less than 100 nm.

Since a combination of sputtering and CVD process is troublesome and the layers are easily contaminated, this method is not desirable for the preparation of films of semiconductor grade. The utilization of 3C-SiC prepared by carbonization as a buffer layer is very attractive because the

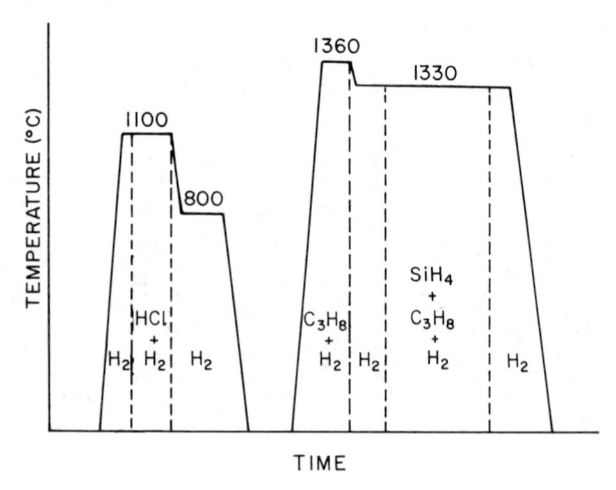

Fig. 7.5. Temperature program for reproducible single-crystal growth of 3C-SiC on Si.

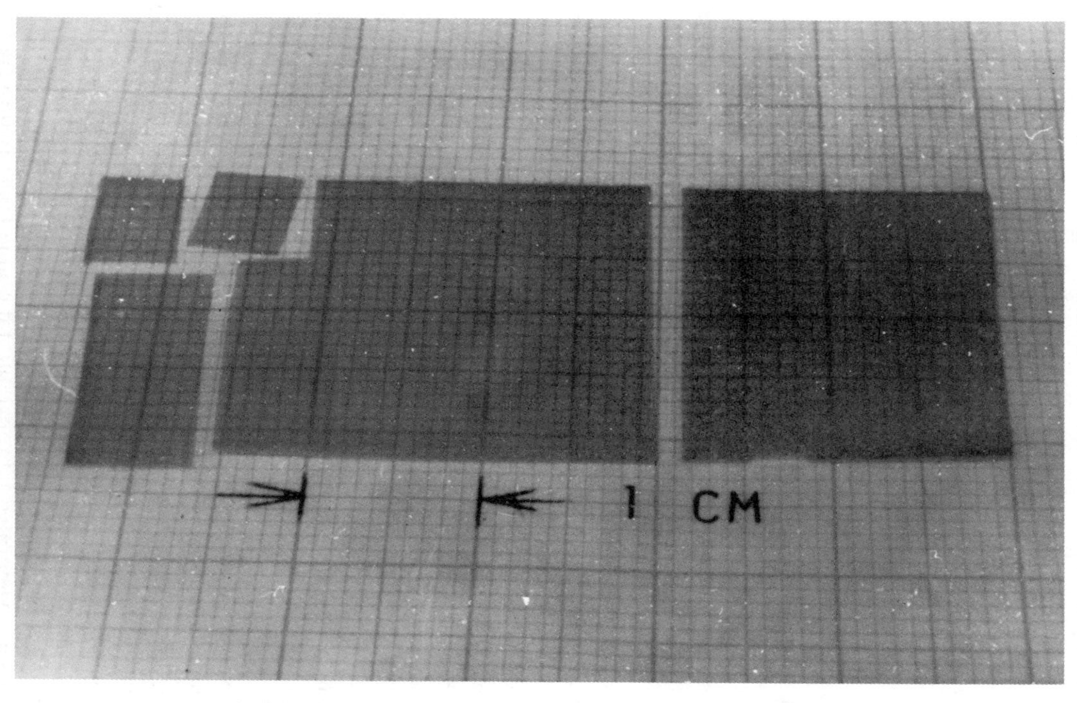

Fig. 7.6. Films of 3C-SiC with large area. (From Nishino *et al.* [43]).

buffer layer formation and CVD process can be carried out in the same reaction chamber sequentially. Although this combination was studied at an early stage, improvement of crystal quality was not observed [37]. By the detailed study of buffer layer formation using Auger and RED analyses, the optimum condition was determined in the author's laboratory for obtaining good epitaxial layers of 3C-SiC on Si by a subsequent CVD process [42]. Using the temperature program shown in Fig. 7.5, a reproducible preparation method for cubic SiC was found. Films of 3C-SiC with large areas were obtained as shown in Fig. 7.6 [43]. Control of conduction type by doping impurities during CVD was demonstrated for the first time [44].

IV. FILM FORMATION BY PHYSICAL AND REACTIVE VAPOR DEPOSITION

Films of SiC were prepared by vacuum evaporation onto cleaved synthetic mica (001) substrates in a high-vacuum vessel [45]. Substrates were taken off by dissolving in HF solution. Formation of SiC films has been achieved by volatilization of Si in the presence of accetylene (C_2H_2) [46]. Vapors of Si were provided both by dc sputtering and by thermal evaporation. In the case of sputtering, transparent films exhibiting yellow color were obtained at a substrate temperature of 900°C under a C_2H_2 pressure of 3.99×10^{-4} Pa. The deposits were impervious to standard silicon etchants. Reflection electron diffraction yielded a pattern that was characteristic of 3C-SiC. In the case of evaporated Si with an impingement rate of 3 Å/s in a C_2H_2 partial pressure of 3.99×10^{-4} Pa, 3C-SiC films were obtained at 900°C. The properties of such films were similar to those obtained by reactive sputtering. Quite different results were obtained for C_2H_2 partial pressures exceeding 1.33×10^{-3} Pa. Diffraction patterns of Si were observed indicating a lack of reaction. At such high pressures, the conversion of Si to SiC was virtually arrested.

Epitaxial deposits on 3C-SiC substrates were obtained by reactive deposition involving the volatilization of Si in C_2H_2 at temperatures substantially lower than by CVD [47]. Substrates of 3C-SiC were prepared by chemical conversion of Si wafers heated in a hydrocarbon atmosphere. If the ratio *r*, which is the rate of impingement of C_2H_2 at the substrate to that of arrival Si atoms, is greater than 30, only 3C-SiC grows epitaxially with a weaker polycrystalline component. Both the converted SiC substrates and the reactively evaporated films exhibit noticeable {111} twinning. At higher substrate temperatures (~1100°C), 3C-SiC was completely epitaxial, and the crystallographic quality was comparable to that of the converted substrates. The low epitaxial temperature, as opposed to those generally observed with CVD, may be largely an effect of the deposition rate. In CVD, the rates usually exceeded 0.1 μm/min, and the increased rate generally led to increased epitaxial temperatures.

Films of SiC were deposited on Si (111) and Mo plates by rf sputtering of a Si target in a mixture of Ar and CH_4 at temperatures of 500–1200°C [48]. The sputtering was carried out at a pressure of 7.98×10^{-1} Pa using 13.6 MHz. Films of 300–700 Å were obtained with a deposition rate of 10 Å/min. Epitaxial SiC films were obtained on Si substrates at temperatures of 850–1200°C. In a narrow temperature range between 900 and 940°C, 2H-SiC exists with a considerable amount of (0001) stacking disorder. Below this range, random stacking predominates, and above it diffraction spots of 3C-SiC appear in its (111) stacking layers parallel to the (0001) plane of 2H-SiC. The amount of 3C-SiC increases with temperature and only 3C-SiC was found above 1100°C. On Mo substrates, 2H-SiC films or 2H-SiC films containing a small amount of 3C-SiC were obtained at about 1000°C.

Thin films of SiC were deposited on pyrex glass, sapphire, and Si substrates by rf sputtering of a SiC target, and the structure and mechanical properties of sputtered SiC films were studied [49]. The RED pattern of the film showed an amorphouslike halo when the substrate temperature was lower than 500°C. The film was crystalline when the substrate temperature was higher than 700°C. Sputtering onto Si substrates yielded amorphous SiC, according to the IR spectrum, which showed a maximum absorption band at about 800 cm^{-1} corresponding to the lattice vibration of bulk SiC. The RED pattern of sputtered crystalline films of 0.3 μm thickness on sapphire (001) substrates at 800°C suggested polycrystalline α-SiC with the hexagonal symmetry similar to the reactively sputtered SiC films [48]. The observed α-SiC phase was unstable, and it transformed into the β-phase by annealing in vacuum at 1100°C for 1 h. Films of SiC of the same polytype were even obtained by using a 3C-SiC target. This suggests that the SiC target surface is mostly decomposed into Si and C atoms. Sputtered particles with a mean energy of 10 eV recombine with each other on the substrate and form SiC. The polytypes of the sputtered SiC films are determined mainly by the substrate temperature.

Single crystals of 3C-SiC were grown on Si substrates by rf sputtering by using 3C-SiC powder in a quartz tray as a target material by our group [50]. Small crystals of 3C-SiC were prepared by melting Si in a graphite crucible and were ground into powder. Films of 1500-Å thickness were obtained by sputtering for 60 min. The surface of the film was yellowish green, mirrorlike, and smooth. Triangular shapes as shown in Fig. 7.7 were observed in the deposited films or at the boundary between Si and the deposit. When the substrate was kept at high temperature during sputtering, the IR reflectivity of the film showed a sharp peak as shown in Fig. 7.8. For the films prepared at 1200°C, the RED pattern indicated single crystals without diffused spots.

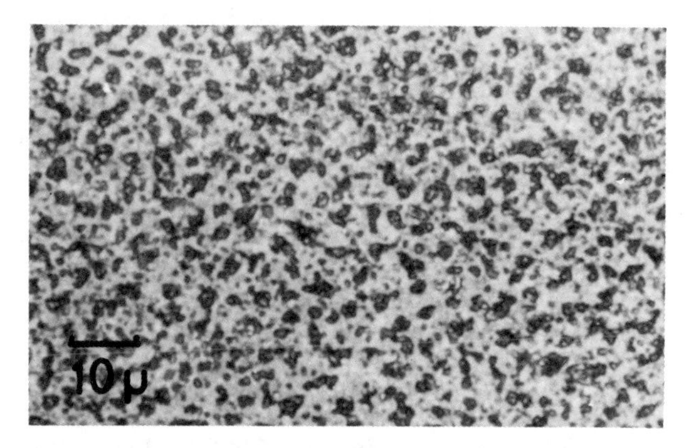

Fig. 7.7. Triangular shapes appearing in 3C-SiC films on Si. (From Nishino *et al.* [50].)

X-ray diffraction of the film prepared at temperatures greater than 1000°C showed only one strong peak corresponding to the (111) plane of 3C-SiC. Films deposited at room temperature showed a broad IR spectrum centered at 12.6 μm. After annealing at a temperature of 1000°C for 3 h in a 1.33-Pa vacuum, this band became narrow, and the peak height increased, which implies an increase in Si–C bonds. On examination of the x-ray diffraction

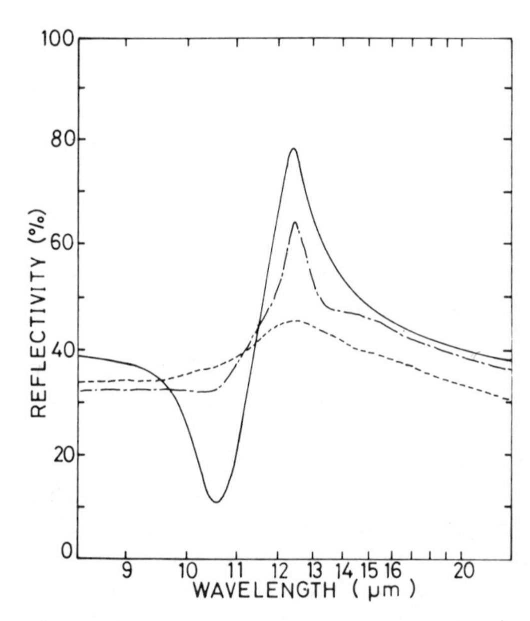

Fig. 7.8. Change in IR reflectivity of sputtered 3C-SiC films with substrate temperature; - - -, 800°C; — - —, 1000°C; and ——, 1200°C. (From Nishino *et al.* [50].)

pattern, however, the films did not show any peak; i.e., they were amorphous.

Effects of target materials on the structural properties of deposited SiC films on Si substrates were examined [51]. Films were deposited at 200–740°C by using a sintered 6H-SiC target and a Si single crystal target. For the SiC target, sputtering was carried out in an Ar atmosphere at a pressure of 2.7 Pa with a deposition rate of 0.3 μm/h by applying rf power of 2 W/cm^2. For the Si target, the sputtering atmosphere was a mixture of Ar and hydrocarbon (CH_4 or C_2H_2) at a pressure of 2.7 Pa. The partial pressure of CH_4 or C_2H_2 was varied. The rf power density was 3 W/cm^2. At low partial pressures the deposition rate was independent of partial pressure, whereas above a critical pressure the deposition rate decreased with increasing partial pressure. The critical pressure for C_2H_2 was much lower than for CH_4. The chemical composition of the film was examined by Auger electron spectroscopy using a SiC bulk single crystal as a standard. The Si/C ratio of the films using the SiC target was not significantly affected by the deposition rate or substrate temperature. The Si/C ratio of the films using the Si target varied with the hydrocarbon partial pressure and the deposition rate. Stoichiometric films were obtained at some partial pressures. The film composition was governed by the ratio of arrival rate of hydrocarbon molecules to that of Si atoms at the substrate. Since C_2H_2 has a higher chemical reactivity than CH_4, the partial pressure needed to obtain stoichiometric films was lower than that of CH_4. By using RED patterns, the SiC films deposited from the SiC target were found to be crystallized more easily than those from the Si target. This was attributed to the higher chemical activity and kinetic energy of C atoms ejected from the SiC target than the C atoms from hydrocarbon. Films of SiC were prepared by rf ion plating [52]. Growth mechanism and film structure are different from reactive evaporation.

V. LIQUID-PHASE GROWTH OF THIN FILMS

Liquid-phase epitaxial (LPE) growth has the following advantages: (1) a gettering effect of the solution on many impurities and (2) the possibility of doping certain impurities that are difficult to incorporate in vapor-phase growth or difficult to obtain in a volatile form. However, little work has been carried out on LPE growth of SiC for the lack of a suitable solvent that dissolves a large amount of C [53].

In early stages, LPE growth of SiC was attempted on SiC substrates floating on a Cr melt in a SiC crucible [54]. To reduce Cr vaporization and supersaturation, He gas was blown on the Cr melt, and the epitaxial layer of the same polytype as the substrate was obtained. Chromium was found to be incorporated in the epitaxial layer as an acceptor.

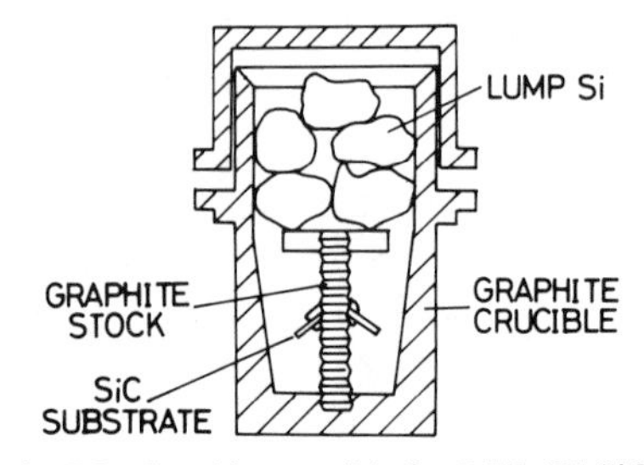

Fig. 7.9. Graphite crucible for LPE 6H-SiC.

In order to avoid contamination by the third constituent, Si was employed as a solvent and melted in a graphite crucible as shown in Fig. 7.9 [55]. To solve the problem of low solubility of C in Si (0.06 at. % at around 1700°C) [56] a temperature gradient was set up in the crucible. Carbon was dissolved from the wall of the crucible maintained at higher temperature and carried to the SiC substrates by thermal convection. Good epitaxial layers were obtained with a growth rate of 0.5 μm/min at 1650°C. The effect of different dopants was studied and the segregation coefficient of *p*-type dopants such as Al or B was found to be much less than unity. Since N has a segregation coefficient larger than unity and is incorporated into the SiC layer easily (and

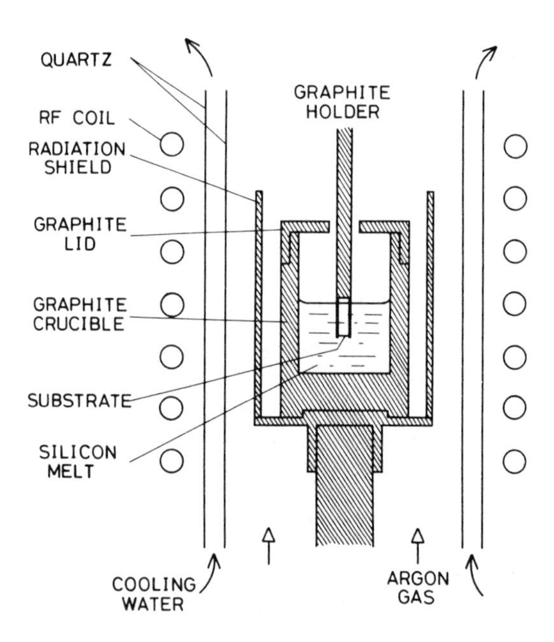

Fig. 7.10. Dipping technique for LPE 6H-SiC. (From Suzuki *et al.* [57]).

works as a donor), it should be carefully removed to obtain *p*-type conduction. In this method, the SiC substrate should be supported on a stock with a certain carefully determined inclination angle. By using this method, however, SiC epitaxial layers are often cracked due to stress when the Si melt is solidified. Moreover, grown crystals are taken out by cutting the graphite crucible and subsequent etching of solidified Si.

In order to solve these problems, a dipping technique has been developed in the author's laboratory [57]. The solvent Si was melted in a highly pure graphite crucible by rf induction heating in the system shown in Fig. 7.10. To prevent vaporization of Si, a graphite lid with a center hole was used. After the crucible was heated to a growth temperature of around 1650°C, the SiC substrate was dipped into the Si melt and allowed to stand for several hours, and then pulled up before the melt was cooled. A growth rate of 5–12 μm/h was obtained at 1650°C. Unintentionally doped layers showed *n*-type

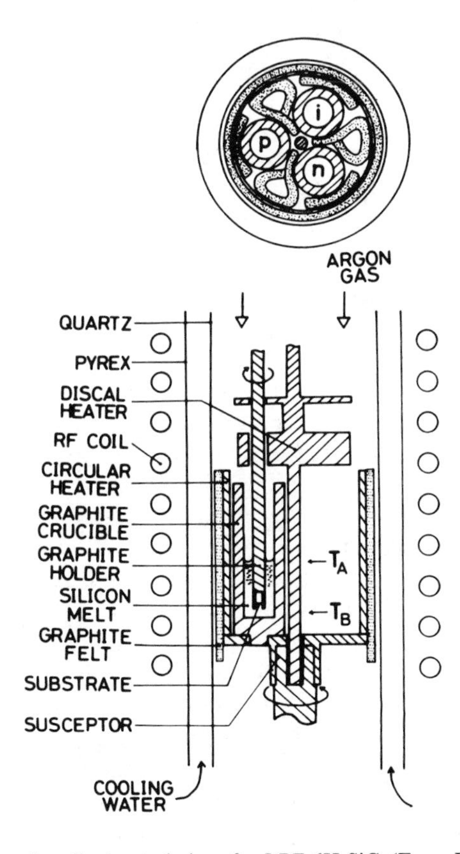

Fig. 7.11. Rotation dipping technique for LPE 6H-SiC. (From Ikeda *et al.* [59]).

conduction, and doping the Si melt with Al gave *p*-type conduction in the grown layer. With this method, grown crystals were free from stress since the crystals were pulled out before the Si melt was solidified. The crucible and the Si melt could be used repeatedly. A similar dipping method using alloys of Si with some transition metals was reported at the same time [58]. High-pressure He gas ($\sim$20 atm) was used to avoid vaporization of Si and hence the growth system was very large.

A rotation dipping technique was proposed by our group to enable *p–n* junction formation in one process [59]. Separate doping for *p*- and *n*-type conduction was carried out using the growth system shown in Fig. 7.11. The system is composed of three crucibles containing melts for *p*- and *n*-layers and for rinsing. The substrate holder can be rotated and moved vertically. After the crucibles were heated up to the growth temperature, the SiC substrate was dipped into the *p*-crucible and then pulled up. Then, the crucible holder was rotated and the substrate was dipped into the *i*-crucible to take out the attached melt for *p*-layer growth. Using a similar procedure the *n*-layer was grown on the *p*-layer. Hence, abrupt *p–n* junctions can be formed sequentially. Impurity doping was extensively studied as well.

As alternative methods for film formation from the liquid phase, the traveling solvent method [60–62] and the traveling heater method [63], which use a very thin molten metal zone such as Cr for LPE growth, were also reported.

VI. THIN FILMS OF AMORPHOUS SiC

Studies on amorphous SiC films were started in 1968 using a diode sputtering method [64]. Structure and electrical conduction were studied. Conductivity was irreducibly reduced by annealing in the temperature range of 200 to 600°C owing to the annihilation of structural defects. Amorphous SiC thin films were deposited on refractory metal strips by a pyrolytic technique. Electrical characteristics of the films were found to be similar to those of granulated SiC crystals [65]. Optical and electrical properties of sputtered amorphous SiC films were studied extensively [66]. Similar to other tetrahedrally coordinated amorphous semiconductors, the amorphous SiC films have generic features: relative insensitivity to impurity incorporation, a shallow optical absorption edge at higher energies, and large annealing effects. As to electrical conduction, alternative models other than phonon-assisted hopping among localized states were required.

Compositional disorder of sputtered amorphous SiC films was studied by Raman spectroscopy [67]. Both graphitelike and diamondlike homonuclear C–C bonds were observed in the Raman spectra. The existence of graphite-

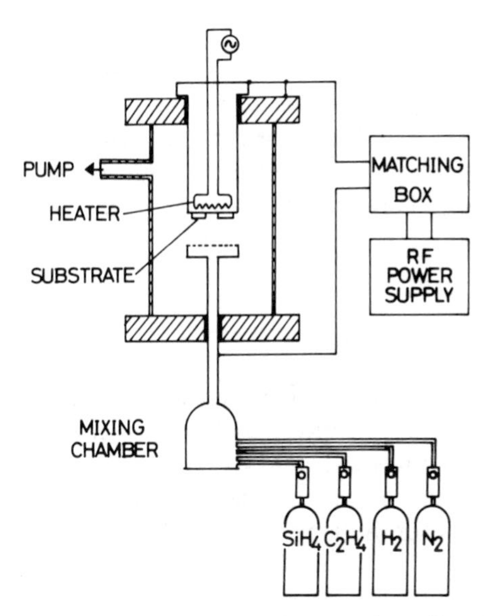

Fig. 7.12. Glow discharge system for amorphous $Si_{1-x}C_x$: H.

like bonds suggested a degree of short-range compositional order inconsistent with the modified continuous random network model.

Based on the success in film preparation of amorphous Si and Ge including hydrogen, a glow discharge method was applied for the preparation of amorphous Si–C films using a similar apparatus schematically shown in Fig. 7.12. Amorphous $Si_{1-x}C_x$: H films were prepared with SiH_4 and C_2H_4 using rf inductive coupling [68]. The film deposition onto the substrate was dependent on the gas flow rate, the overall pressure in the system, the particular gases used, and the intensity of the discharge. The volume ratio of the reactant gases need not reflect the final composition, since the gases may not dissociate at the same rate in the discharge. By this method, it was possible to deposit amorphous Si–C at compositions other than stoichiometric SiC. Optical and electrical properties were investigated extensively for the different film compositions.

Infrared absorption spectra of glow-discharge-deposited amorphous $Si_{1-x}C_x$ films were studied in detail [69]. A random distribution of Si and C atoms, without chemical ordering, in a tetrahedral network terminated at its open points by H atoms successfully accounted for the principal features of the IR spectra. Only one H atom was bonded to a Si atom, whereas H was bonded to a C atom predominantly in the form of CH_2 or CH_3. The bonded H content in the film was estimated to be of the order of 20–30 at. %.

Amorphous $Si_{1-x}C_x$ films including H atoms were also deposited by simultaneous rf reactive sputtering of Si and graphite targets in a H_2–Ar gas mixture [70]. Film composition was determined by electron spectroscopy for chemical analysis (ESCA), Rutherford back scattering (RBS), and thermal evolution of hydrogen. The difference in the properties between sputtered films and the glow-discharge-deposited films were described in the light of chemical bonding.

VII. APPLICATIONS FOR SiC FILMS

As described in the introduction, films of SiC are needed for a wide variety of present and expected applications [71]. In this section, recent applications using epitaxial layers prepared at rather low temperatures are described, together with applications of sputter-deposited SiC films and glow-discharge-deposited amorphous $Si_{1-x}C_x:H$ films.

A. Light-Emitting Diodes

Visible light emission by current flow was discovered in SiC many years ago before the mechanism of electroluminescence was elucidated [1]. Silicon carbide has attracted much interest as a material for visible light-emitting diodes (LEDs). However, the SiC LED has not been used in a practical way because of (1) the requirement of high temperature for junction preparation, (2) difficulty in the supply of substrate crystals, and (3) rather low external quantum efficiency.

Since red, orange, yellow and green LEDs are well developed using GaP and GaAsP at present, much attention for developing blue LEDs is given to SiC single crystals. Blue LEDs were developed using the LPE method with a crucible shown in Fig. 7.9 and *p–n* diode characteristics and the mechanism of light emission were analyzed [54, 72, 73]. On *p*-SiC substrates doped with Al, *n*-SiC layers were grown using the Si melt doped with Al and N. A brightness of 100 fL for forward bias at room temperature was reported, together with no failure for longer than 15,000 h through the test with a current flow in the temperature range between room temperature and 400°C [72, 73].

The dipping method was applied to prepare *p–n* junctions to develop a convenient fabrication technique for blue LEDs [59, 74]. The overcompensation method was developed for *p–n* junction preparations utilizing the dipping technique. Sequential growth of double epitaxial layers was carried out by introducing N_2 gas into the growth system for the *n*-layer growth after the *p*-layer was grown on the *p*-type 6H-SiC substrate dipped in the Si melt doped with Al. The external quantum efficiency was 1×10^{-5} at room temperature.

In order to prepare an abrupt *p–n* junction, the rotation dipping technique was proposed and blue LEDs together with yellow and orange LEDs were fabricated [59]. The mechanism of light emission in electroluminescence was studied extensively, and the optimum doping level was presented.

The influence of the fabrication method on the characteristics of blue LEDs was studied using LEDs made by LPE and vapor-phase epitaxy VPE [75]. Blue LEDs made by LPE seemed to have better external quantum efficiency than those prepared by VPE. In our experiments, however, the VPE method seemed well suited for fabrication of blue LEDs if the Al doping was well controlled [76].

B. Transistors

Junction-type field-effect transistors (J-FETs) were fabricated using *p–n* junctions of 6H-SiC prepared by diffusion at high temperatures [71]. The mutual transconductance of 90 μS was not changed even at 520°C, but the drain current decreased. There are no reports on J-FETs using epitaxial *p–n* junctions prepared at rather low temperatures.

Bipolar transistors were fabricated for the first time using 6H-SiC prepared by the VPE method [77]. The base region was prepared epitaxially using the $SiCl_4$–C_6H_{14}–H_2 system on an *n*-SiC substrate which worked as a collector, and then the emitter *n*-region was grown on the base region. The collector junction has a leakage current of 300 nA at 40 V and a breakdown voltage of 50 V. Although the current amplification factor of the transistor is as low as 4–8, the transistor characteristics can be improved by optimization of the dimensions.

C. Devices for High Frequency and High Power

Electronic properties of 6H-SiC recently reported promise to bring about a new field for SiC application [78]. The interesting feature is a large value of the saturation drift velocity for electrons in 6H-SiC. A value of 2×10^7 cm/s was obtained at an electric field of 5×10^5 V/cm, which is two times larger than in Si. Thus high-frequency and high-power devices will be available if the processing is well developed. The other interesting feature is related to the large value of the breakdown field in VPE 6H-SiC [79]. Step junctions of n^+-SiC on *p*-SiC (hole concentration of 10^{17}–$2 \times 10^{18}/cm^3$) show values of $(2–3) \times 10^6$ V/cm, which indicate the possibility of devices working with high electric field, such as avalanche transit time devices.

D. Other Applications

Highly reliable thermistors were developed by using rf-sputtered SiC thin films [80]. Silicon carbide thin film thermistors are superior to the conventional ceramic thermistors because of high reliability, high accuracy, and good reproducibility, and are useful for temperature sensing, temperature control, and flame detection.

Silicon carbide films are used as the wear layer on the thermal printing head utilizing its hardness and good thermal conductivity [81]. Chemical vapor deposited SiC works as a unique optical material well suited to synchrotron radiation beam lines [82]. The CVD technology has provided highly polishable materials in large sizes, for example, mirrors up to 7.5 cm^2 with surface rms roughness less than 8 Å. The total integrated scattering rises less rapidly in the vacuum ultraviolet for these SiC materials, more so than for metal-coated glass mirrors.

As an application of amorphous SiC films, a-$Si_{1-x}C_x$: H is widely used as a window material for amorphous solar cells [83]. Visible electroluminescent devices are also proposed [84, 85].

VIII. SUMMARY

Thin film deposition methods for SiC were reviewed together with structure and physical properties of the deposited films. Vapor-phase and liquid-phase grown epitaxial films of SiC can be applied for active electronic devices such as LEDs and transistors. Films of SiC are easily deposited by sputtering at rather low temperatures, and these materials can be used as thermistors, wear layers, and so on. Preparation and application of amorphous $Si_{1-x}C_x$: H films was also described.

ACKNOWLEDGMENTS

The author shows many thanks to Dr. S. Nishino for help in preparation of the manuscript.

REFERENCES

1. O. V. Lossev, *Wireless Telegraphy Telephony* **18**, 61 (1923).
2. J. R. O'Connor and J. Smiltens, eds., *Silicon Carbide; A High Temp. Semicond. Proc. Conf. 1960*, Pergamon, Oxford, 1960.
3. H. K. Henisch and R. Roy, eds., *Mater. Res. Bull.* **4**, (1969).
4. R. C. Marshall, J. W. Faust, Jr., and C. E. Ryan, eds., *Silicon Carbide, Proc. Int. Conf., 3rd, 1973*, University of South Carolina Press, Columbia, South Cardina, 1974.
5. J. A. Lely, *Ber. Dtsch. Keram. Ges.* **32**, 229 (1955).

6. W. Spielmann, *Z. Angew. Phys.* **19**, 93 (1965).
7. V. J. Jennings, A. Sommer, and H. C. Chang, *J. Electrochem. Soc.* **113**, 728 (1966).
8. R. B. Campbell and T. L. Chu, *J. Electrochem. Soc.* **113**, 825 (1966).
9. A. Todkill and R. W. Brander, *Mat. Res. Bull.* **4**, S293 (1969).
10. W. Spielmann and K. Brack, *Z. Angew. Phys.* **18**, 321 (1965).
11. J. M. Harris, H. C. Gatos, and A. F. Witt, *J. Electrochem. Soc.* **118**, 335, 339 (1971).
12. S. Minagawa and H. C. Gatos, *Jpn. J. Appl. Phys.* **10**, 1680 (1971).
13. B. Wessels, H. C. Gatos, and A. F. Witt, *Silicon Carbide, Proc. Int. Conf., 3rd, 1973*, p. 25, University of South Carolina Press, Columbia, South Carolina, 1974.
14. G. Gramberg and M. Königer, *Solid-State Electron.* **15**, 285 (1972).
15. W. von Münch and I. Pfaffeneder, *Thin Solid Films* **31**, 39 (1976).
16. S. Nishino, H. Matsunami, and T. Tanaka, *Jpn. J. Appl. Phys.* **14**, 1833 (1975).
17. H. Matsunami, S. Nishino, M. Odaka, and T. Tanaka, *J. Cryst. Growth* **31**, 72 (1975).
18. S. Nishino, H. Matsunami, and T. Tanaka, *J. Cryst. Growth* **45**, 144 (1978).
19. P. Rai-Choudhury and N. P. Formigoni, *J. Electrochem. Soc.* **116**, 1440 (1969).
20. R. W. Brander and A. L. Boughey, *Br. J. Appl. Phys.* **18**, 905 (1967).
21. J. M. Harris, H. C. Gatos, and A. F. Witt, *J. Electrochem. Soc.* **116**, 380 (1969).
22. W. von Münch and I. Pfaffeneder, *J. Electrochem. Soc.* **122**, 642 (1975).
23. S. Minagawa and H. C. Gatos, *Jpn. J. Appl. Phys.* **10**, 844 (1971).
24. B. J. Isherwood and C. A. Wallace, *J. Appl. Crystallogr.* **1**, 145 (1968).
25. S. Yoshida, E. Sakuma, S. Misawa, and S. Gonda, *J. Appl. Phys.* **55**, 169 (1984).
26. W. G. Spitzer, D. A. Kleinman, and C. J. Frosh, *Phys. Rev.* **113**, 133 (1959).
27. N. C. Tombs, J. J. Comer, and J. F. Fitzgerald, *Solid-State Electron.* **8**, 839 (1965).
28. H. Nakashima, T. Sugano, and H. Yanai, *Jpn. J. Appl. Phys.* **5**, 874 (1966).
29. I. H. Khan and R. N. Summergrad, *Appl. Phys. Lett.* **11**, 12 (1967).
30. K. E. Haq and A. J. Learn, *J. Appl. Phys.* **40**, 431 (1969).
31. I. H. Khan, *Mater. Res. Bull.* **4**, S285 (1969).
32. J. Graul and E. Wagner, *Appl. Phys. Lett.* **21**, 67 (1972).
33. C. J. Mogab and H. J. Leamy, *Silicon Carbide, Proc. Int. Conf., 3rd, 1973*, p. 58, 64, University of South Carolina Press, Columbia, South Carolina, 1974; *J. Appl. Phys.* **45**, 1075 (1974).
34. A. J. Learn and I. H. Khan, *Thin Solid Films* **5**, 145 (1970).
35. D. M. Jackson, Jr., and R. W. Howard, *Trans. Metall. Soc. AIME* **233**, 468 (1965).
36. K. A. Jacobson, *J. Electrochem. Soc.* **118**, 1001 (1971).
37. K. Kuroiwa and T. Sugano, *J. Electrochem. Soc.* **120**, 138 (1973).
38. J. J. Rohan and J. L. Sampson, *J. Phys. Chem. Solids, Suppl.* **1**, 523 (1967).
39. I. H. Khan and A. J. Learn, *Appl. Phys. Lett.* **15**, 410 (1969).
40. H. Matsunami, S. Nishino, and T. Tanaka, *J. Cryst. Growth* **45**, 138 (1978).
41. S. Nishino, Y. Hazuki, H. Matsunami, and T. Tanaka, *J. Electrochem. Soc.* **127**, 2674 (1980).
42. H. Matsunami, S. Nishino, and H. Ono, *IEEE Trans. Electron Devices* **ED-28**, 1235 (1981).
43. S. Nishino, J. A. Powell, and H. A. Will, *Appl. Phys. Lett.* **42**, 460 (1983).
44. S. Nishino, H. Suhara, and H. Matsunami, *Ext. Abstr. Conf. Solid-State Devices Mat., 15th, Tokyo, 1983*, 317 (1983).
45. Y. Onuma, *Jpn. J. Appl. Phys.* **8**, 401 (1969).
46. A. J. Learn and K. E. Haq, *J. Appl. Phys.* **40**, 430 (1969).
47. A. J. Learn and K. E. Haq, *Appl. Phys. Lett.* **17**, 26 (1970).
48. S. Matsumoto, H. Suzuki, and R. Ueda, *Jpn. J. Appl. Phys.* **11**, 607 (1972).
49. K. Wasa, T. Nagai, and S. Hayakawa, *Thin Solid Films* **31**, 235 (1976).
50. S. Nishino, H. Matsunami, M. Odaka, and T. Tanaka, *Thin Solid Films* **40**, L27 (1977).
51. T. Tohda, K. Wasa, and S. Hayakawa, *J. Electrochem. Soc.* **127**, 44 (1980).

52. Y. Murayama and T. Takao, *Thin Solid Films* **40**, 309 (1977).
53. R. C. Marshall, *Mater. Res. Bull.* **4**, S73 (1969).
54. W. F. Knippenberg and G. Verspui, *Philips Res. Rep.* **21**, 113 (1966).
55. R. W. Brander and R. P. Sutton, *J. Phys. D* **2**, 309 (1969).
56. R. I. Scace and G. A. Slack, *J. Chem. Phys.* **30**, 1551 (1959).
57. A. Suzuki, M. Ikeda, N. Nagao, H. Matsunami, and T. Tanaka, *J. Appl. Phys.* **47**, 4546 (1976).
58. P. W. Pellegrini and J. M. Feldman, *Silicon Carbide, Proc. Int. Conf. 1973*, p. 161, University of South Carolina, Columbia, South Carolina, 1974; *J. Cryst. Growth* **27**, 320 (1974).
59. M. Ikeda, T. Hayakawa, S. Yamagiwa, H. Matsunami, and T. Tanaka, *J. Appl. Phys.* **50**, 8215 (1979).
60. L. B. Griffiths and A. I. Mlavsky, *J. Electrochem. Soc.* **111**, 805 (1964).
61. M. A. Wright, *J. Electrochem. Soc.* **112**, 1114 (1965).
62. M. Kumagawa, M. Ozeki, and S. Yamada, *Jpn. J. Appl. Phys.* **9**, 1422 (1970).
63. K. Gillessen and W. von Münch, *J. Cryst. Growth* **19**, 263 (1973).
64. C. J. Mogab and W. D. Kingery, *J. Appl. Phys.* **39**, 3640 (1968).
65. T. E. Hartman, J. C. Blair, and C. A. Mead, *Thin Solid Films* **2**, 79 (1968).
66. E. A. Fagen, *Proc. Int. Conf. Amorphous Liquid Semicond., 5th, Garmish-Partenkirchen, 1973*, p. 601; *Silicon Carbide Proc. Int. Conf. 1973*, p. 542, University of South California, Columbia, South Carolina, 1974.
67. M. Gorman and S. A. Solin, *Solid State Commun.* **15**, 761 (1974).
68. D. A. Anderson and W. E. Spear, *Philos. Mag.* **35**, 1 (1977).
69. H. Wieder, M. Cardona, and C. R. Guarnieri, *Phys. Status. Solidi B* **92**, 99 (1979).
70. T. Shimada, Y. Katayama, and K. F. Komatsubara, *J. Appl. Phys.* **50**, 5530 (1979).
71. R. B. Campbell and H. C. Chang, *in* "Semiconductors and Semimetals" (R. K. Willardson and A. C. Beer, eds.), Vol. 7, p. 625, Academic Press, New York, 1971.
72. R. W. Brander, *Mat. Res. Bull.* **4**, S187 (1969).
73. R. W. Brander, *Proc. IEE* **116**, 329 (1969).
74. H. Matsunami, M. Ikeda, A. Suzuki, and T. Tanaka, *IEEE Trans. Electron Devices* **ED-24**, 958 (1977).
75. W. von Münch, W. Kürzinger, and I. Pfaffeneder, *Solid State Electron.* **19**, 871 (1976).
76. S. Nishino, A. Ibaraki, H. Matsunami, and T. Tanaka, *Jpn. J. Appl. Phys.* **19**, L353 (1980).
77. W. von Münch and P. Hoeck, *Solid State Electron.* **21**, 479 (1978).
78. W. von Münch and E. Pettenpaul, *J. Appl. Phys.* **48**, 4823 (1977).
79. W. von Münch and I. Pfaffeneder, *J. Appl. Phys.* **48**, 4831 (1965).
80. K. Wasa, T. Tohda, Y. Kasahara, and S. Hayakawa, *Rev. Sci. Instrum.* **50**, 1084 (1979).
81. M. Sakamoto, Y. Kajiwara, T. Itano, H. Kataniwa, T. Ohkubo, and H. Katoh, *NEC Res. Dev.* **64**, 71 (1982).
82. V. Rehn and W. J. Choyke, *Nuc. Instrum. Mater.* **177**, 173 (1980).
83. Y. Tawada, K. Tsuge, M. Kondo, H. Okamoto, and Y. Hamakawa, *J. Appl. Phys.* **53**, 5273 (1982).
84. H. Munekata and H. Kukimoto, *Appl. Phys. Lett.* **42**, 432 (1983).
85. H. Matsunami, M. Yoshimoto, Y. Fujii, and J. Saraie, *J. Non-Cryst. Solids* **59-60**, 569 (1983).

8 CHARACTERIZATION OF THIN FILMS BY X-RAY DIFFRACTION

Armin Segmüller
Masanori Murakami

IBM Thomas J. Watson Research Center
Yorktown Heights, New York

I. INTRODUCTION

Thin films deposited onto substrates are widely used for electronic device materials. For optimal performance, these materials are required to possess specific electrical, mechanical, or chemical properties that are strongly affected by the choice of materials and their microstructure. Therefore, characterization of thin film microstructure is very important to improve device quality to the level acceptable to the design engineers.

Thin-film microstructure is strongly influenced by the substrate material and by the conditions prevailing during film preparation, such as substrate temperature, deposition rate, residual gas pressure, and angle of incidence of the vapor beam. Thus, nondestructive *in situ* characterization techniques are essential in order to obtain meaningful information on the microstructure of thin films. Most of the techniques developed for characterization of bulk material are difficult to apply to thin films due to their small dimensions perpendicular to the surface. Therefore, surface physics techniques are often applied to thin films, and several unique analytical techniques have been developed for thin films. Among them, x-ray diffraction is a very powerful and suitable technique for characterizing the microstructure of thin films. It is nondestructive and noncontact, and it provides a variety of information, such as presence and composition of phases, film thickness, grain size and orientation, strain state, interfaces, and solute concentration distribution between different layers. In addition, *in situ* studies can be carried out in various environments. In comparison to the other important structural method, electron diffraction, information is obtained on larger volumes, the precision of the lattice spacing measurement is better, and intensities are measured directly if diffractometers are used. For a comparison of both techniques in the study of thin films see, for instance, Mader [1].

Since the early 1970s, x-ray diffractometers have increasingly been automated by use of on-line mini- or microcomputers. The quality of the diffraction data has been improved drastically by using high-power x-ray sources like rotating-anode tubes or synchrotons, by selecting narrow spectral bands of x radiation with single-crystal monochromators, by step-scanning with very long counting periods of minutes or even hours to decrease the statistical error of data obtained from weak scatterers like thin films, and by using monitor counters to correct for intensity changes of the x-ray source. The availability of computers to process the diffraction data enables the experimentalist to determine very precise Bragg angles, total intensities, and line profiles.

The purpose of this chapter is to review x-ray diffraction techniques that are commonly used to characterize the microstructure of thin films deposited onto rigid substrates. Since an excellent review on this topic has been published recently by Vook [2], we emphasize subjects not covered extensively by this article. We assume the reader to be familiar with the basic principles of x-ray diffraction as found in textbooks, e.g., by Buerger [3], Klug and Alexander [4], Cullity [5], Taylor [6], Guinier [7], Barrett and Massalski [8], or Warren [9]. In Section II we discuss x-ray diffraction techniques that are applicable to thin-film studies. Specific experiments carried out using these techniques to obtain information on thin-film microstructure are reviewed in Section III. The limitations and possible extensions of these techniques are discussed in Section IV.

II. X-RAY DIFFRACTION INSTRUMENTATION FOR THIN-FILM STUDIES

Numerous x-ray diffraction methods can be applied to the study of thin films. With the progress made in the last decade in computer control and high-power x-ray sources, diffractometers equipped with counters to measure diffracted intensity directly have become very popular and widely used, and we discuss them first. However, this does not imply that photographic methods to record the diffracted intensity are obsolete. Often, they provide fast, preliminary information, and they are superior to powder diffractometers in recording two-dimensional data that can be used to establish orientational relationships between substrate and film. One of the advantages of x-ray diffraction is that the measurements can be carried out in various environments. We discuss in the last section *in situ* experiments to study growth, recrystallization, and phase transformations in thin films.

A. Diffractometer Methods

Due to the particular shape and dimensions of thin films, diffractometry is especially suited for their study. Figure 8.1 shows the principles of the Bragg–Brentano diffractometer. So-called parafocusing is achieved by placing the sample in a position tangential to the focusing circle, enabling the use of a large sample area. The focusing circle or, in three dimensions, the focusing cylinder is determined by three points (or lines): the tube focus, the diffractometer axis, and the receiving slit. Its radius is a function of the Bragg angle θ. Perfect focusing would be obtained if the sample were bent into an arc, thus adhering to the focusing cylinder. However, this is not possible for thin films deposited on rigid substrates. The deviation of the flat film from the focusing cylinder causes an asymmetric broadening of the diffracted intensity profile shifting it to smaller angles. A similar effect is caused by the vertical divergence (parallel to the diffractometer axis) of the incoming and diffracted beam that can be minimized by use of Soller slits. Since the parafocusing arrangement projects an image of the x-ray source, i.e., the focal

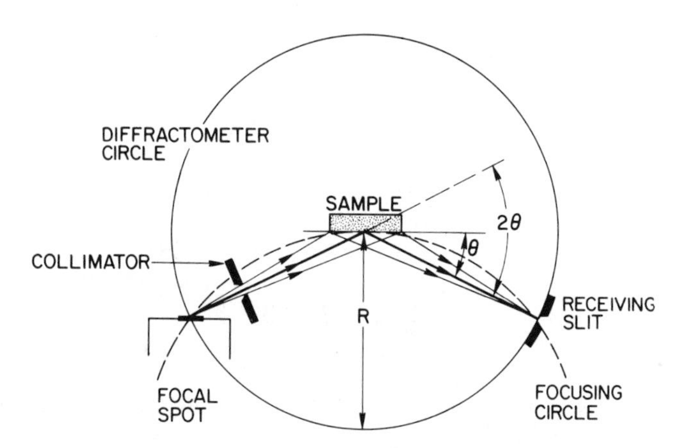

Fig. 8.1. Focusing geometry of the Bragg–Brentano diffractometer.

spot on the anode of the tube, onto the receiving slit of the detector, the line focus is generally used in order to minimize the line broadening. These effects are discussed in detail by Klug and Alexander [4] and Parrish [10]. In order to maintain the parafocusing condition during an angular scan, the sample is rotated with one-half of the angular velocity of the detector. During a scan the sample's surface area irradiated by the incident beam changes in proportion to $1/\sin\theta$. In thin films with a thickness $t \ll 1/\mu$, where μ is the linear absorption coefficient, this effect causes a change of the diffracting volume and thereby of the diffracted intensity with the same proportionality, whereas in bulk samples the change of irradiated surface area is compensated by an exactly reciprocal change of the penetration depth perpendicular to the surface ($\propto \sin\theta/\mu$), thus keeping the diffracting volume constant. Jenkins and Paolino [11] addressed the problem of the varying diffracting volume in thin films with the Theta Compensating Slit that changes the divergence of the incident beam with the Bragg angle θ keeping the irradiated area on the sample constant.

As we have seen, a large diffracting volume can be obtained from thin films at a small angle of incidence, which, on the Bragg–Brentano diffractometer, equals the diffraction or Bragg angle. One way to obtain small Bragg angles is to use x radiation of short wavelength, e.g., Mo–Kα radiation. Another more effective way is to use a different focusing arrangement offered by the Seemann–Bohlin diffractometer and shown schematically in Fig. 8.2. Here, the diffractometer circle, i.e., the circle along which the receiving slit of the

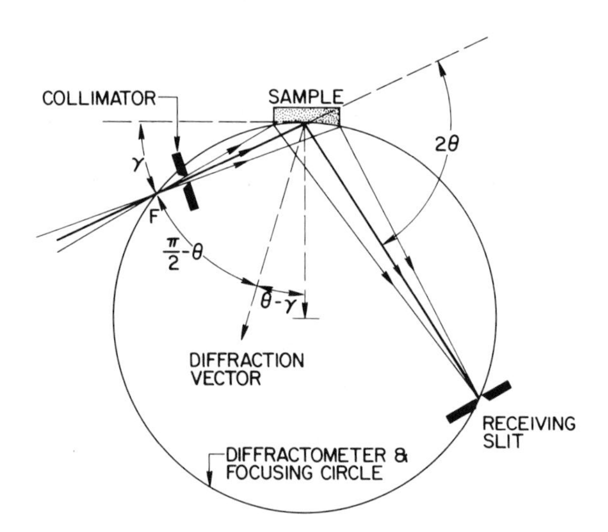

Fig. 8.2. Focusing geometry of the Seemann–Bohlin diffractometer. Here F represents the tube focal spot or image thereof.

detector is moving, coincides with the focusing circle. The sample is mounted tangentially to the diffractometer circle and the angle of incidence, which may now differ from the Bragg angle, can be chosen as small as $\sim 5°$. As x-ray source F either the tube's line focus or its image projected by a focusing monochromator is chosen. The angle along the diffractometer or focusing circle is defined in terms of 4θ. Designs of Seemann–Bohlin diffractometers have been published by Wassermann and Wiewiorowsky [12], King *et al.* [13], and especially for thin-film work with glancing-angle incidence by Weiner [14], and Feder and Berry [15]. Parrish *et al.* [16] described a Seemann–Bohlin linkage for the Norelco x-ray diffractometer. Although the Seemann–Bohlin focusing principle can be applied to diffraction in reflection and, if a focusing monochromator is used, in transmission, only the former mode is possible with thin films deposited on thick substrates. Since the sample is mounted stationary on the perimeter of the focusing circle, vacuum sample chambers for low and high temperature work can easily be adapted to the Seemann–Bohlin diffractometer. A Seemann–Bohlin focusing camera with a focusing Johannson monochromator, a so-called Guinier camera, with counter attachment based on Weiner's [14] design is manufactured by Huber.*

The diffraction vector, perpendicular to the reflecting planes and parallel to the bisectrix of incident and diffracted beam, defines the direction along which structural information, such as lattice parameter or crystallite size, is obtained. On the Seemann–Bohlin diffractometer, it changes its direction with Bragg angle θ, whereas on the Bragg–Brentano diffractometer it is fixed and perpendicular to the sample surface. Only the crystallites of the thin film that are oriented with a specific set of lattice planes perpendicular to the bisectrix or diffraction vector contribute to the intensity diffracted from these planes. Therefore, only polycrystalline films with a random or nearly random distribution of grain orientations may diffract x rays from all major lattice planes on both diffractometers. In this case, the Seemann–Bohlin diffractometer has a clear advantage, since the angle of incidence γ (Fig. 8.2) can be made very small ($\sim 5°$) yielding a diffracting volume ($\propto 1/\sin\gamma$) that is much larger than the one obtained on the Bragg–Brentano diffractometer ($\propto 1/\sin\theta$). However, many thin films have a strong preferred orientation of crystallites. In many cases a low-indexed set of lattice planes is parallel to the surface, e.g., in epitaxial films or in films with a fiber texture. In order to get sufficient diffracted intensity from these planes we must choose an angle of incidence $\gamma \sim \theta$, eliminating the advantage of the Seemann–Bohlin diffractometer over the Bragg–Brentano setup. An important difference between the

* Robert Huber Diffraktionstechnik, distributed by Blake Industries Inc., Scotch Plains, New Jersey 07076.

two instruments lies in the fact that on the Seemann–Bohlin diffractometer the distance between the sample and the receiving slit changes with the Bragg angle θ, thus causing a change of the absorption in air and the angular receptance of the detector; also sample absorption and displacement corrections differ. Correction factors for the Seemann–Bohlin geometry that should be applied to quantitative measurements of total intensities and line profiles are given by Segmüller [17], Kunze [18], Mack and Parrish [19], and King *et al.* [13]. Peak shift corrections have been discussed by Gillham [20] and Gillham and King [21].

As we have seen, both the Bragg–Brentano and the Seemann–Bohlin diffractometer provide us with information along the diffraction vector that is either parallel to the film surface normal or inclined to it by an angle less than 90°. Marra *et al.* [22] devised a method to obtain information from a very shallow surface layer in a direction parallel to it that will be discussed in Section III.F.1.

For orientation studies of epitaxial films and films with a fiber texture, a pole figure* goniometer for the determination of preferred orientation of technical materials, such as cold-rolled sheets, may be used in reflection mode, the so-called Schulz [23] technique. However, due to the finite thickness of the film the scattering volume changes when the sample is tilted around the bisectrix perpendicular to the diffraction vector, thus making intensity corrections necessary [24]. Only (hkl) poles within a cone with $0 \leqslant \alpha \lesssim 70°$, where α is the angle of the pole with the film normal, can be recorded in the reflection mode. To record a complete pole figure ($0 \leqslant \alpha \leqslant 90°$) the transmission technique of Decker *et al.* [25], yielding information in an angular range of $50 \lesssim \alpha \leqslant 90°$, would have to be applied. For most films this is not possible because of the absorption of the x rays in the substrate. Brine and Young [26] discussed the use of the Single-Crystal Orienter† attachment to study the orientation of epitaxial crystals. In the same way any four-circle diffractometer can be used.

B. Photographic Methods

Photographic film methods are also available for x-ray diffraction characterization of thin films. In an *in situ* study, Rühl [27] mounted the thin-film sample stationary under an angle of incidence of ~20°. On the surrounding x-ray film many reflections were recorded simultaneously, but only the reflections with $\theta \sim 20°$ gave sharp lines owing to parafocusing (Fig. 8.1). In

* The (hkl) pole figure is a representation of the density of normals of (hkl) planes or poles in the stereographic projection of the file with the film normal in the center.

† Two-circle (x, ψ) goniometer was marketed originally by General Electric Company.

our laboratory we have used this glancing-angle technique for many years: The thin film/substrate is mounted in a Debye–Scherrer camera under a glancing angle of ~15–20° to the incident beam and it is held stationary during exposure [28]. This method gives first and fast information on crystallinity and preferred orientation of the film. Read and Altman [29] and Read and Hensler [30] used a wide-film Debye–Scherrer camera* to characterize tantalum films with the glancing-angle technique. Glancing-angle diffraction photographs with much smaller angles of incidence, under nonfocusing conditions and oscillating by 5°, were used by Jackson *et al.* [31] to study films of Si on single-crystal Si substrates. Glancing-angle photographs, usually taken with polychromatic radiation, often show two features: Debye–Scherrer rings diffracted from the polycrystalline film with characteristic radiation and Laue spots diffracted from the single-crystal substrate with radiation selected from the bremsstrahlung spectrum. As already mentioned earlier, a Guinier camera for thin film studies in reflection mode, with glancing-angle incidence from ~0 to 10° and providing focusing over the entire angular range, is commercially available from Huber.

For determination of the orientation of thin epitaxial films the polychromatic Laue back-reflection method has been used by Hall and Thompson [32]. Also, monochromatic moving-film methods, described in the textbook by Buerger [3], especially the Weissenberg and the Buerger precession camera, can be used to project reciprocal-lattice planes of film and substrate in a two-dimensional picture. We shall discuss these methods in Section III.D.

C. *In Situ* Methods

Both photographic and diffractometer methods have been applied to numerous *in situ* studies of the growth and annealing behavior of thin films. Rühl [27] designed a vacuum x-ray camera with movable photographic film in which tin films could be evaporated at a gas pressure of $\sim 5 \times 10^{-4}$ Pa and subsequently annealed at temperatures between 20 and 380 K. Keith [33] designed an all-glass x-ray vacuum camera based on the Seemann–Bohlin focusing geometry thus allowing evaporation of Cu films under a gas pressure $\lesssim 10^{-5}$ Pa and annealing at temperatures between 90 K and room temperature. Two Bragg reflections, (200) and (311), could be focused through thin borosilicate windows onto photographic film outside. Vook and Schoening [34] designed an all-metal vacuum chamber with a Be window for the incident and diffracted beam that could be baked at 440 K to attain a gas pressure of $\sim 2 \times 10^{-7}$ Pa. The Huber–Guinier camera can be

* The "Read" camera is marketed by Blake Industries, Inc., Scotch Plains, New Jersey 07076.

equipped with a high-temperature stage and a motor-driven cylinder to move the photographic film axially for real-time studies. It was used by Vandenberg and Hamm [35] to study the phase diagram of Au–Al thin-film couples between 308 and 623 K. With a rotating Cu anode the intensity diffracted from the bilayers (less than 400 nm thick) was sufficient to continuously record the 2-mm-high diffraction pattern on a film cylinder moving with an axial speed of 0.06 mm/min, while the temperature was raised with a rate of 0.38 K/min.

Nowadays, several designs and commercial products are available for cryostats and high-temperature chambers, with built-in evaporation source and ultra-high vacuum (UHV), that can be attached to an all-purpose diffractometer, e.g., the four-circle diffractometer manufactured by Huber. Sample chambers used in other surface physics studies can be adapted to x-ray diffraction [36]; Marra [37] has described such a chamber.

III. CHARACTERIZATION OF THIN FILMS BY X-RAY DIFFRACTION

Several experimental quantitites can be measured from x-ray diffraction data: the angular position, total intensity, linewidth, and line profile of Bragg reflections. Each of these quantities provides certain information on the microstructure of the thin film. In this section we shall discuss each of these basic measurements together with the structural properties that can be derived from them. We shall also discuss methods to determine orientational relationships between film and substrate. Observation of small-angle interferences from thin films and multilayers is not diffraction in the strict sense, i.e., reflection by lattice planes, but it can be viewed as diffraction from interfaces of a thin-film–substrate system providing information on thicknesses and average electron densities of the film layers, and, therefore, it will also be presented here. Since the crystal structure of the film components and possible phases are usually known, the emphasis lies on deviations from the ideal crystal structure.

A. Measurement of Bragg Angles

From the angular position θ of the diffraction lines, lattice or d-spacings, characteristic of the symmetry and dimensions of the unit cell, can be determined by using Bragg's equation

$$d = \lambda/2 \sin \theta, \tag{1}$$

where λ is the wavelength of the x radiation. On photographic films line positions can be measured with a ruler and light box. A microdensitometer allows us to digitize the film data for computer processing. On diffractometer strip-chart recordings line positions again can be measured with a ruler. With an expanded θ scale and using the midchord method [38, 39] good precision can be obtained. Most efficient is, of course, the use of a computer controlled diffractometer to collect the diffraction data in digital form. Digital data, either taken directly with a diffractometer or read from a photographic film with a microdensitometer, can be processed easily on a computer in a number of ways. Parrish *et al.* [40] discussed various strategies for peak search. The centroid or center of gravity of a diffraction line

$$\langle\theta\rangle = \int \theta[I(\theta) - I_B]\, d\theta \Big/ \int [I(\theta) - I_B]\, d\theta, \tag{2}$$

where $I(\theta)$ is the intensity profile and I_B the background intensity, and the higher moments, e.g., the variance, can easily be programmed for a computer. The denominator in Eq. (2) is called the total or integrated intensity. The use of these quantities has been extensively discussed by Parrish and Wilson [41] and Wilson [42]. A least-squares method to compute derivatives and, from their sign change, peak positions has been described by Segmüller [43].

X-ray diffraction data is always subjected to statistical and systematic errors. The former can be minimized by long counting or exposure times and by the use of least-squares methods. The most common cause for systematic error in the Bragg angles θ is the offset of the sample center from the focusing circle of the diffractometer. According to Fig. 8.3 an offset ΔR of the sample causes a parallel shift of the diffracted beam in the Bragg–Brentano geometry of

$$\Delta B = 2\,\Delta R \cos\theta, \tag{3a}$$

which causes a shift (in rad) of the measured peak position of [19, 42, 44]

$$\Delta 2\theta = \Delta B/R = 2(\Delta R/R)\cos\theta. \tag{3b}$$

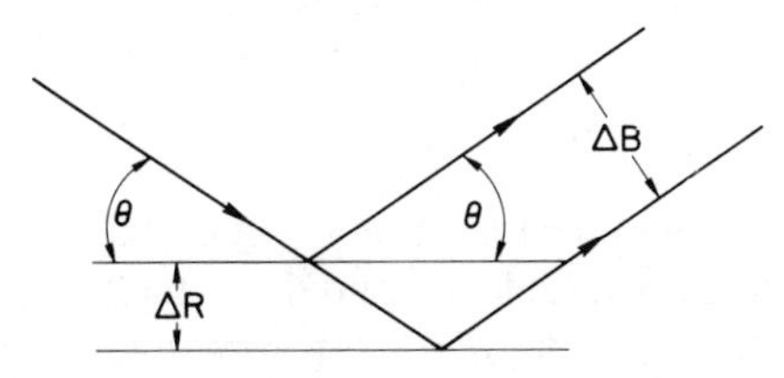

Fig. 8.3. Diffracted beam shift ΔB caused by displacement of the sample from diffractometer center by ΔR.

For the Seemann–Bohlin geometry the displacement error is given by [17, 19]

$$\Delta 4\theta = (\Delta R/R)\sin 2\theta/[\sin\gamma \sin(2\theta - \gamma)]. \quad (3c)$$

Other errors in θ due to the film flatness and the vertical divergence can be kept small.

The d-spacing measured by x-ray diffraction may be affected by a state of strain in the film. Such strains can be caused by different thermal expansion of film and substrate or by lattice misfit between epitaxial layer and substrate. If the strain tensor is known, the measured d-spacing can be corrected.

1. Identification of Phases

From the d-spacings, phases can be identified in a film, using the JCPDS Powder Diffraction File (JCPDS International Centre for Diffraction Data, 1601 Park Lane, Swarthmore, Pennsylvania 19081), and the reflections can also be indexed with Miller indexes. Since thin films often have a preferred orientation, measured peak intensities may not agree with the data given in the JCPDS file. Also a margin of error must be allowed for the d-spacings determined from the experimental data. The proportional amount of phases present can be determined from total intensities as discussed in Section III.B.

2. Determination of Lattice Parameters and X-Ray Density

After the measured reflections have been identified as belonging to a specific phase and have been assigned Miller indexes, using the JCPD file discussed in the previous section, lattice parameters can be obtained by applying least-squares methods [45, 46], as discussed in detail by the textbooks referred to in Section I. We will outline it briefly as follows: The best-fitting lattice parameters are obtained by minimizing the function

$$R_0 = \sum \Delta_i^2 W_i, \quad (4)$$

where the summation is executed over all observations i, and where Δ_i is the residual, i.e., the deviation of the observation from the calculation and W_i the weight of the observation. The residual is defined by

$$\Delta_i = (4/\lambda^2)\sin^2\theta_i - Q_i(A, B, C, \ldots) - \Phi_i(\Delta R/R), \quad (5)$$

where Q_i is a linear function of the scalar products of the reciprocal lattice vectors, the so-called quadratic form, and Φ_i a function of the Bragg angle θ, relating the displacement error $\Delta R/R$ to the change of $(\sin\theta)^2$. In Table 8.1, the quadratic forms Q are listed for the cubic, tetragonal, orthorhombic, and hexagonal system. For monoclinic and triclinic symmetry the reader is

TABLE 8.1

Cubic Form Q and Unit Cell Volume V for Four Symmetry Systems[a]

Symmetry	Orthorhombic ($a \neq b \neq c$) Tetragonal ($a = b \neq c$) Cubic ($a = b = c$)	Hexagonal ($a = b \neq c$)
Q	$(h^2/a^2) + (k^2/b^2) + (l^2/c^2)$	$\frac{4}{3}(h^2 + hk + k^2)/a^2 + l^2/c^2$
V	abc	$a^2c\sqrt{3}/2$

[a] Lattice parameters are a, b, and c.

referred to the textbooks listed in Section I. For an orthorhombic crystal the quadratic form is given by

$$Q_i = h_i^2 A + k_i^2 B + l_i^2 C, \tag{6a}$$

where h_i, k_i, and l_i are the Miller indexes of the observation i. The parameters to be varied in the least-squares method, A, B, C, are given by

$$A = 1/a^2, \qquad B = 1/b^2, \qquad C = 1/c^2, \tag{6b}$$

where a, b, c are the lattice parameters of the orthorhombic crystal. Also varied is the parameter $\Delta R/R$, the displacement error. Its coefficient Φ_i is given by

$$\Phi_i = (8 \sin \theta_i \cos^2 \theta_i)/\lambda^2 \tag{6c}$$

for the Bragg–Brentano geometry, and by

$$\Phi_i = \sin^2 2\theta_i/[\lambda^2 \sin \gamma \sin(2\theta_i - \gamma)] \tag{6d}$$

for the Seemann–Bohlin geometry. Other parameters can be added to Eq. (5), e.g., a zero error $\Delta\theta_0$, however, too many parameters may sometimes give dubious results.

Equal weight should be given to the individual observation of the Bragg angle [47]. This amounts to weighting the quantity Δ_i^2 in Eq. (4) by

$$W_i = 1/\sin^2 2\theta_i. \tag{7}$$

The algorithm to minimize R_0 of Eq. (4) has been outlined by Cruickshank [48]. It can easily be programmed for a computer, and programs are also available.

In films with cubic structure a graphic method can be applied since only two parameters, the lattice parameter a and the displacement ΔR, are to be varied. The lattice parameters obtained from the various reflections are plotted versus a specific function $f(\theta)$ of the Bragg angle θ [49]. A straight line is drawn through the data points. Its slope is determined by the displacement ΔR and its intercept with the ordinate axis ($\theta = 90°$) equals the lattice parameter corrected for the displacement error. Instead of the Nelson–Riley function, used for Debye–Scherrer films, we use

$$f(\theta) = \cos^2\theta/\sin\theta \tag{8a}$$

for the Bragg–Brentano diffractometer, and

$$f(\theta) = \cos^2\theta/[\sin\gamma\sin(2\theta - \gamma)] \tag{8b}$$

for the Seemann–Bohlin diffractometer. However, this method is weighting the lattice parameters with equal weight and not the angle measurements.

From the lattice parameters the volume V of the unit cell can be determined. Table 8.1 also lists V as a function of the lattice parameters for four symmetry systems. The x-ray density ρ_x of the phase can then be determined by using the relation

$$\rho_x = \sum n_i A_i/(VN), \tag{9}$$

where n_i and A_i are the number per unit cell and the atomic weight, respectively, of the atom i, and $N = 6.0249 \times 10^{23}$/mole, Avogadro's number. The summation is executed over all atomic species i in the unit cell.

3. Determination of Phase Composition

In a binary system with complete miscibility of the components, e.g., Cu–Ni, the composition of the film sample often can be determined from the lattice parameter using Vegard's law [50] for a first approximation or from published lattice parameters of compounds. Pearson [51, 52] has compiled lattice parameters for metals and alloys. If the film is strained the measured lattice parameter must be corrected for strain before it can be used to determine composition.

4. Determination of Strain

A deviation of the measured d-spacing d from that of a strain-free bulk material d_0, as calculated from the literature lattice parameters, indicates the presence of strain. The component of the strain tensor in the direction of the diffraction vector (hkl) is given by

$$\varepsilon_{hkl} = (d_{hkl} - d_{0,hkl})/d_{0,hkl}\,. \tag{10}$$

The strain tensor can be determined by measuring d-spacings of the film in several directions off the surface normal. For details we refer the reader to the review article by Segmüller and Murakami [53] that discusses the measurement of strains and stresses in thin films by x-ray diffraction.

5. Determination of Film Thickness

In the diffraction pattern of epitaxial films and films with a strong fiber texture secondary maxima are often observed close to the Bragg reflection. These are caused by the finite size of the crystallites along the film normal due to the restriction imposed by the film thickness. The film thickness t is related to the spacing $\Delta\theta$ of the maxima by

$$t = \lambda/(2\,\Delta\theta \cos\theta). \tag{11a}$$

Croce *et al.* [54] and Vook and Witt [55] measured the thickness of Au films deposited on glass with a strong (111) fiber texture. Vook *et al.* [56] applied the method to highly oriented Sn films deposited on glass, and they explained the maxima in the framework of the kinematical theory. Stacy and Janssen [57], using dynamical theory, derived the relation

$$t = \lambda \sin(2\theta - \gamma)/(\Delta\theta \sin 2\theta). \tag{11b}$$

to determine the thickness of thin epitaxial garnet layers from the spacings of the maxima. For symmetric reflection ($\gamma = \theta$) Eq. (11b) is identical to Eq. (11a). Often, the secondary maxima are called Pendellösung (pendulum solution) fringes because they are a consequence of the two interfering wavefields that, according to dynamical theory, exist inside the crystal with the energy oscillating between them [58].

B. Measurement of Total Intensities

The total or integrated intensity diffracted by a polycrystalline thin film for a specific reflection is obtained from the intensity I versus θ plot as the area, above background intensity I_B, enclosed by the diffraction line. It is given by

$$I_{tot} = \int [I(\theta) - I_B]\, d\theta = I_0 C\, \mathrm{LP}\, B A A_a F_{hkl}^2 n_{hkl} p_{hkl}, \tag{12}$$

where I_0 is the intensity of the incident x rays, C a constant determined mainly by the scattering power of one electron for x rays polarized perpendicular to the scattered beam, LP the Lorentz–polarization factor, B the fraction of the entire diffraction cone received by the detector, A a factor determined by the size of the scattering volume and by the absorption of x rays in it, A_a the absorption of x rays between source and detector outside the sample, F_{hkl} the absolute value of the structure factor of the (hkl) reflection (including the Debye–Waller factor), n_{hkl} the number of equivalent planes (hkl) in the crystal lattice, and p_{hkl} the fraction of crystals with their reciprocal lattice vector (hkl) parallel to the scattering vector in comparison with a random sample for which $p = 1$. The Lorentz–polarization factor is given by

$$\mathrm{LP} = (1/\sin\theta)(1 + \cos^2 2\theta \cos^2 2\theta_{\mathrm{M}})/(1 + \cos^2 2\theta_{\mathrm{M}}), \tag{13}$$

where θ_{M} is the Bragg angle of the incident-beam monochromator ($\theta_{\mathrm{M}} = 0$, if no monochromator used) diffracting kinematically. The use of perfect crystals, diffracting dynamically, is not common in thin-film work because of intensity reasons. By using a diffracted-beam monochromator, which is very common on Bragg–Brentano diffractometers, the denominator of the second fraction in Eq. (13) is replaced by 2. The fraction of the entire diffraction cone received by the detector B is normally included in LP. Since it differs for the two geometries we give it separately as

$$B = H/(2\pi R \sin 2\theta) \tag{14a}$$

for the Bragg–Brentano geometry, and

$$B = H/[2\pi R \sin 2\theta \sin(2\theta - \gamma)] \tag{14b}$$

for the Seemann–Bohlin geometry, where H is the length of the receiving slit. The use of Soller slits to limit the height divergence of the diffracted beam on a Seemann–Bohlin diffractometer imposes a maximum condition on B at small Bragg angles [17]. The absorption factor A is given by

$$A = \frac{F_i}{\mu \sin\gamma} \frac{\sin\gamma \sin(2\theta - \gamma)}{\sin\gamma + \sin(2\theta - \gamma)} \left\{1 - \exp\left[-\mu t\left(\frac{1}{\sin\gamma} + \frac{1}{\sin(2\theta - \gamma)}\right)\right]\right\} \tag{15a}$$

for the Seemann–Bohlin geometry, where F_i is the cross section of the incident beam, μ the linear absorption coefficient, and t the thickness of the thin film. Its values for the Bragg–Brentano geometry is obtained by simply setting $\gamma = \theta$. For thin films with $t \ll 1/\mu$, A reaches the limit

$$\lim_{t \to 0} A = \frac{F_i t}{\sin\gamma}, \tag{15b}$$

which is simply the volume irradiated by the incident x-ray beam. For bulk material with $t \gg 1/\mu$, A reaches the limit

$$\lim_{t \to \infty} A = \frac{F_i \sin(2\theta - \gamma)}{\mu[\sin\gamma + \sin(2\theta - \gamma)]}, \tag{15c}$$

which assumes the well-known value $F_i/2\mu$ for $\gamma = \theta$. It is noted that Eqs. (15) are valid only if the cross section of the incident beam is kept constant and is entirely received by the sample at all angles. On the Seemann–Bohlin diffractometer the distance from the sample to the receiving slit changes with the Bragg angle. Therefore, the absorption in air A_a changes according to

$$A_a = \exp\{-2R\mu_a[\sin\gamma + \sin(2\theta - \gamma)] + T\mu_a\}, \tag{16}$$

where μ_a is the linear absorption coefficient of air and T the distance from the focal line of the tube to the tube window. For the Bragg–Brentano diffractometer A_a is, of course, constant. Mack and Parrish [19] found excellent agreement between measurement and computation by Eqs. (14–16) of the total intensity ratios for the two geometries using a Seemann–Bohlin diffractometer with variable aperture of the receiving slit that pointed towards the sample at all angles θ.

Values for the multiplicity factor n_{hkl} are published in textbooks and in "International Tables for X-Ray Crystallography," Vol. I (1952). The parameter p is a measure for any preferred orientation of grains in the thin-film sample.

1. Determination of Film Thickness

According to Eqs. (12) and (15b), the total intensity diffracted from a thin film with a thickness $t \ll 1/\mu$ is proportional to the irradiated or scattering volume of the thin film. In a one-phase system the scattering volume is determined by the film thickness. Therefore, the total intensity is proportional to the film thickness. The proportionality factor is determined by calibration with standard samples with thicknesses determined by other methods, e.g., Rutherford back-scattering (RBS) (see, for instance, Chu *et al.* [59]). However, the comparison of two films is only possible if both films have the same preferred orientation.

In bilayer films or in films with a crystalline substrate the attenuation by the top layer of x rays diffracted from the bottom layer or substrate can be used to determine the thickness t of the top layer. The attenuation is given by

$$A_t = \exp(-2\mu t/\sin\theta). \tag{17}$$

The total intensity is a measure of the so-called mass thickness of the film, i.e., the number of atoms per film area. To determine the true thickness optical or x-ray interference methods must be applied. We shall discuss the latter in Section III.E.1.

2. Estimate of Preferred Orientation

If we measure the total intensities of several reflections of a polycrystalline film, we can estimate the amount of preferred orientation of the grains. After dividing the total intensities by the factors given in Eqs. (13)–(16), by the square of the structure factor F_{hkl}, and by the multiplicity factor n_{hkl}, the values of p are obtained, multiplied by a constant that is independent of θ. The ratio of the p value for two different reflections (hkl), e.g., p_{111}/p_{200}, can be used to characterize the preferred orientation. When using a Seemann–Bohlin diffractometer the p values are defined for different directions. Vook and Witt [55, 60] estimated the preferred orientation of Cu and Au films on glass from the total intensities.

The preferred orientation can be determined more quantitatively by measuring the total intensity for one reflection (hkl) in different directions, as discussed in Section III.D.

3. Determination of Phase Proportions

If the film consists of several phases the total intensity diffracted from the different phases is a measure for their proportion in the film. Again, we must be aware of possible preferred orientation.

C. Measurement of Linewidths and Profiles

The width of a diffraction line is primarily determined by the crystallite size in a reciprocal relation. Imperfections of the crystal structure, such as dislocations, stacking faults, or microstrains, cause an additional broadening. The linewidth can be defined in several ways. On a strip-chart recording the half-peak width is measured as the length of a chord drawn horizontally across the profile halfway between peak and background intensity. Other definitions, more suitable to mathematical treatment, are the integral width

$$W_{\mathrm{I}} = I_{\mathrm{tot}}/(I_{\mathrm{p}} - I_{\mathrm{B}}), \tag{18a}$$

where I_{p} is the peak intensity, and the variance

$$W_\theta = \langle(\theta - \langle\theta\rangle)^2\rangle = \int(\theta - \langle\theta\rangle)^2[I(\theta) - I_{\mathrm{B}}]\,d\theta/I_{\mathrm{tot}}, \tag{18b}$$

where the centroid $\langle\theta\rangle$ is given by Eq. (2).

1. Determination of Grain Size and Microstrains

The linewidth W, measured in radians and in terms of θ by any of the three methods discussed earlier, allows us to determine the mean grain size D parallel to the diffraction vector by use of Scherrer's equation

$$D = k\lambda/(2W\cos\theta), \tag{19}$$

where k is a shape-characteristic constant of the order of 1. Often, the grain size perpendicular to the film surface is given by the film thickness t. If values of D, measured for two or more orders of one reflection along the same direction of the film, agree reasonably well, the finite grain size is indicated as primary cause of the line broadening. If the value of D increases considerably with the order of reflection, then the presence of microstrains is strongly indicated. Microstrains or rms strains change from grain to grain and also within a grain. Macrostrains, i.e., strains homogeneous over the entire film, cause a shift of the Bragg angle according to Eq. (10), but no line broadening. For a certain microstrain distribution with rms strain $\langle \varepsilon \rangle = \langle \Delta d/d \rangle$ the contribution to the line broadening is given by

$$W_\varepsilon = \langle \varepsilon \rangle \tan \theta, \tag{20}$$

as can be verified by differentiating Bragg's equation [Eq. (1)].

Sometimes, the average grain size of a large-grain thin film can be estimated by counting diffraction spots along a certain arc of the Debye–Scherrer ring on a photographic film and dividing their number into the diffracting volume [6, 8] or by measuring the spot dimensions [6]. Vook and Witt [55, 60] used the latter method to estimate the grain size in recrystallized Cu and Au films and compared the results with those obtained by profile analysis.

Line profile analysis by Fourier methods can be applied to the profile to determine mean grain size and rms strain and possibly distributions of sizes and strains. The application of this method, developed by Warren and Averbach [61], to thin film has been reviewed by Segmüller and Murakami [53], and we refer the reader to this article. Some academic discussion has arisen in the past as to whether the contributions to the line broadening by grain size and microstrain distribution can be separated by analyzing only one order of reflection. If only one reflection can be measured, as it happens often with thin films, the peak-to-background intensity ratio is probably rather low and the statistical error high. The application of these highly mathematical methods to not-so-well-defined experimental data often seem not very meaningful physically. If, on the other hand, several reflections or several orders can be measured, then they should be used for proper separation of the two effects.

2. Determination of Solute Concentration Profile

By analyzing the shape of the diffraction intensity band, Houska [62, 63] developed a technique to determine the concentration profiles in films with diffused concentration gradients perpendicular to the film surface. Knowledge of the concentration profile is necessary to determine interdiffusion coefficients during homogenization of the layered film.

The method can be applied to binary systems that form a single solid solution. The lattice parameters of the pure components should be sufficiently different, and there should be a unique relation between the composition and the lattice parameter of the solid solution. The starting sample, a bilayer of the pure components (each layer $\sim 1\ \mu m$ thick), has a diffraction pattern consisting of two sharp lines at Bragg angles corresponding to the lattice parameters of the pure components. With progressing interdiffusion, the two lines decrease in intensity and the intensity in the range between the two lines increases forming an intensity band $P(\theta)$. Assuming a reasonable concentration profile, the shape of the intensity band can be calculated. The Bragg angle θ is a function of the concentration, and the intensity $P(\theta)$ depends on the structure factor, which is also a function of the concentration, and on the attenuation of the incident and diffracted beam in the film between the surface and the layer with the specific concentration. The shape of the intensity band $P(\theta)$ is very sensitive to a change of the concentration profile. Thus, the shape of the intensity band can be simulated by varying the concentration profile. An extensive review of this method, also called single intensity profile analysis (SIPA), is prepared by Murakami *et al.* [64].

3. Study of Imperfections in Single-Crystalline Films

Lighty *et al.* [65] used a photographic divergent-beam back-reflection technique developed by Imura *et al.* [66] to study imperfections in single-crystalline Cu films prepared by electrodeposition. In this method, x rays diffracted from one set of lattice plates (*hkl*) form a cone around the (*hkl*) normal (analogous to the Kossel cones generated from a divergent source within the crystal) intersecting the photographic film in an ellipselike curve. From the sharpness of these curves, manifesting itself in the separation of the concentric $K\alpha_1$ and $K\alpha_2$ curves and their deviation from an ideal ellipselike figure, information can be obtained on imperfections, polygonization boundaries, polycrystalline deposits, etc. From the diameters of these curves precise lattice parameters can be determined, and, since each curve represents a different direction in the crystal, the strain tensor can also be obtained from one exposure.

D. Measurement of Film Orientation

The Laue back-reflection method was used by Hall and Thompson [32] for the determination of the orientation of thin epitaxial films of Cu, Ag, Pd, and Au deposited on LiF and mica single crystals. If substrate and film have the same structure, their reflections coincide or are very close together. A slight separation may be due to a strain in the epitaxial film to accommodate the misfit, thus causing corresponding planes in film and substrate to be tilted slightly against each other. This has been discussed in detail by Segmüller and Murakami [53]. If in addition to the bremsstrahlung, the x radiation used for the Laue photograph has a strong characteristic component, e.g., Cu–$K\alpha$ when a copper tube is used, it sometimes becomes possible to identify spots diffracted with this component as belonging to the film or to the substrate of equal structure but different lattice parameter [67]. Different degrees of film perfection and substrate structure may cause the Laue spots to have different shapes and sharpnesses. Wallace and Ward [68, 69] have used a cylindrical texture camera for Laue back-reflection photographs of epitaxial films and have drawn a Greninger chart for this method to evaluate the photographic films. If the structures of film and substrate are equal and the misfit of the system is large enough the reflections of film and substrate can be separated by using the moving-film methods described in the textbook by Buerger [3]. Bettini and Brandt [70] used the Buerger precession camera to study epitaxy of CdS and CdTe on GaAs (111). The Weissenberg camera was used by Wallace and Ward [69] in an oscillatory, inclined-beam mode with stationary film and without layer screen to study the orientation of epitaxial films. Lo *et al.* [71] used a Weissenberg camera to study orientation of CdTe on Si substrates.

By using a pole-figure goniometer attachment to a diffractometer, Witt *et al.* [72] recorded a complete pole figure for an Au film. Since it was deposited on a glass slide sufficiently thin to be penetrated by x rays without strong attenuation, the transmission method of Decker *et al.* [25] could be applied with modified intensity corrections, in addition to the Schulz [23] reflection method. Clarke and Court [24] compared the Schulz [23] technique and a photographic technique with a forward-reflection film camera for the characterization of ZnO and Au films.

E. Small-Angle Interference Measurements

Total reflection of x rays from thin films can be observed for glancing angles of the incident beam below the critical angle for total reflection ϕ_c. Even slightly above ϕ_c, x rays are specularly reflected with considerable intensity. Kiessig [73] observed interference of a monochromatic x-ray beam specularly reflected from the surface of a thin Ni film with the beam penetrating through the film and specularly reflected from the interface film/substrate. Interference fringes are observed in specular reflection as a function of the reflection angle above the critical angle.

1. Determination of Film Thickness

Figure 8.4 shows the interference pattern of a thin film of amorphous Ge deposited on a sapphire substrate. Total reflection is observed below the critical angle for Ge at ~0.3°. Above the critical angle numerous interferences can be observed. The positions of the extrema are given by

$$\theta_m^2 = 2\delta + (m + k)^2(\lambda^2/4t^2), \tag{21}$$

where m is a whole number (the order of the extremum), $\delta = \mathrm{Re}(1 - n)$, n the refractive index of the film, and $k = 0$ or $\frac{1}{2}$, for minima or maxima, respectively. The quantity δ is proportional to the electron density N_e and given by

$$\delta = r_e N_e \lambda^2/2\pi, \tag{22}$$

where $r_e = 2.818 \times 10^{-13}$ cm is the classical electron radius. The definition of k holds for $\delta_s < \delta_f$, i.e., the substrate's electron density is smaller than that of the film. If $\delta_s > \delta_f$ then $k = 0$ or $\frac{1}{2}$, for maxima or minima, respectively. For details see, for instance, Segmüller [74]. Hink and Petzold [75] have applied least-squares methods to determine the film thickness t and the quantity δ from the positions of the extrema. For the Ge film of Fig. 8.4 we obtained $t = (70.37 \pm .35)$ nm and $\delta = (14.0 \pm 0.4) \times 10^{-6}$.

Similar to optical interference methods, the true geometric thickness of the film is determined, whereas the mass thickness is obtained by measuring the total intensity (Section III.B.1.). The method does not depend on crystallinity or random orientation distribution of the grains, it can be applied to amorphous films, it requires extremely smooth surfaces, and it provides an

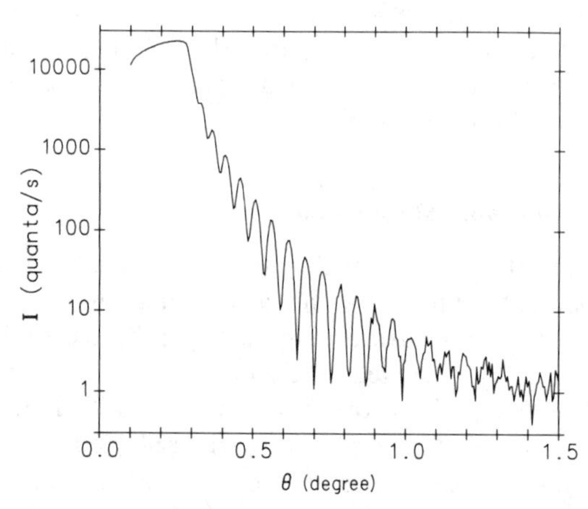

Fig. 8.4. Small-angle interference pattern of a film of amorphous germanium (70.4 nm thick) deposited on sapphire.

estimate of the surface roughness as will be discussed in Section III.E.3. The precision of the thickness measurement is much higher than that of other noninterferometric methods with a probable error often less than 1%.

2. Structural Studies on Thin-Film Multilayers

Structural information on thin-film multilayers can be obtained from small-angle interference studies. Parratt [76] adapted optical dispersion theory to specular x-ray reflection and developed a recursive algorithm to compute small-angle interference patterns of model structures consisting of N stratified thin layers each characterized by its thickness t_j and its refractive index $n_j = 1 - \delta_j - i\beta_j$. According to Eq. (22), δ_j is proportional to the average electron density of layer j, and the imaginary part of the refractive index is related to the average linear absorption coefficient of layer j by

$$\beta_j = \mu_j \lambda / 4\pi. \tag{23}$$

Structural details on an atomic scale cannot be resolved by this method. Equation (21), which determines the position of the interference maxima is closely related to Bragg's equation. Interfaces between the layers now assume the role of lattice planes to reflect x rays. Computer-model calculations of small-angle interferences have been used to study semiconductor superlattices [77, 78], periodic Langmuir–Blodget multilayers of Mn stearate [79], and near-periodic multilayers used for soft-x-ray mirrors [80]. The sensitivity is good enough to detect one monomolecular layer of Mn stearate ~2.5-nm thick. Least-squares refinement of the parameters t_j and n_j has been carried out by Segmüller [81] by using the nonlinear simplex method.

3. Estimate of Surface Roughness

If the film surface is not smooth, but rather has a certain roughness, the intensity of the interference maxima will be increasingly reduced with increasing reflection angle. Croce *et al.* [82] (see also Névot and Croce [83]) introduced a Debye–Waller factor for surface roughness

$$D = \exp(-16\pi^2 \sin^2\theta \langle z^2 \rangle / \lambda^2), \tag{24}$$

where $\langle z^2 \rangle$ is the mean square deviation of the real surface from an ideal plane perpendicular to the surface. It was introduced into the model computations referred to in the last section as an additional parameter to be varied.

F. Diffraction Studies of Ultrathin Films

In the last decade, high-power x-ray sources have increasingly become available that, together with new experimental techniques, have enabled the scientist to apply x-ray diffraction to the study of ultrathin films and surface-adsorbed layers that, up to then, could be detected and studied only with such specific surface techniques as low-energy electron diffraction (LEED) or reflection high-energy electron diffraction (RHEED). These studies demonstrate the progress x-ray diffraction has undergone in the past decade in its application to thin films.

1. Study of Films and Surfaces by Grazing-Incidence X-Ray Diffraction

As discussed in the preceding section, x rays can be specularly reflected from surfaces at angles close to the critical angle for total reflection. Marra *et al.* [22] have demonstrated that strong diffracted x-ray beams can also be observed under these conditions. In their experiment, the sample was mounted on a Bragg–Brentano diffractometer with its surface perpendicular to the diffractometer axis as depicted in Fig. 8.5. The primary x-ray beam, monochromated and focused at the sample by a vertically bent graphite monochromator, was incident on the sample under an angle close to the critical angle for total reflection ($\lesssim 0.5°$), thus enabling intense specular reflection. Executing a $2\theta - \theta$ scan, and thereby rocking the sample around its surface normal, they also observed strong diffracted beams exiting from the surface under very shallow angles with diffraction vectors parallel to the surface. The penetration depth, typically some 10 nm perpendicular to the surface, could be changed by varying the angle of incidence. From the diffraction angles the two-dimensional lattice of the epitaxial layer parallel to the interface can be determined, but no information is obtained perpendicular to the interface. From the linewidth we can determine the size of coherently scattering domains which, in the case of epitaxial misfit, is related to the spacing of misfit dislocations. By using this extremely surface-sensitive method, Marra *et al.* [22] studied the interface between the epitaxial Al film

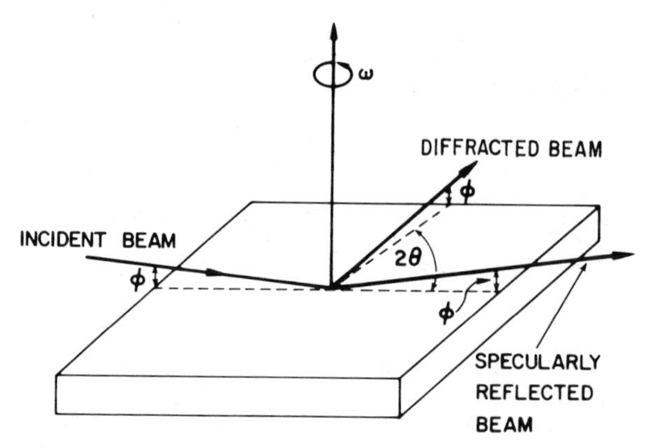

Fig. 8.5. Geometry of the grazing incidence diffraction method.

and the (001) GaAs substrate; Weng *et al.* [84, 85] studied the Al–Ge(001) interface; Eisenberger and Marra [86] studied the reconstruction of the Ge (001) surface in UHV; Marra *et al.* [87] studied the melting of Pb monolayers on Cu (110) surfaces in UHV; and Robinson [88] studied the reconstruction of the Au (110) surface in UHV. The first two experiments and the last one were carried out with a rotating-anode x-ray source, the fourth one with a synchroton source, and the third one with both sources for comparison. The UHV experiments were carried out in a sample chamber designed by Marra [37]. An x-ray diffraction chamber could be decoupled without breaking the vacuum from the main system in which the sample was prepared and which was also equipped with a LEED–Auger electron optics system.

The grazing-incidence diffraction method is very well suited to determine orientation relationships between the substrate and an ultrathin epitaxial film. The detector is set to the diffraction angle 2θ for a set of lattice planes (hkl) perpendicular to the interface, and the sample is then rotated around the surface normal in an azimuthal scan, thus giving the (hkl) pole density on the equator of the pole figure.

2. X-Ray Diffraction Studies on Kr and Xe Monolayers on Graphite

Research on the structure, thermodynamics, and phase transitions of simple gases adsorbed on graphite has intensified. Horn *et al.* [89] have used exfoliated ZYX graphite* that presents a large (00.1) surface area and a strong preferred orientation of the (00.1) planes to study the liquid–solid phase transition in submonolayers of krypton by x-ray diffraction experiments with Cu–$K\alpha$ radiation from a rotating anode tube and monochromated by a vertically bent graphite crystal in the incident beam. A survey of these studies on the structure and phase transition of monolayers of krypton and xenon was published by Birgeneau *et al.* [90]. With the availability of high-power x radiation from synchrotons, high-resolution diffraction experiments were made possible [91]. Very recently, D'Amico *et al.* [92] reported x-ray diffraction by Kr monolayers on single-crystal graphite.

IV. SUMMARY

Rapidly increasing demand of thin films for use as electronic device materials has triggered the development of various techniques for the characterization of thin films. Since the film microstructure is very sensitive

* Union Carbide Corporation.

TABLE 8.2

Survey of X-Ray and Alternative Methods
to Characterize Various Characteristic Properties of the Film Microstructure

Film property	X-ray method	Section	Alternative
Thickness	Total intensity	III.B.1	Interferometry
	Pendellösung fringes	III.A.5	Ellipsometry
	Small-angle interferences	III.E.1	RBS
Grain size	Linewidth	III.C.1	TEM
Density	Bragg angles	III.A.2	Weighting
	Small-angle interferences	III.E.1	—
Macrostrain	Bragg angles	III.A.4	Strain gauge
Microstrain	Linewidth	III.C.1	—
	Line profile	III.C.1	—
Crystal perfection	Linewidth	III.C.1	TEM
	Line profile	III.C.1	
Preferred orientation	Intensity	III.B.2	TEM
	Pole figure	III.D	
Surface roughness	Small-angle interferences	III.E.3	SEM, TEM replica
Phase identification	Bragg angles	III.A.1	TEM
Phase composition	Bragg angles	III.A.3	Microprobe
Concentration profile	Line profile	III.C.2	STEM, RBS

to the preparation conditions, *in situ* characterization of thin films is desirable. X-ray diffraction is one of the *in situ* analytical techniques. The present chapter reviewed the x-ray diffraction techniques applicable to the characterization of thin films.

Characteristics of the film microstructure that can be obtained by x-ray diffraction techniques are summarized in Table 8.2. The most appropriate alternative analytical techniques for obtaining these characteristics are also listed in the last column of the table. An excellent reference for alternative techniques is the "Handbook of Thin Film Technology" [93]. The x-ray diffraction techniques have an important advantage over other techniques, because information can be obtained without destroying or contacting the thin film samples. However, the x-ray diffraction techniques have several limitations: (1) the information obtained is averaged over a macroscopic area irradiated by the x rays, (2) the sample depth examined is limited by the x-ray penetration depth, (3) the information is obtained along the direction parallel to the x-ray diffraction vector, and (4) much less information can be obtained on amorphous thin films. Therefore, use of both techniques, x-ray and alternative, is recommended to obtain detailed microstructural information on thin films.

REFERENCES

1. S. Mader, *in* "Handbook of Thin Film Technology" (L. I. Maissel and R. Glang, eds.), pp. 9-1–9-34, McGraw-Hill, New York, 1970.
2. R. W. Vook, *in* "Epitaxial Growth, Part A" (J. W. Matthews, ed.), pp. 339–364, Academic Press, New York, 1975.
3. M. J. Buerger, "X-Ray Crystallography," Wiley, New York, 1942.
4. H. P. Klug and L. E. Alexander, "X-Ray Diffraction Procedures," Wiley, New York, 1954.
5. B. D. Cullity, "Elements of X-Ray Diffraction," Addison-Wesley, Reading, Massachusetts, 1956.
6. A. Taylor, "X-Ray Metallography," Wiley, New York, 1942.
7. A. Guinier, "X-Ray Diffraction," Freeman, San Francisco, 1963.
8. C. S. Barrett and T. B. Massalski, "Structure of Metals," McGraw-Hill, New York, 1966.
9. B. E. Warren, "X-Ray Diffraction," Addison-Wesley, Reading, Massachusetts, 1969.
10. W. Parrish, *in* "Instruments and Measurements," *Proc. Int. Instrum. Measur. Conf., 5th, September 13–16, Stockholm, 1960*, Vol. 1, Chemical Analysis, pp. 346–359, Academic Press, New York, 1960. [Reprinted in "Advances in X-Ray Diffractometry and X-Ray Spectroscopy" (W. Parrish, ed.), pp. 1–18, Centrex, Eindhoven, 1962].
11. R. Jenkins and F. R. Paolini, *Norelco Rep.* **21**, 9–14 (1974).
12. G. Wassermann and J. Wiewiorowsky, *Z. Metallkd.* **44**, 567–570 (1953).
13. H. W. King, C. J. Gillham, and F. G. Huggins, *Adv. X-Ray Anal.* **13**, 550–577 (1970).
14. K. L. Weiner, *Z. Kristallogr.* **123**, 315–319 (1966).
15. R. Feder and B. S. Berry, *J. Appl. Crystallogr.* **3**, 372–379 (1970).
16. W. Parrish, M. Mack, and I. Vajda, *Norelco Rep.* **14**, 56–59 (1967); W. Parrish and M. Mack, *Acta Crystallogr.* **23**, 687–692 (1967).
17. A. Segmüller, *Z. Metallkd.* **48**, 448–453 (1957).
18. G. Kunze, *Z. Angew. Phys.* **17**, 412–421, 522–534; **18**, 28–37 (1964).
19. M. Mack and W. Parrish, *Acta Crystallogr.* **23**, 693–700 (1967).
20. C. J. Gillham, *J. Appl. Crystallogr.* **4**, 498–506 (1971).
21. C. J. Gillham and H. W. King, *J. Appl. Crystallogr.* **5**, 23–27 (1972).
22. W. C. Marra, P. Eisenberger, and A. Y. Cho, *J. Appl. Phys.* **50**, 6927–6933 (1979).
23. L. G. Schulz, *J. Appl. Phys.* **20**, 1030–1033 (1949).
24. R. N. Clarke and I. N. Court, *Electrocomponent Sci. Technol.* **5**, 107–112 (1978).
25. B. F. Decker, E. T. Asp, and D. Harker, *J. Appl. Phys.* **19**, 388–392 (1948).
26. D. A. Brine and R. A. Young, *Natl. Symp. Vac. Technol. Trans., 7th, 1960*, 250–259, Pergamon, New York (1961).
27. W. Rühl, *Z. Phys.* **138**, 121–135 (1954).
28. J. Angilello, private communication, 1984.
29. M. H. Read and D. H. Altman, *Appl. Phys. Lett.* **7**, 51–52 (1965).
30. M. H. Read and D. H. Hensler, *Thin Solid Films* **10**, 123–135 (1972).
31. D. M. Jackson, J. B. Newkirk, and M. J. Urban, *J. Appl. Phys.* **33**, 2301–2304 (1962).
32. M. J. Hall and M. W. Thompson, *Br. J. Appl. Phys.* **12**, 495–498 (1961).
33. H. D. Keith, *Proc. Phys. Soc., London, B* **69**, 180–192 (1956).
34. R. W. Vook and F. R. L. Schoening, *Rev. Sci. Instrum.* **34**, 792–793 (1963).
35. J. M. Vandenberg and R. A. Hamm, *J. Vac. Sci. Technol.* **19**, 84–88 (1981).
36. P. Eisenberger and L. C. Feldman, *Science* **214**, 300–305 (1981).
37. W. C. Marra, Ph.D. dissertation, Stevens Institute of Technology, Hoboken, New Jersey, 1981.
38. J. A. Bearden, *Phys. Rev.* **38**, 2089–2098 (1931).
39. J. A. Bearden, *Phys. Rev.* **43**, 92–97 (1933).

40. W. Parrish, J. Taylor, and M. Mack, *Adv. X-Ray Anal.* **7**, 66–85 (1964).
41. W. Parrish and A. J. C. Wilson, "International Tables for X-Ray Crystallography," Vol. II, pp. 216–232, Kynoch, Birmingham, England, 1959.
42. A. J. C. Wilson, "Mathematical Theory of X-Ray Powder Diffractometry," Philips Tech. Lib., Centrex Publ. Eindhoven, 1963.
43. A. Segmüller, *Adv. X-Ray Anal.* **13**, 455–467 (1970).
44. A. J. C. Wilson, "Elements of X-Ray Crystallography," Addison-Wesley, Reading, Massachusetts, 1970.
45. M. U. Cohen, *Rev. Sci. Instrum.* **6**, 68–74 (1935).
46. M. U. Cohen, *Rev. Sci. Instrum.* **7**, 155; *Z. Kristallogr.* **94**,288, 306 (1936).
47. J. B. Hess, *Acta Crystallogr.* **4**, 209–215 (1951).
48. D. W. J. Cruickshank, "International Tables for X-Ray Crystallography," Vol. II, pp. 84–95, Kynoch, Birmingham, England, 1959.
49. J. B. Nelson and D. P. Riley, *Proc. Phys. Soc., London* **57**, 160–177 (1945).
50. L. Vegard, *Z. Phys.* **5**, 17 (1921).
51. W. B. Pearson, "A Handbook of Lattice Spacings and Structures of Metals and Alloys," Pergamon, New York, 1958.
52. W. B. Pearson, "A Handbook of Lattice Spacings and Structures of Metals and Alloys," Vol. 2, Pergamon, Oxford, 1967.
53. A. Segmüller and M. Murakami, *IBM Research Report* RC 1077. Treatise on Materials Science and Technology," "Analytical Techniques for Thin Films" (K. N. Tu and R. Rosenberg, eds.), Academic Press, New York. To be published.
54. P. Croce, G. Devant, M. Gandais, and A. Marraud, *Acta Crystallogr.* **15**, 424 (1962).
55. R. W. Vook and F. Witt, *J. Vac. Sci. Technol.* **2**, 243–249 (1965).
56. R. W. Vook, T. Parker, and D. Wright, *in* "Surfaces and Interfaces, I. Chemical and Physical Characteristics" (J. Burke, N. Reed, and V. Weiss, eds.), pp. 347–358, Syracuse Univ. Press, Syracuse, New York, 1967.
57. W. T. Stacy and M. M. Janssen, *J. Cryst. Growth* **27**, 282–286 (1974).
58. B. W. Batterman and G. Hildebrandt, *Acta Crystallogr.* **A24**, 150–157 (1968).
59. W. K. Chu, J. W. Mayer, and M. A. Nicolet, "Backscattering Spectrometry," Academic Press, New York, 1978.
60. R. W. Vook and F. Witt, *J. Vac. Sci. Technol.* **2**, 49–57 (1965).
61. B. E. Warren and B. L. Averbach, *J. Appl. Phys.* **21**, 595–599 (1950).
62. C. R. Houska, *High Temp.-High Pressures* **4**, 417–429 (1972).
63. C. R. Houska, *in* "Treatise on Materials Science and Technology," Vol. 19A, 1980; "Experimental Methods, Part A" (H. Herman, ed.), pp. 63–105, Academic Press, New York, 1980.
64. M. Murakami, A. Segmüller, and K. N. Tu, "Treatise on Materials Science and Technology," "Analytical Techniques for Thin Films" (K. N. Tu and R. Rosenberg, eds.), Academic Press, New York. To be published.
65. P. E. Lighty, D. Shanefield, S. Weissmann, and A. Shrier, *J. Appl. Phys.* **34**, 2233–2239 (1963).
66. T. Imura, S. Weissmann, and J. J. Slade, Jr., *Acta Crystallogr.* **15**, 786–793 (1962).
67. V. L. Lambert, *J. Appl. Phys.* **46**, 2303–2305 (1975).
68. C. A. Wallace and R. C. C. Ward, *J. Appl. Crystallogr.* **8**, 255–260 (1975).
69. C. A. Wallace and R. C. C. Ward, *J. Appl. Crystallogr.* **8**, 545–556 (1975).
70. M. Bettini and G. Brandt, *J. Appl. Phys.* **50**, 869–873, 6938–6941 (1979).
71. Y. Lo, R. N. Bicknell, T. H. Myers, J. F. Schetzina, and H. H. Stadelmaier, *J. Appl. Phys.* **54**, 4238–4240 (1983).
72. F. Witt, R. W. Vook, and M. Schwartz, *J. Appl. Phys.* **36**, 3686–3687 (1965).
73. H. Kiessig, *Ann. Phys.* [5] **10**, 769–788 (1931).

74. A. Segmüller, *Thin Solid Films* **18**, 287–294 (1973).
75. W. Hink and W. Petzold, *Z. Angew. Phys.* **10**, 135–138 (1958).
76. L. G. Parratt, *Phys. Rev.* **95**, 359–369 (1954).
77. L. L. Chang, A. Segmüller, and L. Esaki, *Appl. Phys. Lett.* **28**, 39–41 (1976).
78. A. Segmüller, P. Krishna, and L. Esaki, *J. Appl. Crystallogr.* **10**, 1–6 (1977).
79. M. Pomerantz and A. Segmüller, *Thin Solid Films* **68**, 33–45 (1980).
80. E. Spiller and A. Segmüller, *Ann. New York Acad. Sci.* **342**, 188–200 (1980).
81. A. Segmüller, *in* "Modulated Structures—1979" (J. M. Cowley, J. B. Cohen, M. B. Salamon, and B. J. Wuensch, eds.), *AIP Conf. Proc.* **53**, 78–80, American Institute of Physics, New York (1979).
82. P. Croce, L. Névot, and B. Pardo, *Nouv. Rev. d'Opt. Appl.* **3**, 37–50 (1972).
83. L. Névot and P. Croce, *J. Appl. Crystallogr.* **8**, 304–314 (1975).
84. S. L. Weng, A. Y. Cho, and P. Eisenberger, *J. Vac. Sci. Technol.* **16**, 1134 (1979).
85. S. L. Weng, A. Y. Cho, W. C. Marra, and P. Eisenberger, *Solid State Commun.* **34**, 843–846 (1980).
86. P. Eisenberger and W. C. Marra, *Phys. Rev. Lett.* **46**, 1081–1084 (1981).
87. W. C. Marra, P. H. Fuoss, and P. E. Eisenberger, *Phys. Rev. Lett.* **49**, 1169–1172 (1982).
88. I. K. Robinson, *Phys. Rev. Lett.* **50**, 1145–1148 (1983).
89. P. M. Horn, R. J. Birgeneau, P. Heiney, and E. M. Hammonds, *Phys. Rev. Lett.* **41**, 961–964 (1978).
90. R. J. Birgeneau, E. M. Hammonds, P. Heiney, P. W. Stephens, and P. M. Horn, *in* "Ordering in Two Dimensions," (S. K. Sinha, ed.), pp. 29–38, Elsevier, North-Holland Publ., New York, 1980.
91. P. A. Heiney, P. W. Stephens, R. J. Birgeneau, P. M. Horn, and D. E. Moncton, *Phys. Rev. B* **28**, 6416–6434 (1983).
92. K. L. D'Amico, D. E. Moncton, E. D. Specht, R. J. Birgeneau, S. E. Nagler, and P. M. Horn, *Phys. Rev. Lett.* **53**, 2250–2253 (1984).
93. L. I. Maissel and R. Glang, eds., "Handbook of Thin Film Technology," McGraw-Hill, New York, 1970.

INDEX

D

E

F

G

H

I

J

K

L

M

Y

Z